WILLIAM F. MAAG LIBRARY
YOUNGSTOWN STATE UNIVERSITY

Very High Speed Integrated Circuits: Gallium Arsenide LSI

SEMICONDUCTORS AND SEMIMETALS
Volume 29

Semiconductors and Semimetals

A Treatise

Edited by R. K. Willardson　　*Albert C. Beer*
　　ENICHEM AMERICAS, INC.　BATTELLE COLUMBUS LABORATORIES
　　PHOENIX, ARIZONA　　　　COLUMBUS, OHIO

Very High Speed Integrated Circuits: Gallium Arsenide LSI

SEMICONDUCTORS
AND SEMIMETALS

Volume 29

Volume Editor

TOSHIAKI IKOMA

CENTER FOR FUNCTION-ORIENTED ELECTRONICS
INSTITUTE OF INDUSTRIAL SCIENCE
UNIVERSITY OF TOKYO
TOKYO, JAPAN

ACADEMIC PRESS, INC.
Harcourt Brace Jovanovich, Publishers

Boston   San Diego   New York
London   Sydney   Tokyo   Toronto

This book is printed on acid-free paper. ∞

COPYRIGHT © 1990 BY ACADEMIC PRESS, INC.
ALL RIGHTS RESERVED.
NO PART OF THIS PUBLICATION MAY BE REPRODUCED OR
TRANSMITTED IN ANY FORM OR BY ANY MEANS, ELECTRONIC
OR MECHANICAL, INCLUDING PHOTOCOPY, RECORDING, OR
ANY INFORMATION STORAGE AND RETRIEVAL SYSTEM, WITHOUT
PERMISSION IN WRITING FROM THE PUBLISHER.

ACADEMIC PRESS, INC.
1250 Sixth Avenue, San Diego, CA 92101

United Kingdom Edition published by
ACADEMIC PRESS LIMITED
24-28 Oval Road, London NW1 7DX

The Library of Congress has cataloged this serial title as follows:

Semiconductors and semimetals.—Vol. 1—New York: Academic Press, 1966-

   v.: ill.; 24 cm.

  Irregular.
  Each vol. has also a distinctive title.
  Edited by R. K. Willardson and Albert C. Beer.
  ISSN 0080-8784 = Semiconductors and semimetals

    1. Semiconductors—Collected works. 2. Semimetals—Collected works.
I. Willardson, Robert K. II. Beer, Albert C.
QC610.9.S48   621.3815'2—dc19   85-642319
                                                                                               AACR 2 MARC-S

Library of Congress   [8709]
ISBN 0-12-752129-1 (v. 29)

Printed in the United States of America
90 91 92 93   9 8 7 6 5 4 3 2 1

Contents

LIST OF CONTRIBUTORS	vii
PREFACE	ix

Chapter 1 Active Layer Formation by Ion Implantation
M. Kuzuhara and T. Nozaki

List of Symbols	1
I. Introduction	2
II. Furnace Annealing	3
III. Rapid Thermal Annealing	25
References	58

Chapter 2 Focused Ion Beam Implantation Technology
H. Hashimoto

I. Introduction	63
II. Implanter	64
III. FIB Implants	71
IV. Compositional Disordering of GaAs–AlGaAs	82
V. FIB-Doping MBE Growth	85
VI. Prospect of FIB Technology	93
References	95

Chapter 3 Device Fabrication Process Technology
T. Nozaki and A. Higashisaka

I. Introduction	97
II. Fundamental Process Technologies	99
III. Self-Aligned MESFET Technology	114
IV. Device Technology Effective for Improving Characteristics in Submicron Gate MESFETs	131
V. Summary	153
References	155

Chapter 4 GaAs LSI Circuit Design
M. Ino and T. Takada

I.	Introduction	160
II.	Device Analysis	160
III.	FET Model	170
IV.	Other Elements	187
V.	Basic Circuit Configurations	190
VI.	LSI Design	202
VII.	Testing Technology	223
VIII.	Summary	230
	References	230

Chapter 5 GaAs LSI Fabrication and Performance
M. Hirayama, M. Ohmori, and K. Yamasaki

I.	Introduction	234
II.	Characterization of Semi-insulating GaAs Wafers for IC Application	238
III.	TEG Selection for LSI Fabrication Process and Chip/Wafer Layout	245
IV.	FET Fabrication and Performance for LSIs	251
V.	Circuit Elements and Interconnection	276
VI.	Assembly Techniques and Packaging	283
VII.	LSI Fabrication and Performance	287
VIII.	Summary and Future Trends	300
	References	301

INDEX	305
CONTENTS OF PREVIOUS VOLUMES	310

List of Contributors

Numbers in parentheses indicate the pages on which the authors' contributions begin.

H. HASHIMOTO, *Fujitsu Laboratories, Ltd., 10-1, Morinosato Wakamiya, Atsugi, Kanagawa 243-01, Japan* (63)

A. HIGASHISAKA, *Microelectronics Research Laboratories, Nippon Electric Company, Ltd., 4-1-1, Miyazaki, Miyamae-ku, Kawasaki, Kanagawa 213, Japan* (97)

M. HIRAYAMA, *Atsugi Electrical Communication Laboratory, Nippon Telegraph and Telegraph Company, Ltd., 3-1, Morinosato Wakamiya, Atsugi, Kanagawa 243-01, Japan* (233)

M. INO, *Atsugi Electrical Communication Laboratory, Nippon Telegraph and Telephone Company, Ltd., 3-1, Morinosato Wakamiya, Atsugi, Kanagawan 243-01, Japan* (159)

M. KUZUHARA, *Microelectronics Research Laboratories, Nippon Electric Company, Ltd., 4-1-1, Miyazaki, Miyamae-ku, Kawasaki, Kanagawa 213, Japan* (1)

T. NOZAKI, *Optoelectronics Research Laboratory, Nippon Electric Company, Ltd., 4-1-1, Miyazaki, Miyamae-ku, Kawasaki, Kanagawa 213, Japan* (1) and (97)

M. OHMORI, *Electric Materials and Components Laboratories, Nippon Mining Company, Ltd., 3-17-35, Niizominami Toda, Saitama 335, Japan* (233)

T. TAKADA, *Atsugi Electrical Communication Laboratory, Nippon Telegraph and Telephone Company, Ltd., 1839 Ono, Atsugi, Kanagawa 243-01, Japan* (159)

K. YAMASAKI, *NTT Research and Development Headquarters, Yamato Seimei Bldg., Uchisaiwaicho, Chiyoda-ku, Tokyo 100, Japan* (233)

Preface

Very high speed integrated circuits are important for supercomputers and for digital communications (e.g., integrated digital networks and mobile land communications). Although silicon integrated circuit technologies are progressing into a submicron range of the scaling rule and the nanosecond range of operational speed, silicon itself has a limit of low mobility as compared with gallium arsenide and with some of the other III–V semiconductors. Moreover, the existence of the gallium arsenide semi-insulating substrate allows the reduction of parasitic circuit elements in integrated circuits; hence, gallium arsenide integrated circuits have an advantage over a p-n junction isolation used in silicon LSIs. Gallium arsenide LSIs are thus expected to be superior to silicon LSIs in terms of speed and power dissipation, and they have been developed intensively.

The realization of gallium arsenide LSIs requires an integrated knowledge of thin-film growth, ion implantation and subsequent annealing, fabrication processes such as etching and metalization, as well as design and testing of circuits. In this volume, technologies necessary to develop gallium arsenide LSIs are described.

In Chapter 1, the formation of active layers in a gallium arsenide metal-semiconductor field effect transistor (MESFET) by ion implantation is described. Ion implantation continues to be a standard technology for fabricating gallium arsenide MESFETs and LSIs. A most important feature is precise control of the threshold voltage of MESFETs, which is closely related to the activation efficiency of implanted species. This activation is achieved by post-implantation annealing, the condition of which is very important for obtaining activation efficiency with good reproducibility. Practical technologies for the formation of active layers are also described. Chapter 2 describes a new technology of focused ion beam implantation. This technology is not used for formation of active layers in the present MESFETs, but it is very important for maskless processing of III–V semiconductors and also for lithographic processing of very small devices. The operational principle of a focused ion beam implanter and some advanced applications combined with molecular beam epitaxy are also described.

Chapter 3 deals with device fabrication technologies, such as Ohmic and Schottky contact formation and dry etching. Self-aligned processes are essential for very high speed devices and several of the proposed methods are summarized in this chapter. Also covered are methods for suppressing short channel effects in submicron MESFETs.

Chapter 4 is devoted to LSI design, including device modeling and circuit considerations. Two advantages of gallium arsenide LSIs are low dissipation power and high-speed operation, which contradict one another. To utilize these advantages, an appropriate selection of circuit configurations is important. In this chapter, a variety of circuit configurations and their features is presented; testing is also described. Chapter 5 describes the GaAs LSI performances that have been attained so far and the fabrication processes. This chapter also deals with practical technologies for gallium arsenide LSIs and will be useful for practicing engineers.

Because it presents state-of-the-art technologies of gallium arsenide high speed LSIs, this volume will be very valuable not only to researchers and engineers in this field, but also to engineers working in telecommunications and high-speed computers.

I am indebted to the contributors and editors of this treatise for making this volume possible.

TOSHIAKI IKOMA

CHAPTER 1

Active Layer Formation by Ion Implantation

M. Kuzuhara
T. Nozaki

MICROELECTRONICS RESEARCH LABORATORIES
NEC CORPORATION
KAWASAKI, JAPAN

		Page
	LIST OF SYMBOLS	1
I.	INTRODUCTION	2
II.	FURNACE ANNEALING	3
	1. *Historical Background*	3
	2. *Encapsulating Material Requirements*	4
	3. *Commonly Employed Encapsulants*	5
	4. *Annealing with SiO_xN_y Encapsulants*	10
	5. *Annealing under Controlled Atmosphere*	23
III.	RAPID THERMAL ANNEALING	25
	6. *General Considerations*	25
	7. *n-Type Channel Implants*	27
	8. *High-Dose n^+ Implants*	34
	9. *p-Type Implants*	42
	10. *Minimization of Anneal-Induced Defects*	51
	11. *Device Application and Future Subjects*	54
	REFERENCES	58

List of Symbols

AES	Auger electron spectroscopy	LSI	Large-scale integration
C–V	Capacitance–voltage	LSS	Lindhard Scharff Shiott
CVD	Chemical vapor deposition	MESFET	Metal-semiconductor field effect transistor
DLTS	Deep-level transient spectroscopy		
FET	Field effect transistor	NAA	Neutron activation analysis
FIB	Focused ion beam	PL	Photoluminescence
IC	Integrated circuit	RBS	Rutherford backscattering
IR	Infrared	RI	Refractive index
LEC	Liquid-encapsulated Czochralski	RTA	Rapid thermal annealing
LM	Liquid-metal	SIMS	Secondary ion mass spectrometry

I. Introduction

Ion implantation into a semi-insulating GaAs substrate is currently an established technology to introduce desired impurities into near-surface regions of the substrate and to produce high-quality active regions for the fabrication of GaAs integrated circuits (ICs). In this chapter, the important parameters that affect the electrical properties of ion-implanted GaAs active regions will be discussed, with particular emphasis on the post-implantation annealing procedures including rapid thermal annealing.

Ion implantation doping has some inherent advantages over other doping techniques such as diffusion and epitaxial techniques. These include independent control of doping levels and doped layer thicknesses, excellent uniformity and reproducibility in doping, and the availability of selective doping without degrading planarity. The last advantage is further improved by the use of focused ion beam implantation. These advantages clearly offer considerable promise of developing as a high-yield integrated circuit technology.

One of the critical problems in ion implantation doping is the need for high-temperature heat treatment after implantation to anneal out the radiation-induced defects and to electrically activate the implanted dopants. There are several differences in annealing characteristics for GaAs, compared with those for Si. The most notable of these is the dissociation of GaAs during high-temperature annealing because of the preferential evaporation of As atoms. Such evaporation is reported to take place above $657 \pm 10°C$ for $\langle 100 \rangle$ GaAs (Foxon *et al.*, 1973). Since the activation of ion-implanted GaAs has been conventionally accomplished using a furnace in the 800–900°C temperature range, the annealed materials have been encapsulated with a suitable protective layer or processed in an arsenic pressure controlled atmosphere (Immorlica and Eisen, 1976; Malbon *et al.*, 1976; Kasahara *et al.*, 1979a, 1979b).

In Part II, a number of applications of furnace annealing to the formation of channel and n^+ contact regions for GaAs FETs are presented. Commonly employed encapsulating films are reviewed in view of the desirable material requirements for GaAs annealing encapsulants. Emphasis is laid upon the activation of Si implants in GaAs, since Si implantation is currently being employed toward developing high-speed GaAs ICs throughout the world. Descriptions of furnace annealing characteristics for group VI dopants, such as S, Se, and Te, have been presented in a number of review articles by Eisen (1980), Donnelly (1981), Morgan *et al.* (1981), and Stolte (1984). The conditions required to achieve maximum electrical activation for Si-implanted channel regions are discussed in view of the type of encapsulants, encapsulating film thickness, and annealing temperature. Some methods for capless annealing under an arsenic pressure are also described, in which electrical activation dependence on arsenic partial pressure controlled by an AsH$_3$ flow is discussed.

Recent advanced GaAs IC technology has year by year demanded shallower channel regions and more heavily doped n^+ contact regions. These increasing demands have accelerated the progress of rapid thermal annealing both in processing development and in apparatus improvement.

In Part III, the advantages of rapid thermal annealing over conventional furnace annealing are discussed, together with a comparison with pulse annealing using Q-switched lasers or pulsed electron beams. The application of rapid thermal annealing to low-dose channel implants and high-dose n^+ implants is separately described, since optimum annealing conditions are somewhat different from each other. The formation of p^+ regions by rapid thermal annealing is also presented for the fabrication of fully ion-implanted GaAs junction FET ICs (Zuleeg *et al.*, 1978; Troeger *et al.*, 1979; Kasahara *et al.*, 1981). The p^+ implant activation by rapid thermal annealing has recently received increasing interest for the fabrication of planar GaAs/AlGaAs heterojunction bipolar transistor ICs (Asbeck *et al.*, 1982), although the heterojunction device technology is beyond the scope of this chapter. An effective method to minimize rapid thermal annealing–induced defects is presented. This method provides slip–free surface morphology, which is essential for uniform activation over the entire water. The potential advantages of rapid thermal annealing are experimentally confirmed not only for channel implant activation but also for n^+ contact implant activation. Finally, some applications of rapid thermal annealing to GaAs MESFETs are described and the problems to be solved before its practical use as a fabrication tool for GaAs ICs are pointed out.

II. Furnace Annealing

1. HISTORICAL BACKGROUND

Ion implantation doping for GaAs was first demonstrated by Mayer *et al.* (1967) and then by Schroeder and Dieselman (1967). The low temperature ($<600°C$) employed in their annealing, however, required no encapsulating films, since such a low-temperature annealing gave rise to negligible arsenic evaporation during annealing. Roughan and Manchester (1969) reported the first use of an encapsulant. They covered the GaAs surface with an evaporated SiO$_x$ film to suppress As loss at an elevated temperature (650°C). Hunsperger and Marsh (1969) reported high-temperature annealing, up to 900°C, for activating Zn and Cd implants in GaAs using a CVD SiO$_2$ encapsulant. Activation of Se implants in GaAs was also reported using an

SiO$_2$ encapsulant (Foyt *et al.*, 1969). Since then, a variety of annealing encapsulants, including SiO$_2$, Si$_3$N$_4$, SiO$_x$N$_y$, AlN and their combined structures, have been studied and the usefulness of these encapsulants has been discussed.

In 1970, it was observed that an SiO$_2$ encapsulant on GaAs permits Ga outdiffusion during annealing (Gyulai *et al.*, 1970). Therefore, nitride films such as Si$_3$N$_4$ and AlN have been extensively studied as encapsulants for implanted GaAs for the purpose of suppressing Ga loss during annealing. These nitride encapsulants have been successfully applied particularly to group VI dopant activation (Gamo *et al.*, 1977; Eisen *et al.*, 1977; Inada *et al.*, 1978a), since Ga vacancy formation is detrimental to such dopant activation. In contrast with the evident effectiveness of nitride encapsulants for group VI dopant activation, advantages for group IV dopant activation have been less definitely manifested, although excellent controllability of the carrier concentration profile was reported in the use of an AlN encapsulant for Si implant activation (Okamura *et al.*, 1982).

Annealing methods without the use of encapsulating films have also been examined, in which an arsenic ambient was used to protect the GaAs surface from dissociation at an elevated temperature. Various methods have been proposed to implement such a new approach. It was reported that the strict control of an arsenic partial pressure is necessary to realize reproducible annealing characteristics (Kasahara *et al.*, 1979b).

Following this section, requirements for encapsulating-film materials are presented in Section 2. Section 3 provides a brief review of commonly employed encapsulants that have been applied for activation of implanted GaAs. Section 4 focuses on annealing characteristics of Si implants in GaAs activated using SiO$_x$N$_y$ encapsulants, including the film preparation procedure and the characteristics of the deposited films. In Section 5, some applications are described concerning capless annealing techniques.

2. Encapsulating Material Requirements

Properties of ion-implanted GaAs layers activated using various encapsulants are known to exhibit certain differences, depending on the kind and/or the quality of the employed encapsulating materials. It is recognized that the differences can be ascribed to the physicochemical properties of the encapsulating materials. There are several desirable, but sometimes mutually conflicting, properties that an ideal encapsulant should exhibit in a high-temperature annealing process. These requirements have been pointed out by several authors (Morgan *et al.*, 1981; Oberstar and Streetman, 1983).

(1) The encapsulant should not interact chemically with the substrate.

(2) The encapsulant should not permit outdiffusion of substrate host atoms.

(3) The encapsulant constituent atoms should not diffuse into the substrate.

(4) The stress at the encapsulant/substrate interface should be small.

(5) The encapsulant should exhibit great resistance to cracking, blistering and peeling off from the substrate.

(6) The encapsulant should be compositionally homogeneous and should not contain structural defects.

The electrical properties, such as electrical activation or mobility of the implanted layers, can be directly affected whether requirements (1)–(3), (5) and (6) are satisfied or not. Requirement (4) is believed to be responsible for the redistribution of implanted atoms or for the surface accumulation of residual impurities, such as Cr, Fe, Cu and Mn, existing in the substrate. Besides the requirements listed above, the capability for low temperature deposition and availability of selective etching for the substrate are also desirable requirements in view of practical applicability.

3. Commonly Employed Encapsulants

In this section, a brief review of commonly employed encapsulants is given taking into account the requirements described in the previous section.

a. SiO_2

Among the encapsulants employed, SiO_2 has been the most popular material. This material has been prepared by various techniques, including pyrolytic or plasma-enhanced chemical vapor deposition, sputtering, evaporation and spin-on coating. The principal feature of a SiO_2 encapsulant is that this material permits a significant Ga outdiffusion from the GaAs substrate during annealing, which was first demonstrated by Gyulai *et al.* (1970) using Rutherford backscattering spectroscopy. Since then, this phenomenon has been identified by a variety of techniques, including photoluminescence (Harris *et al.*, 1972; Chetterjee *et al.*, 1975), Auger electron spectroscopy (Vaidyanathan *et al.*, 1977; Inada *et al.*, 1978a), neutron activation technique (Ohdomari *et al.*, 1978) and X-ray photoelectron spectroscopy (König and Sasse, 1983).

It is clear that the Ga outdiffusion gives rise to a stoichiometry change (i.e., an As-rich condition) in GaAs. This effect has been reported to be detrimental to the activation of Te- and Se-implanted GaAs (Harris *et al.*, 1972; Inada *et al.*, 1978a). In the contrast, Davies *et al.* (1975) and Antell (1977) have reported that the activation of Si-implanted GaAs is enhanced by SiO_2 capped annealing. These results reflect the characteristics of group IV and VI dopants, where the difference in electrical activation can be ascribed to the different sublattice occupation. Furthermore, Davies and McNally (1984) observed enhanced activation on Zn-implanted GaAs resulting from a SiO_2 capped rapid thermal annealing. The effect of Ga outdiffusion on the activation is discussed in Section 4.

b. Si_3N_4

Since the SiO_2 encapsulant proved to bring about Ga outdiffusion, Si_3N_4 has become popular as an annealing encapsulant in which Ga outdiffusion effects are negligible. This material has been prepared by pyrolytic or plasma-enhanced chemical vapor deposition and sputtering. An important point for a Si_3N_4 deposition is the strict control of contaminants, such as oxygen or hydrogen which are introduced from source gases and/or residual gases in the deposition system. These contaminants can be easily incorporated into Si_3N_4 films and affect the annealing characteristics. Therefore, in order to ensure a reproducible annealing, incorporation of these contaminants should be suppressed to a minimal amount.

One of the effective techniques for Si_3N_4 deposition is rapid pyrolytic chemical vapor deposition from SiH_4 and NH_3, which was first developed by Donnelly et al. (1975). Although this technique required rather high temperature ($\sim 700°C$) compared with sputtering and plasma techniques, the pyrolytic Si_3N_4 permitted reproducible annealing at temperatures up to 950°C. Inada et al. (1978a) employed pyrolytic Si_3N_4 encapsulants for activating Se implants in GaAs. Carrier concentration exceeding 4×10^{18} cm^{-3} was achieved after a 900°C, 15-min annealing using an oxygen-free, pyrolytic Si_3N_4 encapsulant, as shown in Fig. 1. Characteristics of pyrolytically deposited Si_3N_4 on GaAs were reported by Leigh (1982), in which complete suppression of Ga outdiffusion up to 950°C annealing was confirmed by Auger profiling and cathodoluminescence measurements.

A crucial problem in the Si_3N_4 encapsulation is the frequent occurrence of cracking or peeling-off during high-temperature annealing. Inada et al. (1978b) found that the maximum Si_3N_4 thickness that can be deposited without causing cracking is 150 nm, when the film is prepared by pyrolytic chemical vapor deposition at 700°C. It is believed that a large Young's modulus of Si_3N_4 ($E/[1 - v] = 3.9 \times 10^{13}$ dyn/cm^2) and a thermal expansion mismatch ($\alpha_{GaAs} = 6.8 \times 10^{-6}$ deg^{-1}, $\alpha_{Si_3N_4} = 3.2 \times 10^{-6}$ deg^{-1}) are responsible for the cracking or peeling-off problems (Tokuyama et al., 1967).

c. AlN

AlN has been introduced as an annealing encapsulant that exhibits an excellent adherence to GaAs up to high temperature (1000°C), while suppressing Ga outdiffusion during annealing (Pashley and Welch, 1975). Sputtering techniques have been generally employed for the film deposition. The

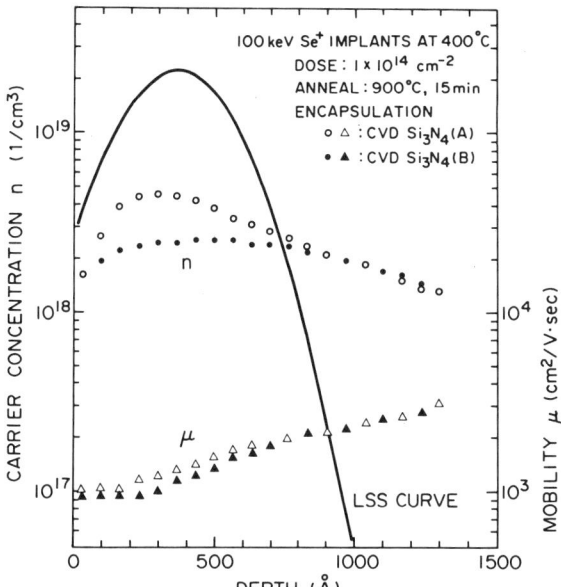

Fig. 1. Carrier concentration and mobility profiles for GaAs implanted with 100 keV, 1×10^{14} cm^{-2} Se ions at 400°C and annealed with two kinds of CVD Si$_3$N$_4$ encapsulants: ○, △, type A (oxygen-free Si$_3$N$_4$); and ●, ▲, type B (oxygen-contaminated Si$_3$N$_4$). Annealing was carried out at 900°C for 15 min. [From Inada et al. (1978a).]

adherence improvement is due to the similar thermal expansion coefficients of AlN and GaAs (i.e., 6.1×10^{-6} deg^{-1} and 6.8×10^{-6} deg^{-1}, respectively). Pashley and Welch (1975) reported that the electrical activation for Te implants was enhanced by using AlN instead of Si$_3$N$_4$. A peak electron concentration of 7×10^{18} cm^{-3} was obtained after a 900°C, 10-min annealing using an AlN encapsulant. Similar improvements in activation have been reported for Se implants (Gamo et al., 1977). AlN encapsulation resulted in 3-4 times higher electron concentration values than Si$_3$N$_4$ encapsulation. Davies et al. (1975), on the other hand, have obtained a different result for S implants, where AlN encapsulation gave the lowest electrical activation when compared with SiO$_2$ and Si$_3$N$_4$ encapsulations. For Si implants, Okamura et al. (1982) have applied an AlN encapsulant to shallow n-type layer formation. Thermal conversion effects, caused by the Cr outdiffusion on Cr-doped semi-insulating substrates, were significantly suppressed by a greatly reduced interfacial stress. Okamura et al. (1982) observed no thermal pits, cracks or peeling-off, even on the 1.2-μm-thick AlN encapsulant, after up to 1000°C annealing, and reported an excellent controllability in carrier concentration profile for Si-implanted GaAs, as shown in Fig. 2.

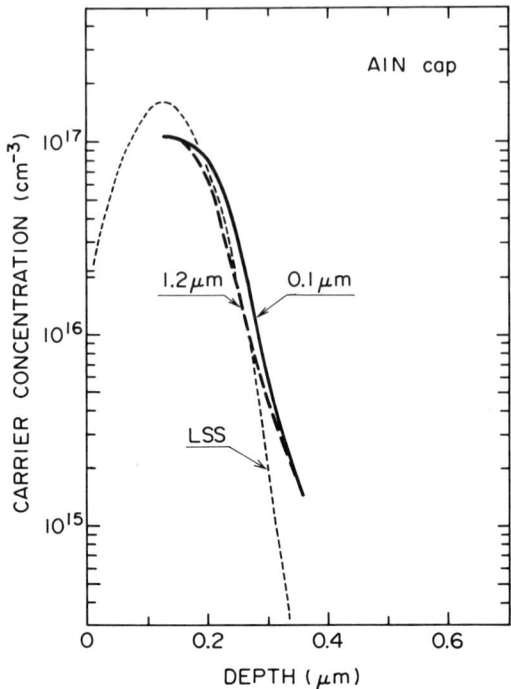

Fig. 2. Carrier concentration profiles for Si-implanted (145 keV, 2.4×10^{12} cm^{-2}) GaAs annealed at 850°C for 20 min with 0.1-μm-thick (solid line) and 1.2-μm-thick (dashed line) AlN encapsulants. [From Okamura et al. (1982).]

d. SiO_xN_y

Since SiO_xN_y is composed of a mixture of SiO_2 and Si_3N_4, a nature intermediate between those of two binary films can be anticipated for the SiO_xN_y encapsulant. SiO_xN_y encapsulants have been prepared, as noted in most of the previous reports, by an accidental oxygen leakage into Si_3N_4 deposition systems (Vaidyanathan et al., 1977; Inada et al., 1978a). Accordingly, annealing behaviors with respect to the SiO_xN_y encapsulation have not been systematically investigated in their work. Kuzuhara and Kohzu (1984) have first studied in detail the SiO_xN_y encapsulation effect on the activation of Si-implanted GaAs and have pointed out its potential feasibility as an annealing encapsulant. A detailed description will be given in Section 4.

e. $SiO_2(Ga)$ and $SiO_2(As)$

Doped silicate glasses, such as Ga-doped SiO_2 ($SiO_2(Ga)$) and As-doped SiO_2 ($SiO_2(As)$), have been examined to minimize the effects of Ga outdiffusion and As evaporation during annealing (Davies *et al.*, 1975). These films have been prepared by spin-on techniques. It has been reported that the $SiO_2(Ga)$ encapsulant is useful for enhancing the activation of S implants in GaAs because of the benefit of the Ga outdiffusion suppression. On the other hand, no enhancement in activation has been seen for the $SiO_2(As)$ encapsulant, indicating that the amount of As loss under the examined annealing temperature (825°C) is not so significant, even if an undoped SiO_2 encapsulant is employed.

Lidow *et al.* (1977, 1978) have proposed an effective application of $SiO_2(As)$, in which $SiO_2(As)$ is used as an overlayer for an Si_3N_4 encapsulant. This $SiO_2(As)/Si_3N_4$ bilayer encapsulant showed extremely stable characteristics against high-temperature annealing up to 1100°C. Figure 3 shows carrier concentration profiles for 120 keV, 1×10^{16} cm^{-2} Se implants in

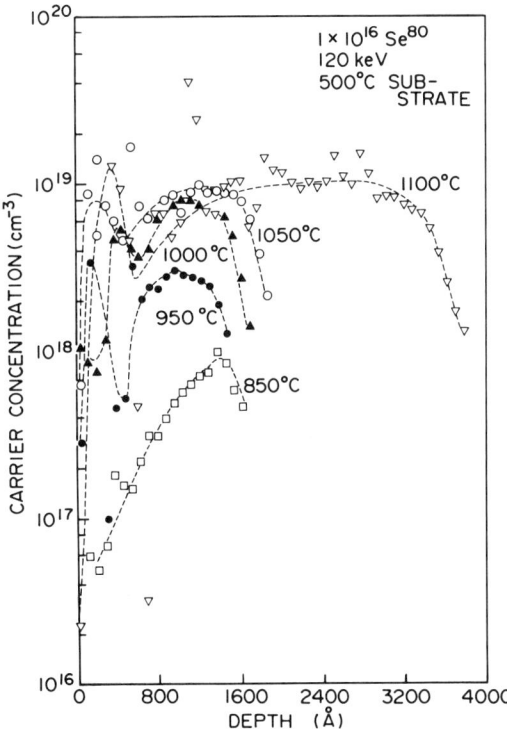

FIG. 3. Carrier concentration profiles for Se-implanted (120 keV, 1×10^{16} cm^{-2}, 500°C substrate) GaAs annealed with As-doped SiO_2 (1 µm)/Si_3N_4 (0.1 µm) double-layered encapsulants at 850°C (□), 950°C (●), 1000°C (▲), 1050°C (○), and 1100°C (▽). [From Lidow *et al.* (1978).]

GaAs annealed with the bilayered encapsulant, in which a maximum electron concentration of 1×10^{19} cm^{-3} was achieved.

4. ANNEALING WITH SiO_xN_y ENCAPSULANTS

Most of the current GaAs ICs are fabricated on *n*-type active layers by implanting Si ions. Despite the amphoteric doping characteristics, Si implantation has been extensively studied for a number of reasons. First, the implanted Si ions in GaAs can be easily activated by a rather low-temperature annealing (750–850°C). Second, an elevated-temperature implantation is not in general required for Si implantation. Third, the atomic mass of Si is rather small compared with that of other dopants, such as Se, Sn, and Te, resulting in reduced lattice damage formation and wide range controllability of the implanted ion depth.

Although a variety of annealing encapsulants have been examined for Si-implanted GaAs, it is still uncertain what film is practically the best for obtaining good electrical properties. In order to determine the optimum encapsulating film, annealing characteristics must be systematically studied for various kinds of encapsulants. In the following, the use of SiO_xN_y as an annealing encapsulant for GaAs is described, including the film deposition procedure, film characterization and excellent annealing results obtained on Si-implanted GaAs.

a. SiO_xN_y deposition procedure

SiO_xN_y films were deposited by the pyrolytic reaction of the SiH_4–NH_3–O_2 system at a substrate temperature of 700°C. The deposition temperature (700°C), which is required to thermally decompose NH_3, is above the As preferential evaporation point. It is therefore of great importance to initiate the film deposition quickly before the As evaporation takes place. For this purpose, infrared radiation from halogen lamps was used as a heat source so that the GaAs substrate could be rapidly heated to the deposition temperature.

A diagram of the SiO_xN_y deposition system is shown in Fig. 4. The SiH_4 (4% in Ar), NH_3 and O_2 (1% in N_2) used were high purity electronic grade gases. The SiH_4 and NH_3 were further diluted with N_2 prior to being introduced into the reaction tube. Substrates were supported on a carbon susceptor, which was preferentially heated by the infrared radiation. The temperature of the carbon susceptor was measured with thermocouples mounted in the susceptor. The readout was delivered to a PID controller to provide a precise and quickly changeable temperature program.

A typical temperature cycle for the SiO_xN_y deposition is shown in Fig. 5. The substrate temperature was raised to 700°C within several seconds to

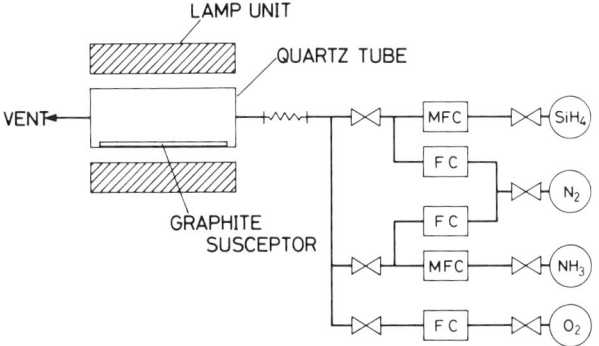

Fig. 4. Pyrolytic SiO_xN_y deposition system.

suppress transition layer formation, as well as to prevent thermal As evaporation before the initiation of film deposition. The source gas flow was stopped 30 sec before the start of cooling, so as to suppress any oxygen-rich surface layer formation in the cooling process.

b. Characterization of SiO_xN_y

The deposited SiO_xN_y films were characterized by various techniques including ellipsometry for the refractive index and film thickness measurements, Auger electron spectroscopy (AES) for determining the film composition, infrared (IR) absorption for evaluating atomic bond structures and the interference fringe technique (the Newton's ring technique) for the film stress estimation.

The SiO_xN_y film composition is very sensitive to the oxygen flow rate. Variations in the refractive index and film thickness with changes in the net

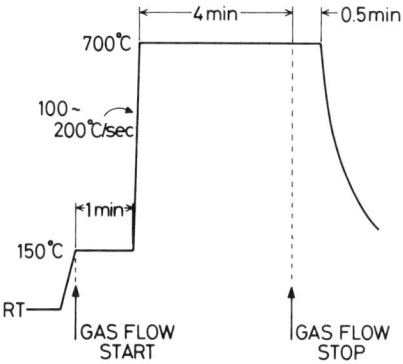

Fig. 5. Temperature cycle for SiO_xN_y deposition.

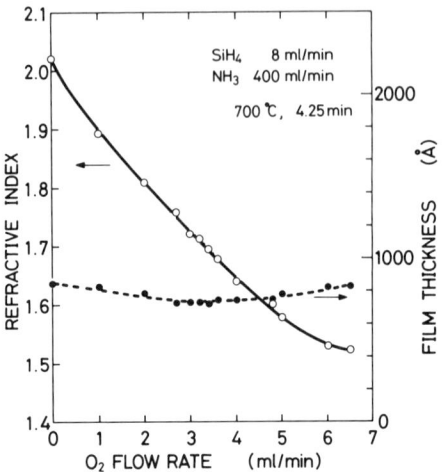

FIG. 6. Variations in refractive index (solid line) and thickness (dashed line) of SiO_xN_y films with changes in net oxygen flow rate. Flow rates for SiH_4, NH_3, and N_2 were fixed at 8, 400, and 9600 ml/min, respectively. Growth time was 4.25 min at 700°C.

oxygen flow rate are shown in Fig. 6. The SiH_4, NH_3, and N_2 carrier gases were introduced at fixed flow rates of 8, 400, and 9600 ml/min, respectively. The growth time (defined as the period between reaching 700°C and stopping the source gas flow) was set for 4.25 min. The refractive index at zero O_2 flow rate was 2.03, which corresponds to SiN_x almost in stoichiometry. Introducing a small amount of O_2 into the reaction system caused a sharp drop in the refractive index. The refractive index saturated near 1.46 for O_2 flow rates over 10 ml/min, indicating that the deposited film is SiO_2 in spite of the presence of NH_3. From these results, it is found that an SiO_xN_y film with any refractive indexes ranging from 1.46 to 2.0 can be prepared by solely controlling the O_2 flow rate in the reaction system employed. The film thickness, on the other hand, was hardly affected by the O_2 flow rate variation and was maintained near 900 Å (i.e., the deposition rate being ~210 Å/min).

The film composition, which was estimated by the atomic ratio of oxygen and nitrogen (i.e., O/N ratio), is plotted against the refractive index in Fig. 7. The O KLL and the N KLL signals in the Auger energy spectra from various SiO_xN_y films were measured by simultaneously sputter-etching the sample surface with a 2 keV Ar-ion beam to obtain the depth distributions of O and N atoms. A 10 keV incident electron beam was used for the measurements. The O/N ratio was derived from the Auger intensities, taking into account the AES sensitivity factors reported by Davis *et al.* (1976). The variation in the O/N ratio over the ~900 Å thickness range is expressed by error bars.

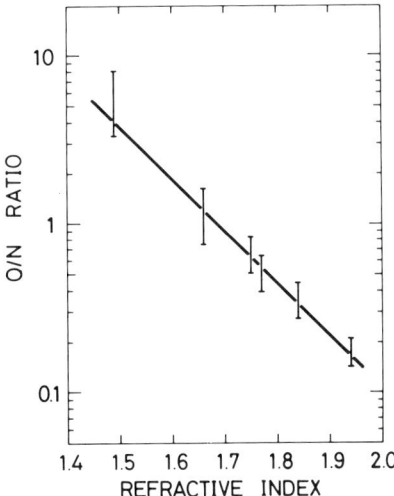

FIG. 7. Relative SiO$_x$N$_y$ film composition (i.e., O/N atomic ratio) versus the film refractive index. O/N ratios were measured by AES.

As shown in Fig. 7, the O/N ratio is seen to gradually decrease with increasing refractive index. This result is consistent with the similar plot reported by Kuiper *et al.* (1983), in which the O/N ratio in the SiO$_x$N$_y$ film prepared from the SiH$_2$Cl$_2$–NH$_3$–N$_2$O system was determined using Rutherford backscattering. It is found, from Fig. 7, that the SiO$_x$N$_y$ film with equal O and N concentrations (i.e., O/N ratio = 1) has a refractive index of ~1.68.

Figure 8 shows IR spectra taken from various SiO$_x$N$_y$ films deposited on Si substrates. Spectra 1, 2, 3, 4, and 5 are for SiO$_x$N$_y$ films having 2.03, 1.89, 1.85, 1.70, and 1.49 refractive indexes, respectively. The spectrum for a pure SiO$_2$ film is plotted by a dashed line. The films were all of approximately the same thickness (4000–5000 Å). IR measurements were carried out using a double beam technique. The four major bands observed in the spectra are due to the absorptions from Si–N (~850 cm^{-1}), Si–O (~1085 cm^{-1}), Si–H (~2200 cm^{-1}), and N–H (~3300 cm^{-1}) vibrations. The strong absorption near 670 cm^{-1} is due to the Si–Si vibration from the Si substrate. With decreasing film refractive index, the spectra in the 850–1100 cm^{-1} region show a peak shift to a higher wavenumber caused by the increase and decrease in the Si–O (~1085 cm^{-1}) and Si–N (~850 cm^{-1}) bands, respectively. The absorption coefficients for the 850 and 1085 cm^{-1} bands are plotted against the SiO$_x$N$_y$ film refractive index in Fig. 9. It is clear that the absorption due to the Si–N band shows a linear increase, while the absorption due to the Si–O band gradually decreases with an increase in the refractive index.

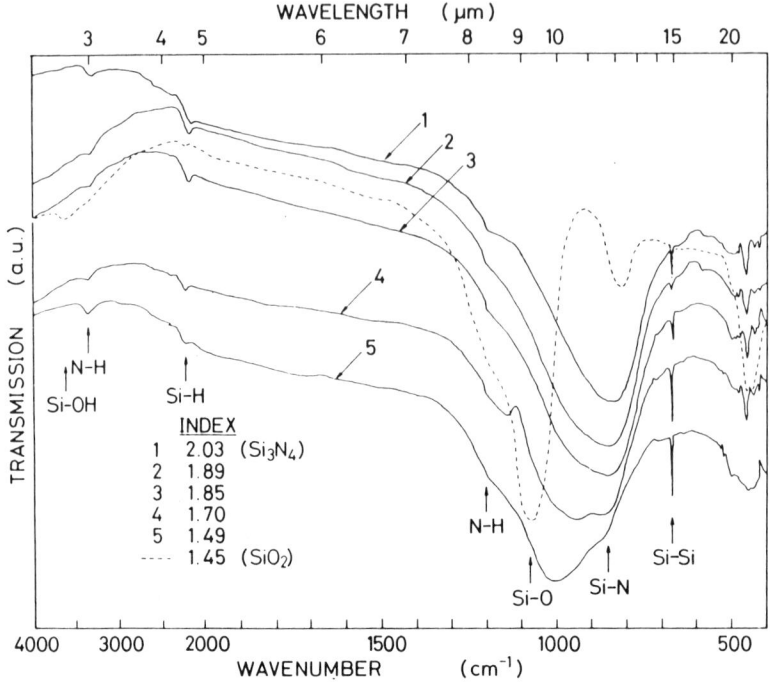

FIG. 8. Infrared absorption spectra for various SiO_xN_y films deposited on Si substrates. Refractive indexes: (1) 2.03 (Si_3N_4); (2) 1.89; (3) 1.85; (4) 1.70; (5) 1.49; (dashed line) 1.45 (SiO_2).

Finally, the stress in the film was evaluated by measuring the curvature of the GaAs substrate on which the SiO_xN_y film was deposited. In general, the film stress is given by the sum of two components, the internal stress and the thermal stress. The separation of these components, however, was not made since the stress measurements were carried out only at room temperature. The radius of curvature for the GaAs substrate was measured by counting the number of Newton rings generated by thallium lamp irradiation ($\lambda = 0.546\ \mu m$). Using the radii of curvature for the substrate with and without the film, i.e., R_f and R_0, respectively, the average stress in the film is given by Eq. (1) (Jaccodine and Schlegel, 1966),

$$\sigma_f = \frac{E_s}{6(1-v_s)} \frac{t_s^2}{t_f} \left(\frac{1}{R_f} - \frac{1}{R_0} \right), \quad (1)$$

where E_s and v_s are the Young's modulus and Poisson's ratio for the substrate, respectively, and t_s and t_f are the substrate and film thicknesses, respectively. For $\langle 100 \rangle$ oriented GaAs, $E_s/(1-v_s) = 1.23 \times 10^{12}\ dyn/cm^2$ (Kirkby et al., 1979).

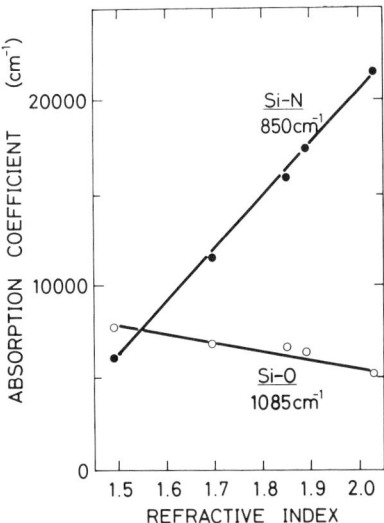

FIG. 9. Absorption coefficients for (●) 850 (Si–N) and (○) 1085 (Si–O) cm^{-1} bands as a function of SiO$_x$N$_y$ film refractive index.

The stress in the SiO$_x$N$_y$ films, calculated from Eq. (1) as a function of the refractive index, is shown in Fig. 10. Films with ~900 Å thicknesses were deposited on 2-inch-diameter GaAs substrates. The stresses in the SiO$_x$N$_y$ films were all tensile and became more tensile with increasing refractive index. A maximum tensile stress of ~1×10^{10} dyn/cm^2 was observed for the Si$_3$N$_4$ film.

In order to estimate the film stress effect on the annealing characteristics, the film stress at high temperature, instead of at room temperature, should be considered. Blaauw (1983) evaluated, at up to 300°C, the stresses in SiO$_2$ and Si$_3$N$_4$ films deposited on GaAs substrates and reported a linear increase in the film stress toward the tensile direction with increasing temperature. This result is consistent with the fact that the thermal expansion coefficient for GaAs is larger than that for SiO$_2$ or Si$_3$N$_4$. Assuming the same stress change for all SiO$_x$N$_y$ films up to high temperature, the SiO$_x$N$_y$ films at an annealing temperature (e.g., at 800°C) are expected to be under more increased tension compared to the stress at room temperature.

c. SiO$_x$N$_y$ capped annealing for Si-implanted GaAs

In the following, the characteristics of Si-implanted GaAs annealed using SiO$_x$N$_y$ encapsulants are described, and the mechanism for the enhanced electrical activation is discussed. Substrates used were ⟨100⟩ Cr-doped

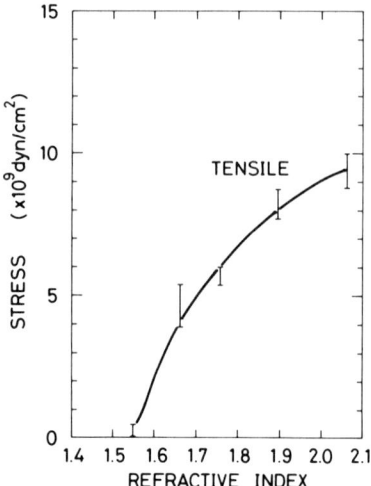

FIG. 10. Stress in SiO_xN_y film deposited on a GaAs substrate versus film refractive index. Film stress was measured by interference fringe technique at room temperature.

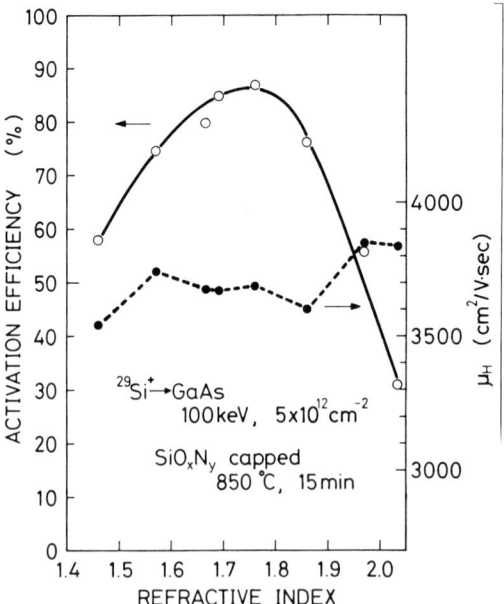

FIG. 11. Dependence of electrical activation (solid line) and Hall mobility (dashed line) on SiO_xN_y film refractive index for 100 keV, 5×10^{12} cm^{-2} Si implants in GaAs annealed at 850°C for 15 min. [From Kuzuhara and Kohzu (1984).]

(~0.5 wt ppm) semi-insulating GaAs crystals. ^{29}Si ions were implanted at room temperature in a nonchanneling direction. The GaAs samples were then encapsulated with ~1000 Å-thick SiO_xN_y films with various oxygen contents. Annealing was carried out, in most cases, at 850°C for 15 min in a flowing N_2 atmosphere. After annealing, Hall effect/sheet resistivity measurements were made using the standard van der Pauw technique. Carrier concentration profiles were evaluated by C–V technique or differential Hall measurements.

Figure 11 represents electrical activation efficiency and sheet Hall mobility as a function of the refractive index of the SiO_xN_y encapsulant. Implantations were carried out at 100 keV with a dose of 5×10^{12} cm^{-2}. Activation efficiencies were determined as the ratio of the measured sheet carrier concentration and the ion dose. Electrical activation reaches 87% for the SiO_xN_y encapsulant with ~1.75 refractive index, while activations for SiO_2 (RI = 1.46) and Si_3N_4 (RI = 2.03) encapsulants are 58% and 31%, respectively. Figure 12 shows carrier concentration profiles for samples annealed with an SiO_xN_y (RI = 1.75) encapsulant and with an SiO_2 (RI = 1.46) encapsulant. A theoretical profile, calculated by LSS theory (Lindhard et al., 1963; Gibbons et al., 1975), is also shown by a dashed line. Excellent

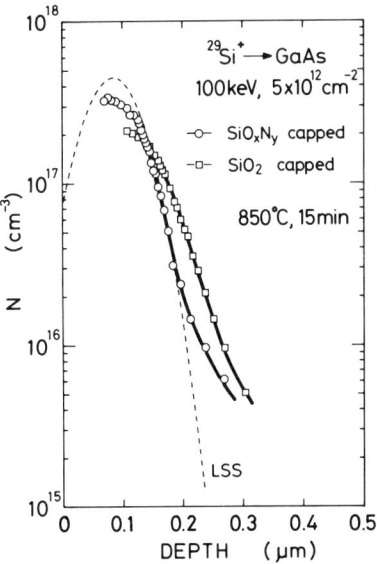

FIG. 12. Carrier concentration profiles for 100 keV, 5×10^{12} cm^{-2} Si implants in GaAs annealed at 850°C for 15 min with an SiO_xN_y (RI = 1.75) encapsulant (○) and with an SiO_2 (RI = 1.46) encapsulant (□).

agreement was obtained between the profile in the sample annealed using the SiO_xN_y encapsulant and the LSS theoretical profile. This fact indicates that the enhanced electrical activation obtained by SiO_xN_y encapsulation is, in fact, due to the activation of implanted Si dopants and not to the excess carriers generated by a thermal conversion of the substrate or by a diffusion of Si atoms from the encapsulant.

As pointed out in Fig. 10, reduced interfacial stress can be expected for the SiO_xN_y encapsulant with 1.75 refractive index compared to the Si_3N_4 encapsulant. No cracks or peeling-off were observed on the surface of the 850°C annealed SiO_xN_y (RI = 1.75) encapsulant with up to 2500 Å thickness, whereas a considerable number of cracks were generated on the annealed Si_3N_4 surface with more than 1500 Å thickness, in accord with the data reported by Inada *et al.* (1978b). Electrical activation values for Si implants (100 keV, 5×10^{12} cm^{-2}), as a function of the thickness of the SiO_xN_y encapsulant with 1.75 refractive index, are shown in Fig. 13. The electrical activations after annealing with SiO_xN_y encapsulants remained almost constant up to 2300 Å film thickness, while retaining high values (~90%) in electrical activation. The result in Fig. 13 further supports the

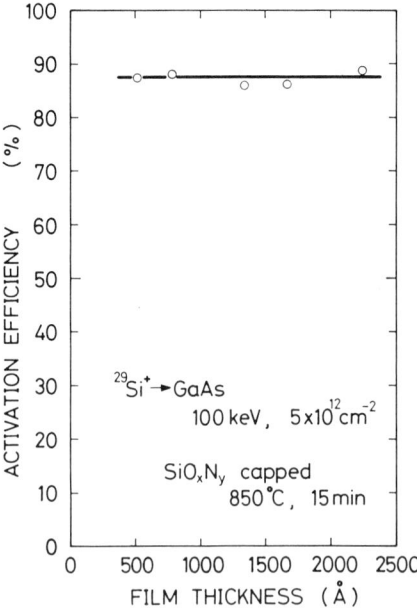

FIG. 13. Electrical activation dependence on SiO_xN_y (RI = 1.75) encapsulant thickness for 100 keV, 5×10^{12} cm^{-2} Si implants in GaAs annealed at 850°C for 15 min. [From Kuzuhara and Kohzu (1984).]

1. ACTIVE LAYER FORMATION BY ION IMPLANTATION

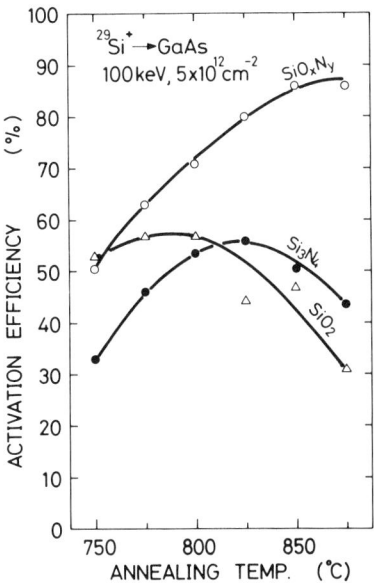

FIG. 14. Electrical activation dependence on annealing temperature for Si-implanted (100 keV, 5×10^{12} cm^{-2}) GaAs annealed with ($\triangle$) a SiO$_2$ encapsulant, ($\bigcirc$) a SiO$_x$N$_y$ (RI = 1.75) encapsulant, and ($\bullet$) a Si$_3$N$_4$ encapsulant.

concept that the stress in the encapsulant is significantly minimized by the use of SiO$_x$N$_y$.

Figure 14 shows electrical activations for three encapsulants, i.e., SiO$_2$, SiO$_x$N$_y$ (RI = 1.75) and Si$_3$N$_4$, at annealing temperatures between 750 and 875°C. Samples were implanted at 100 keV to a dose of 5×10^{12} cm^{-2}. A monotonic increase in activation was seen for the SiO$_x$N$_y$ encapsulant with increasing annealing temperature, whereas a decrease in activation was observed for the SiO$_2$ or Si$_3$N$_4$ encapsulant in the high temperature region. It should be noted that the activation for the SiO$_x$N$_y$ encapsulant is always higher than activations for the other two encapsulants at temperatures above 775°C.

Similar experiments were carried out for high-dose Si implants (150 keV, 7×10^{13} cm^{-2}) with a view to examining the usefulness of SiO$_x$N$_y$ encapsulants for n^+ layer formation. The results of Hall measurements for three encapsulants, SiO$_2$, SiO$_x$N$_y$ (RI = 1.75) and Si$_3$N$_4$, are presented in Table I. The electrical activation for the SiO$_x$N$_y$ encapsulant was about twice as high as that for either the SiO$_2$ or Si$_3$N$_4$ encapsulant. The carrier concentration and Hall mobility profiles for the SiO$_x$N$_y$ and Si$_3$N$_4$ encapsulants are shown in Fig. 15. A maximum electron concentration, 2.5×10^{18} cm^{-3}, was

TABLE I

RESULTS OF HALL MEASUREMENTS FOR 150 keV Si IMPLANTS (7×10^{13} cm^{-2})

Encapsulant	Refractive index	R_s ($\Omega/\square$)	μ_s (cm^2/V·sec)	n_s (cm^{-2})	η (%)
SiO$_2$	1.46	137	2347	1.94×10^{13}	27.7
SiO$_x$N$_y$	1.75	60	2094	4.95×10^{13}	70.7
Si$_3$N$_4$	2.03	105	2225	2.68×10^{13}	38.3

attained after annealing with SiO$_x$N$_y$ encapsulant. This electron concentration exceeds the upper achievable concentration limit, 2×10^{18} cm^{-3}, that has been reported on the activation of high-dose Si implants using a sputtered Si$_3$N$_4$ encapsulant (Tandon *et al.*, 1979; Masuyama *et al.*, 1980; Banwell *et al.*, 1983). These results suggest that the SiO$_x$N$_y$ encapsulation is promising for forming low-resistivity n^+ contact layers in the GaAs device fabrication processes.

In order to examine the reason for the enhanced activation obtained by the use of SiO$_x$N$_y$ encapsulants, Auger depth profiling studies were carried out for various SiO$_x$N$_y$/GaAs structures that were annealed at 850°C for 15 min. Figure 16 represents the Auger profile for the SiO$_x$N$_y$ (RI = 1.49)/GaAs

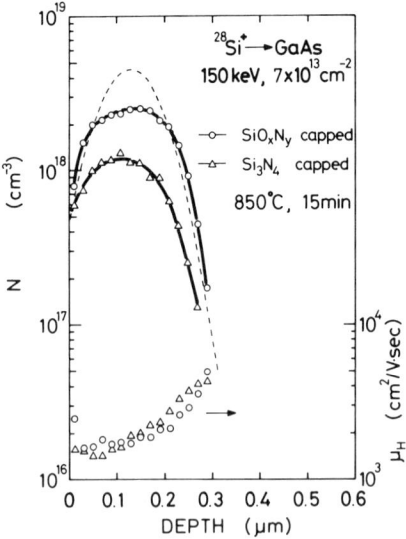

FIG. 15. Carrier concentration (solid line) and Hall mobility profiles for 150 keV, 7×10^{13} cm^{-2} Si implants in GaAs annealed at 850°C for 15 min with an SiO$_x$N$_y$ (RI = 1.75) encapsulant (O) and with an Si$_3$N$_4$ encapsulant (△). [From Kuzuhara and Kohzu (1984).]

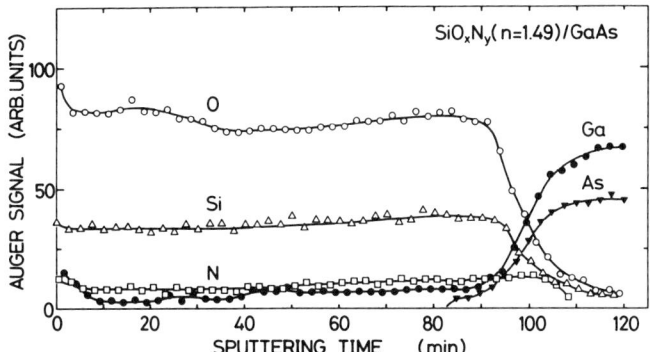

FIG. 16. Auger depth profiles for SiO$_x$N$_y$ (RI = 1.49)/GaAs structure annealed at 850°C for 15 min. [From Kuzuhara and Kohzu (1984).]

structure, in which a Ga signal was clearly detected both at the film surface and within the film layer, indicating the occurrence of a significant Ga outdiffusion from the substrate. The Ga outdiffusion was also observed in the SiO$_x$N$_y$ (RI = 1.53) film. In this case, however, the Ga detected region was limited only within the film layer (see Fig. 17). When the SiO$_x$N$_y$ refractive index became more than 1.59, no Ga signals were detected within the Auger detection limit (~0.5 atmic %). The Auger profile for the SiO$_x$N$_y$ (RL = 1.75)/GaAs structure is depicted in Fig. 18. Results of Auger analyses are summarized in Table II. These results indicate that the amount of Ga outdiffusion is gradually depressed with an increase in refractive index (or a decrease in O/N ratio) in the SiO$_x$N$_y$ encapsulant. Although no Ga

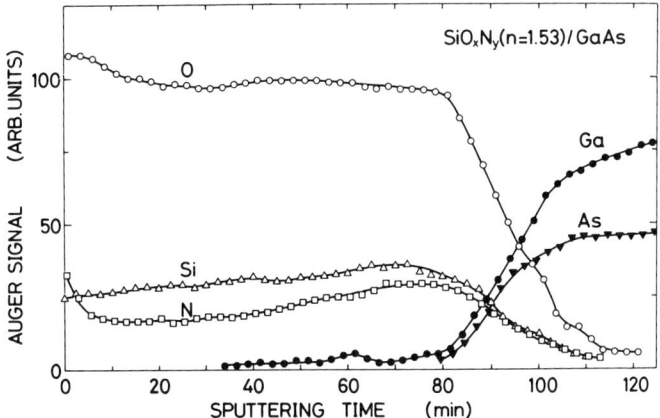

FIG. 17. Auger depth profiles for SiO$_x$N$_y$ (RI = 1.53)/GaAs structure annealed at 850°C for 15 min.

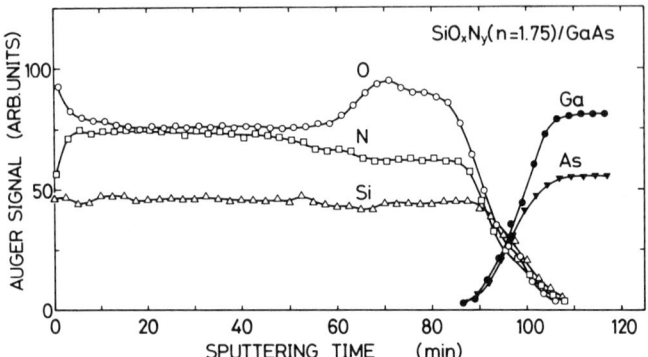

FIG. 18. Auger depth profiles for SiO_xN_y (RI = 1.75)/GaAs structure annealed at 850°C for 15 min. [From Kuzuhara and Kohzu (1984).]

signal was detected in the Auger profile for the SiO_xN_y (RI = 1.75)/GaAs structure, some amount of Ga outdiffusion, while not as excessive as observed in the SiO_2 encapsulant, must have occurred in this structure. This Ga outdiffusion gives rise to a controlled amount of Ga vacancy in the GaAs substrate, which promotes the incorporation of Si dopants into Ga lattice sites. This model can well elucidate the uselessness of SiO_xN_y encapsulants for group VI dopants (Inada *et al.* 1978a), since these dopants occupy As lattice sites. Therefore, the enhanced activation for SiO_xN_y encapsulants, in particular for the SiO_xN_y (RI = 1.75) encapsulant, can be ascribed to the generation of an optimized amount of Ga vacancy caused by the Ga outdiffusion during annealing.

TABLE II

RESULTS OF AES ANALYSES FOR VARIOUS SiO_xN_y/GaAs STRUCTURES

Refractive index	O/N ratio	AES Ga signal
1.46	∞	A, B[a]
1.49	4	A, B
1.53	3	B only
1.59	2	Not detected
1.66	1.2	Not detected
1.75	0.6	Not detected
1.83	0.35	Not detected
2.03	~0	Not detected

[a] A, detected at the film surface; B, detected within the film layer.

5. Annealing under Controlled Atmosphere

Another approach to suppress dissociation of GaAs due to As evaporation during annealing is the introduction of a controlled atmosphere, whereby the need for the critical encapsulation process can be eliminated. Immorlica and Eisen (1976) reported a capless annealing method that utilized a graphite powder saturated with As. The GaAs sample to be annealed was embedded in the graphite powder to provide arsenic overpressure. Se-implanted GaAs samples were annealed at 850°C for 30 min using this method, giving an effective mobility of 5300 cm^2/V·sec and a carrier concentration profile comparable to that obtained with a Si$_3$N$_4$ cap. GaAs MESFETs were fabricated on the Se-implanted, caplessly annealed active layers, where a state-of-the-art device performance (optimum noise figure of 3.4 dB and maximum gain of over 10 dB at 10 GHz) was realized. Despite the good electrical properties obtained, a strong peak at 0.88 μm, believed to be due to As vacancy complexes (Williams and Elliott, 1969), was manifested in the photoluminescence spectrum for the caplessly annealed sample. This suggests that the arsenic partial pressure provided in this method was not sufficient to completely suppress As evaporation.

Malbon et al. (1976) reported a reproducible technique for annealing ion-implanted GaAs layers without the need for encapsulants, in which a flowing high purity hydrogen gas was passed over a liquid Ga solution saturated with As in order to supply arsenic overpressure. The Ga solution also served as a reaction site for the trace amounts of H$_2$O in the annealing atmosphere, thus minimizing erosion of GaAs by H$_2$O. They observed no surface degradation on the annealed GaAs surface up to 800°C.

Kasahara et al. (1979a) employed thermal decomposition of arsine (AsH$_3$) to control arsenic partial pressure in the annealing atmosphere. This method is simple and reproducible, and thus suitable for the annealing process in mass production. No sign of surface degradation was observed even after annealing up to 950°C. Kasahara et al. (1979b) also reported the effects of arsenic partial pressure on activations of ion-implanted GaAs. Sheet carrier concentrations were measured against arsenic partial pressure (P_{AsH_3}) ranging from 0.15 to 10.64 torr for S, Si, and Mg implants in GaAs annealed at 900°C for 15 min. As shown in Fig. 19, an increase in electrical activation was seen for Si or Mg implants with the increase in arsenic partial pressure, whereas an opposite dependence was obtained for S implants. These differences were ascribed to the different pertinent lattice sites, i.e., Ga sites or As sites, for respective implanted impurities. Thermodynamic calculations gave an excellent explanation for the variation in sheet carrier concentration with arsenic partial pressure, in which generation of As vacancy donors was involved in relation to the applied arsenic partial pressure.

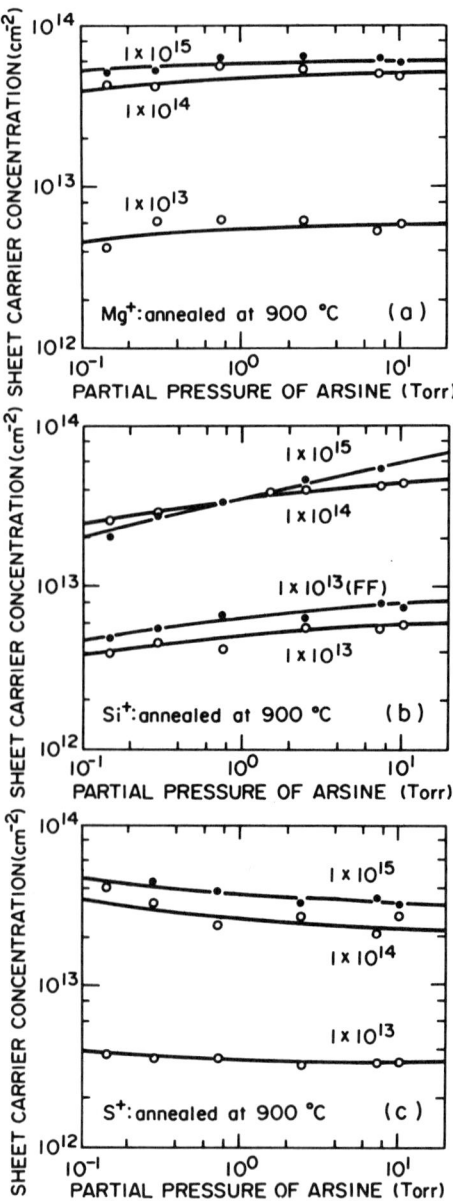

FIG. 19. Sheet carrier concentration dependence on partial pressure of arsine during capless annealing at 900°C for 15 min. Data are shown for (a) 100 keV Mg implants, (b) 130 keV Si implants, and (c) 150 keV S implants. [From Kasahara et al. (1979).]

These results indicate the need for a strict control of arsenic partial pressure, so as to realize reproducible annealing characteristics in the capless annealing systems.

III. Rapid Thermal Annealing

6. GENERAL CONSIDERATIONS

High-speed and large-scale GaAs integrated circuit technology requires the formation of shallower channel layers and more heavily doped n^+ contact layers as a result of the scaling down of device dimensions. In this respect, the conventionally employed furnace annealing process, which involves heat treatment for several tens of minutes, has become inadequate for suppressing both vertical and lateral diffusion of implanted dopants during annealing and for achieving sufficiently high doping levels in n-type dopant activation. With a view to overcoming the above-mentioned problems in furnace annealing, considerable effort has first been concentrated on the development of pulse annealing, in which Q-switched lasers or pulsed electron beams were employed with duration times of 10^{-9}–10^{-3} sec. The application of pulse annealing, however, has met with limited success. Only high-dose implants have been able to be activated, giving electron concentrations exceeding 10^{19} cm^{-3} (Mozzi *et al.*, 1979; Pianetta *et al.*, 1980a, 1980b; Liu *et al.*, 1980). In these cases, however, the mobility values obtained were significantly low. In addition, no activation was generally observed for low-dose ($<10^{13}$ cm^{-2}) n-type implants, because of the formation of compensating defects by pulse annealing itself (Davies *et al.*, 1980; Oraby *et al.*, 1984). This implies the uselessness of pulse annealing for application to FET channel-layer formation.

In recent years, rapid or transient thermal annealing techniques have been introduced for ion-implanted III–V compounds with a view to accomplishing heat treatment within several seconds. The primary difference between rapid thermal annealing and pulse annealing is that the former implies thermal processing in the solid phase, whereas the latter involves processing in the liquid phase. Therefore, the rapid thermal annealing can be regarded as being more similar to furnace annealing from the annealing dynamics standpoint. Various heating sources have so far been explored for realizing such short annealing times, including incoherent light sources, graphite strip heaters and multiply scanned electron beams.

The applicability of rapid thermal annealing to post-implantation annealing in III–V compounds was first confirmed in the work at the University of Surrey (Surridge *et al.*, 1977; Sealy *et al.*, 1979), in which graphite strip heaters were employed on the analogy of the rapid Si$_3$N$_4$ film deposition

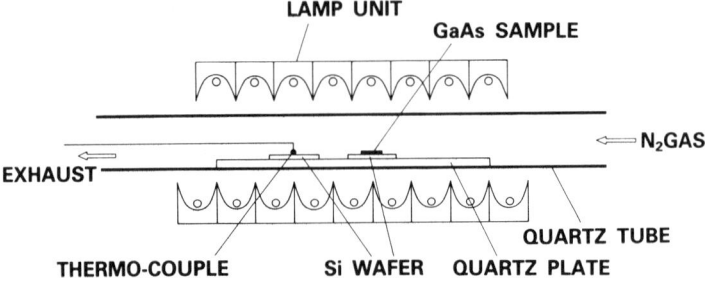

FIG. 20. Cross-sectional view of rapid thermal annealing system using incoherent light sources. [From Kuzuhara et al. (1982).]

developed by Donnelly et al. (1975). However, the graphite heater technique involves a possibility of contamination from the graphite heater itself, since the sample is not spatially isolated from the heater. In addition, the large thermal capacity of the heater is not advantageous for securing rapid temperature cooling after annealing. These disadvantages can be improved by the use of incoherent light sources. Figure 20 illustrates an example of the rapid thermal annealing system utilizing radiation from incoherent light sources (Kuzuhara et al., 1982). Both thermal and spatial isolation between heating sources and samples are established in this configuration, thus not only enabling sufficiently rapid heating and cooling but also permitting a free choice of ambient gases, including an arsenic pressure controlled ambient.

One of the important factors in rapid thermal annealing concerns how the substrate temperatures are determined. Technical difficulties in temperature determination arise principally from the transient nature of the temperature cycle—that is, the substrate temperature is not in general identical with the temperature of either heating sources or ambient gases. An interesting experimental result was reported by Davies et al. (1983), in which the temperature response of Ge during rapid thermal annealing was measured by a thermocouple placed directly on the Ge substrate. The maximum temperature recorded coincided well with that anticipated from the melting point of Ge (937°C), indicating that the substrate temperature during rapid thermal annealing could be measured with sufficient precision by a thermocouple. In practical application, the thermocouple was generally placed on another monitor wafer rather than on an actual sample (see Fig. 20), to prevent the sample from being mechanically damaged and/or contaminated by the thermocouple. A more direct method for temperature monitoring has been established using an optical pyrometer. This method enables contactless-temperature monitoring and thus provides more precise and reproducible temperature readings. In both temperature-monitoring methods, feedback

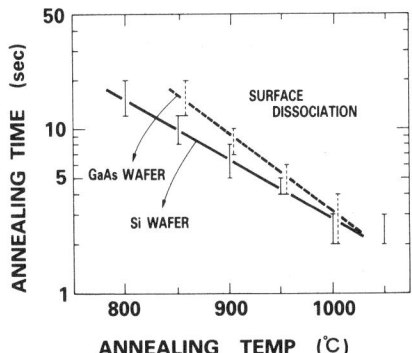

FIG. 21. Critical surface dissociation conditions during rapid thermal annealing in an N_2 ambient with unencapsulated GaAs sample surfaces in close contact with Si wafer susceptors (solid line) and with GaAs wafer susceptors (broken line). [From Kohzu et al. (1983).]

temperature control is indispensable to realize desired temperature–time cycles with sufficient reproducibility.

One of the advantages in rapid thermal annealing is a capless annealing capability. This is primarily due to the very short period of the annealing times involved. Figure 21 represents surface dissociation conditions for GaAs samples, which were annealed with their unencapsulated sample surfaces in close contact with Si wafers or GaAs wafers (Kohzu et al., 1983). Critical surface dissociation conditions are depicted by a solid line for Si wafer susceptors and by a broken line for GaAs wafer susceptors. Under the annealing conditions at the lower portion of each line, no thermal pits were observed on the annealed GaAs surface through a microscope. Effectiveness of the face-to-face capless techniques have been pointed out for n-type channel layer formation (Arai et al., 1981; Kuzuhara et al., 1982; Ito et al., 1983; Badawi and Mun, 1984; Kanber et al., 1985a). On the other hand, since rather high temperatures are required for high dose n^+ implants, various kinds of encapsulants have generally been employed for heavily doped n^+ layer formation (Shah et al., 1981; Davies et al., 1982; Chapman et al., 1982; Bensalem et al., 1983; Ito et al., 1983; Tabatabaie-Alavi et al., 1983a; Kuzuhara et al., 1985).

7. n-Type Channel Implants

Recent GaAs integrated-circuit technology requires uniformly activated, shallow n-type channel-layers with controlled doping levels. Since pulse annealing has failed to activate low dose implants, the means for activation of n-type channel implants has conventionally been restricted to the use of furnace annealing. The long heating times used in furnace annealing, however, caused some undesirable effects, such as substrate-quality degradation and implanted-dopant diffusion, whereby the development of advanced GaAs integrated circuits has been greatly hindered. Potential advantages of rapid thermal annealing over conventional furnace annealing are: (1) dopant diffusion during annealing is minimized; (2) diffusion of background impurities such as Mn and Cr during annealing is minimized; (3) encapsulants may not be necessary, thus eliminating stress at the substrate interface; (4) toxic gases such as AsH_3 may not be necessary for capless annealing; (5) annealing temperatures exceeding 900°C can be easily realized, leading to high electrical activation of implanted dopants; (6) operation time for one annealing cycle is short; and (7) large annealing systems can be constructed by increasing the number and/or the length of heating sources.

A successful low-dose ($<10^{13}$ cm^{-2}) n-type implant activation by rapid thermal annealing was first reported by Shah et al. (1980), in which multiply scanned electron beams were used as heat sources. Rapid scanning by a 30 keV electron beam with a power density of 15 W/cm^2 for an exposure time of 5 sec produced 50–60% activation and 3800 cm^2/V·sec sheet mobility from

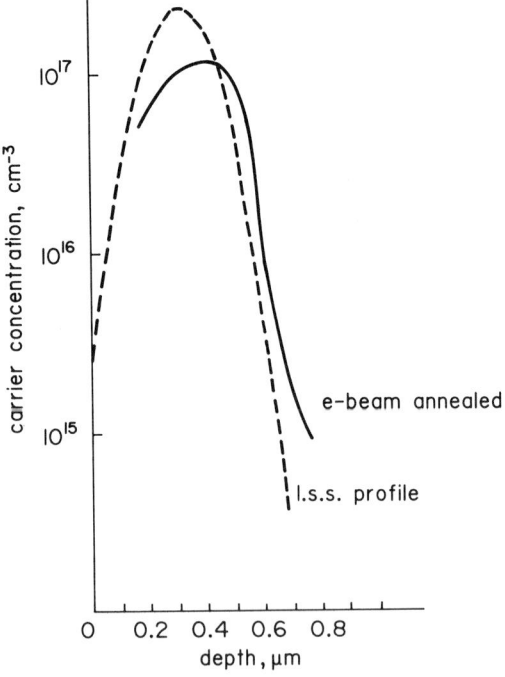

FIG. 22. Carrier concentration profile (solid line) of electron-beam annealed GaAs implanted with Si (360 keV, 6×10^{12} cm^{-2}). [From Shah et al. (1980).]

360 keV, 6×10^{12} cm^{-2} Si implants in GaAs. Figure 22 shows the corresponding carrier concentration profile, in which a peak electron concentration of $\sim 10^{17}$ cm^{-3} is obtained with minimized dopant diffusion. In this work with Si implants, samples could be annealed without encapsulation. However, the subsequent work with Se implants (400 keV, 5×10^{12} cm^{-2}) required an Si$_3$N$_4$ encapsulant to protect the surface from dissociation before significant activation could be obtained (Shah et al., 1981).

The superiority of using rapid thermal annealing for low-dose n-type implant activation was reported by Arai et al. (1981). This is the first report employing incoherent light sources to anneal implanted GaAs. The annealing system with an array of halogen lamps, such as that shown in Fig. 20, was first developed by the same group for annealing implanted Si (Nishiyama et al., 1980). Cr-doped semi-insulating GaAs substrates, which exhibited pronounced thermal conversion to n-type conductivity due to Cr outdiffusion when heat-treated at 850°C for 15 min using conventional furnace annealing, were implanted with 3×10^{12} Si$^+$ cm^{-2} at 70 keV and then annealed under four different conditions: (1) 850°C for 15 min with a Si$_3$N$_4$ encapsulant in N$_2$, (2) 850°C for 10 min with no encapsulation in AsH$_3$, (3) 950°C for 5 sec with a Si$_3$N$_4$ encapsulant in N$_2$, and (4) 950°C for 5 sec with no encapsulation in N$_2$. Rapid thermal annealing corresponds to conditions (3) and (4). The resultant carrier concentration profiles are shown in Figs. 23a

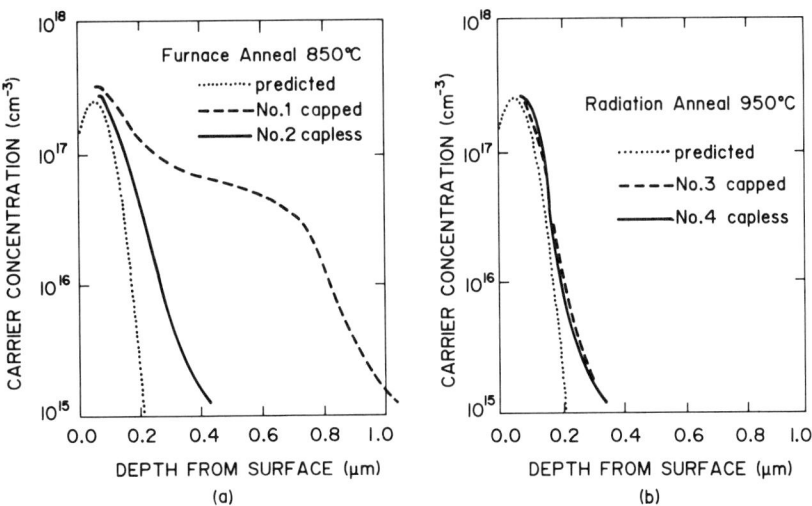

FIG. 23. Carrier concentration profiles for 70 keV, 3×10^{12} cm^{-2} Si-implanted Cr-doped GaAs following (a) furnace annealing at 850°C and (b) rapid thermal annealing at 950°C. A pronounced thermal conversion is observed after furnace annealing, while this effect is avoidable after rapid thermal annealing. [From Arai et al. (1981).]

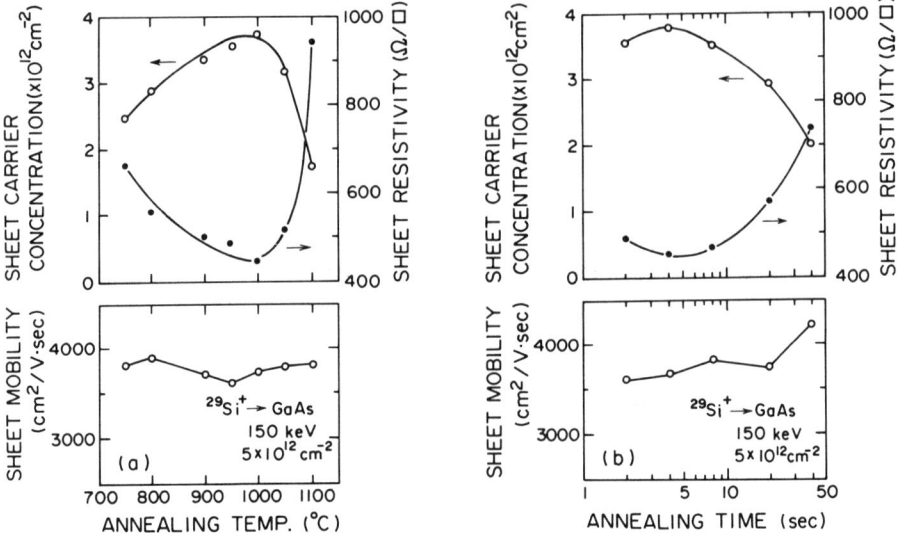

FIG. 24. Dependence of sheet carrier concentration (○, upper graph), sheet resistivity (●, upper graph), and sheet mobility (lower graph) on (a) annealing temperature and (b) annealing time for 150 keV, 5×10^{12} cm^{-2} Si implants in GaAs under unencapsulated conditions. [From Kuzukara et al. (1982).]

and 23b, in which the superiority of rapid thermal annealing in suppressing Cr outdiffusion is dramatically illustrated. Presumably, such serious thermal conversion will not occur on the currently available undoped semi-insulating liquid encapsulated Czochralski (LEC) substrates, since the residual Si donor concentration is extremely low ($<10^{15}$ cm^{-3}). For LEC crystals, rapid thermal annealing with be advantageous to suppress outdiffusion of residual acceptor impurities such as Mn, which sometimes produce p-type layers near the substrate surfaces (Hallais et al., 1976; Zucca, 1976; Klein et al., 1980).

Dependence of electrical properties for rapid thermal annealed Si implants on annealing temperature and time was reported by Kuzuhara et al. (1982). The results are shown in Fig. 24. For optimizing the activation of channel implants (150 keV, 5×10^{12} Si$^+$ cm^{-2}) under unencapsulated conditions, annealing temperature and time should be carefully controlled to ~950°C and within 2–4 sec, respectively. A similar abrupt decrease in activation observed at temperatures above the optimum, as shown in Fig. 24, has been reported for unencapsulated (Yuen et al., 1985) and Si$_3$N$_4$- or AlN-encapsulated (Tiku and Duncan, 1985) samples. Tiku and Duncan (1985) measured 4.2 K photoluminescence (PL) spectra for Si-implanted (180 keV, 1.25×10^{12} cm^{-2} + 50 keV, 0.35×10^{12} cm^{-2}) GaAs samples annealed at various temperatures for a 10 sec annealing time. Figure 25 shows the

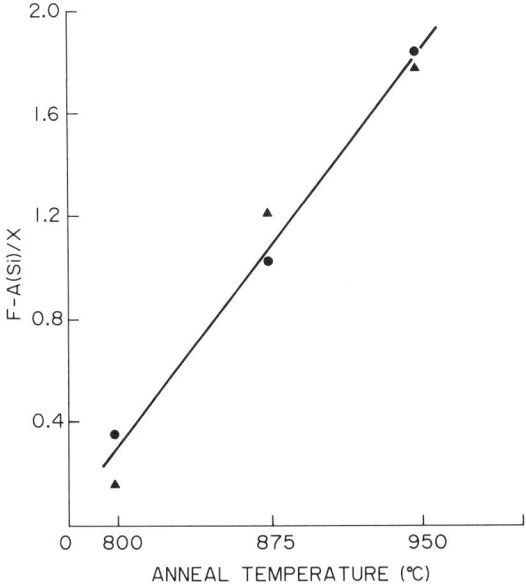

FIG. 25. Free-to-silicon acceptor [F–A(Si)] normalized to bound exciton [X] band intensity versus annealing temperatures for 10 sec. Samples were implanted with Si at 180 keV and 50 keV with doses of 1.25×10^{12} and 0.35×10^{12} cm^{-2}, respectively. [From Tiku and Duncan (1985).]

free-to-Si acceptor band [F–A(Si)] at 1.4843 eV intensity normalized by bound exciton band [X] at 1.515 eV intensity as a function of annealing temperature. The PL intensity, which related to the Si acceptor, showed a monotonic increase with increasing annealing temperature, indicating that the decrease in activation observed in the high-temperature region could be ascribed to self-compensation caused by the amphoteric behavior of Si dopants in GaAs. Kuzuhara *et al.* (1983, 1984) reached a similar conclusion by comparing annealing characteristics of a series of Si and S implants. Figure 26 compares the dependence of sheet carrier concentration on annealing temperature for Si and S implants, which were both implanted at 150 keV and subsequently annealed for 2 sec without encapsulation. Despite the occurrence of appreciable As loss at high temperatures (>1000°C), the activation of S implants never decreased with increasing annealing temperature up to 1100°C, in contrast with the temperature dependence observed for Si implants. These results demonstrate that the decrease in Si implant activation observed in the high-temperature region is predominantly associated with the self-compensation for Si dopants, rather than with surface dissociation of GaAs substrates.

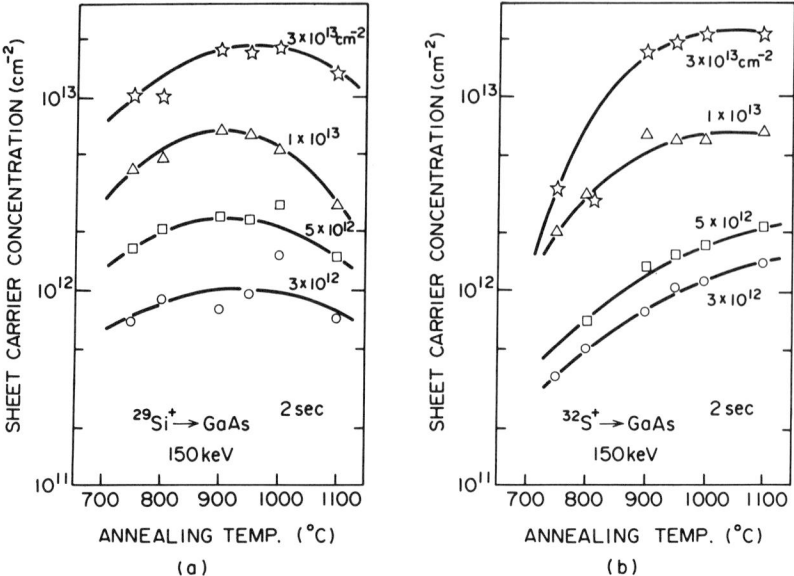

FIG. 26. Sheet carrier concentration versus annealing temperature curves for (a) Si implants (150 keV) and for (b) S (150 keV) implants, both activated under unencapsulated conditions. [From Kuzuhara et al. (1983, 1984).]

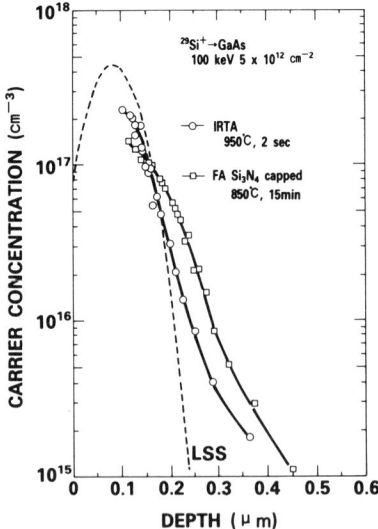

FIG. 27. Comparison of carrier concentration profiles following capless 950°C, 2 sec rapid thermal annealing (O) and Si_3N_4 capped 850°C, 15 min furnace annealing (□). Samples were implanted with 100 keV Si ions to a dose of 5×10^{12} cm^{-2}. [From Kohzu et al. (1983).]

It has been repeatedly reported that rapid thermal annealing yields sharp dopant profiles in channel implants (Kuzuhara *et al.*, 1982; Badawi and Mun, 1984; Rosenblatt *et al.*, 1984; Ezis *et al.*, 1984). Figure 27 shows the carrier concentration profile for 100 keV, 5×10^{12} cm^{-2} Si implants in GaAs followed by a capless 950°C, 2 sec rapid thermal annealing, together with the profile after an Si$_3$N$_4$-encapsulated 850°C, 15 min furnace annealing. Potential advantages of higher electrical activation and sharper dopant profile are clearly manifested in the profile after rapid thermal annealing. The sharp carrier concentration profiles obtained were, in general, explained by a minimized dopant diffusion. Kanber *et al.* (1985a, 1985b), however, attributed these sharp profiles to the difference in the lattice recovery process. Si-implanted LEC GaAs samples were annealed by rapid thermal (900°C, 10 sec, capless) and furnace (850°C, 30 min, capless) techniques and their Si atom concentration profiles were characterized by secondary ion mass spectrometry (SIMS). As shown in Fig. 28, no detectable Si atom

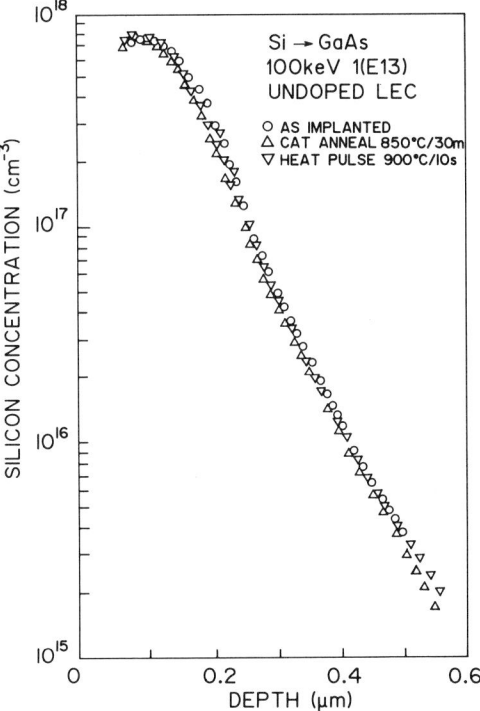

FIG. 28. Si atom concentration profiles measured by SIMS in undoped LEC GaAs after Si implantation (100 keV, 1×10^{13} cm^{-2}) (○), after capless furnace annealing (850°C, 30 min) (△), and after capless rapid thermal annealing (900°C, 10 sec) (▽). [From Kanber *et al.* (1985a).]

redistribution was observed after both annealing procedures. Thus, the observed difference in carrier concentration profiles was ascribed to the different extents of host atom damage removal, since the heating rate for rapid thermal annealing was faster by a factor of 100 compared to that for furnace annealing.

For Se implants, Barrett *et al.* (1984) used a graphite strip heater to anneal 350 keV, 5×10^{12} Se^{2+} cm^{-2} implants in GaAs. A Si$_3$N$_4$-encapsulated annealing at 900°C for 25 sec resulted in 74% electrical activation with about 4000 cm^2/V · sec Hall mobility. No significant Se redistribution was observed in the carrier concentration profile. Channel layer formation by rapid thermal annealing of Se implants was also reported by Sanders *et al.* (1980) and Matsumoto *et al.* (1984).

8. HIGH-DOSE n^+ IMPLANTS

Compared to the case for low-dose ($<10^{13}$ cm^{-2}) implants, it becomes rather difficult to obtain high electrical activations for high-dose ($>10^{14}$ cm^{-2}) Si implants. Achievable doping levels have been limited to below 2×10^{18} cm^{-2}, when samples are processed by conventional furnace anneal-

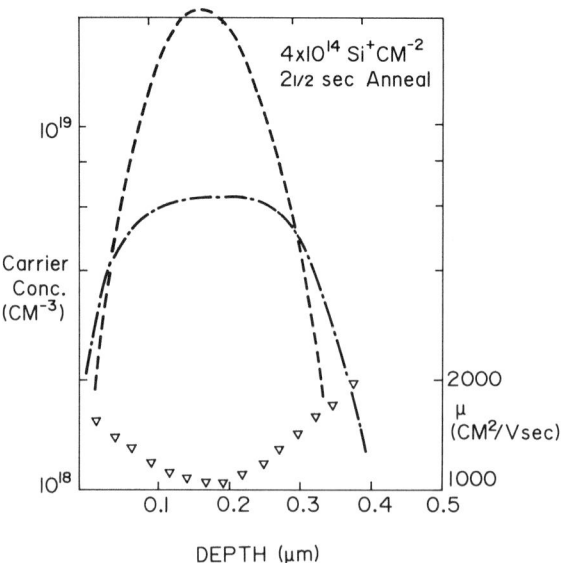

FIG. 29. Carrier concentration and Hall mobility profiles for 200 keV, 4×10^{14} cm^{-2} Si implants in GaAs after Si$_3$N$_4$ capped rapid thermal annealing at ~1000°C for a total irradiation time of 2.5 sec. [From Davies *et al.*, (1982).]

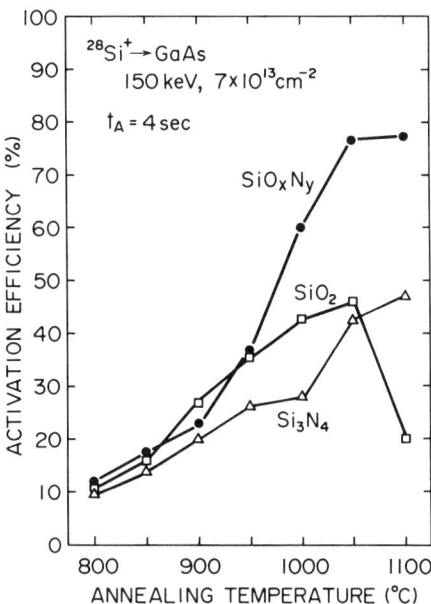

FIG. 30. Electrical activation dependence on annealing temperature for 150 keV, 7×10^{13} cm^{-2} Si implants in GaAs. Three kinds of encapsulants were examined: SiO$_2$ (□), Si$_3$N$_4$ (△), and SiO$_x$N$_y$ (●). All samples were annealed for 4 sec. [From Kuzuhara et al. (1985).]

ing (Tandon et al., 1979; Masuyama et al., 1980; Banwell et al., 1983). To achieve still higher doping levels, annealing temperatures higher than those employed in the furnace annealing are generally required. Davies et al. (1982) reported a peak electron concentration of 6.5×10^{18} cm^{-3} on 200 keV, 4×10^{14} cm^{-2} Si implants, following rapid thermal annealing at ~1000°C for a total lamp irradiation time of 2.5 sec. The corresponding electron concentration profile is shown in Fig. 29, where a CVD Si$_3$N$_4$ layer was used as an encapsulant. Similarly, Tabatabaie-Alavi et al. (1983a) obtained a doping level as high as 7.5×10^{18} cm^{-3} on 200 keV, 4×10^{14} cm^{-2} Si implants after 1160°C, 3 sec annealing using a sputter-deposited SiO$_2$ encapsulant.

The encapsulation effect in rapid thermal annealing was studied by Kuzuhara et al. (1985), in which high-dose Si implants were annealed using various encapsulants prepared by pyrolytic chemical vapor deposition. Figure 30 illustrates the electrical activation dependence on annealing temperature for 150 keV, 7×10^{13} cm^{-2} Si implants in GaAs. All samples were annealed for 4 sec at temperatures ranging from 800 to 1100°C using three kinds of encapsulants: SiO$_2$, Si$_3$N$_4$, and SiO$_x$N$_y$ with 1.75 refractive

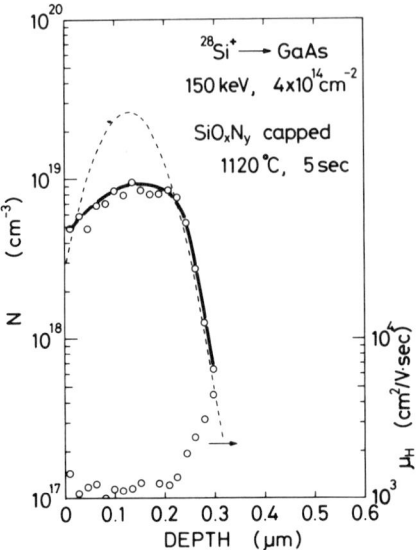

FIG. 31. Carrier concentration (solid line) and Hall mobility profiles for 150 keV, 4×10^{14} cm^{-2} Si implants in GaAs after SiO$_x$N$_y$ (RI = 1.75) capped rapid thermal annealing at 1120°C for 5 sec. [From Kuzuhara et al. (1985).]

index. For each encapsulant, the electrical activation increases with increasing annealing temperature. In particular, the improvement in activation is conspicuous for SiO$_x$N$_y$ capped annealing—that is, the electrical activation is doubled between 950 and 1050°C. Figure 31 shows the depth profile of carrier concentration for 150 keV, 4×10^{14} cm^{-2} Si implants following SiO$_x$N$_y$ capped rapid thermal annealing at 1120°C for 5 sec. A peak electron concentration of 9×10^{18} cm^{-3} is achieved with negligible dopant diffusion.

High-dose Se implants have also been studied in an attempt to achieve high doping levels in GaAs. Chapman et al. (1982) studied an 18% electrical activation and a 25 $\Omega/\square$ sheet resistivity on samples implanted at 300°C with a 1×10^{15} cm^{-2} dose of 400 keV Se$^+$ ions, which were annealed at 1140°C for 10 sec using a graphite strip heater. Barrett et al. (1984) studied carrier concentration profiles for room temperature implanted, 1×10^{14} cm^{-2}, 100–400 keV Se implants, which were annealed with an AlN encapsulant at 1000°C for 2-sec using a graphite strip heater. As shown in Fig. 32, a peak electron concentration of $5-6 \times 10^{18}$ cm^{-3} with about 80% electrical activation is achieved for 400 keV Se implants. The activation energy, which was determined from a plot of sheet carrier concentration versus reciprocal annealing temperature, was calculated to be 1.2 ± 0.1 eV (Chapman et al.,

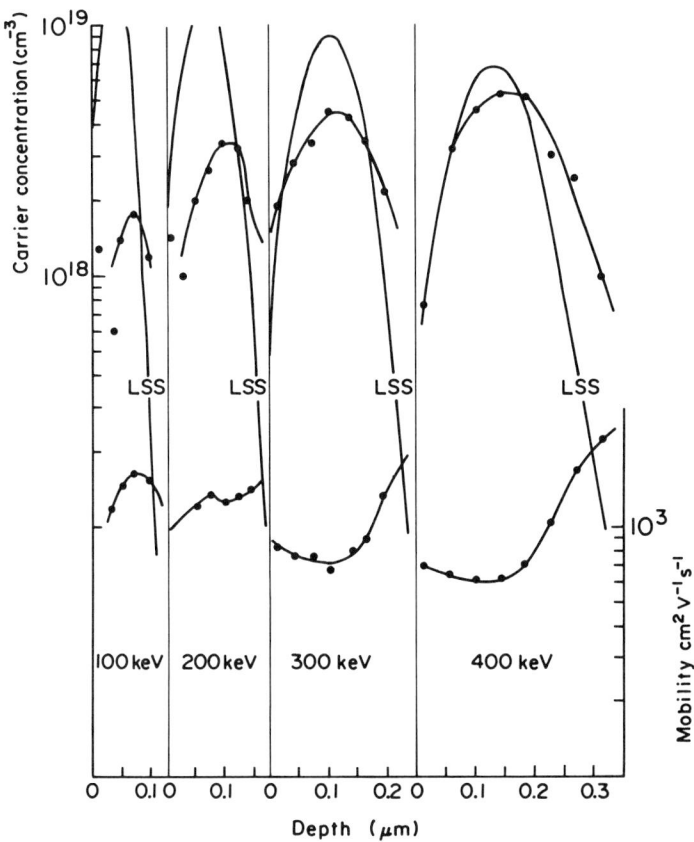

FIG. 32. Carrier concentration and Hall mobility profiles for 1×10^{14} cm^{-2}, 100–400 keV Se implants in GaAs after AlN capped, 1000°C, 2 sec rapid thermal annealing using a graphite strip heater. [From Barrett et al. (1984).]

1982; Barrett et al., 1984). This value is suggested to be associated with splitting up of selenium–gallium vacancy complexes.

Rapid thermal annealing of high-dose S implants in GaAs was reported by Kuzuhara et al. (1983). Room temperature, 150 keV S-implanted GaAs samples were caplessly annealed using incoherent light sources. Electrical depth profiles were measured for S doses of 3×10^{13}, 5×10^{13}, and 1×10^{14} cm^{-2}. Doses less than 3×10^{13} cm^{-2} were activated without causing broadening in carrier concentration profile. However, both in- and out-diffusion became evident for higher doses ($>5 \times 10^{13}$ cm^{-2}). Figure 33 shows a comparison between carrier concentration profiles for 1×10^{14} cm^{-2} S implants following 1000°C, 2 sec rapid thermal annealing and Si$_3$N$_4$

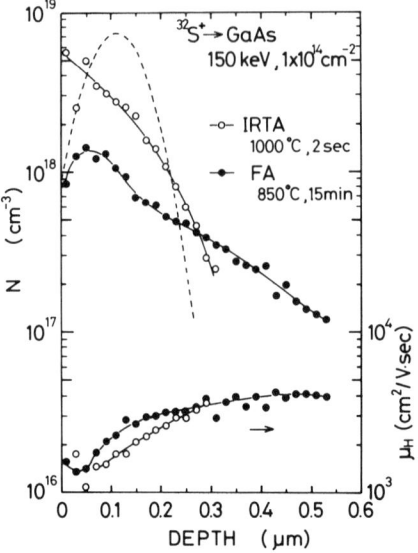

FIG. 33. Comparison of carrier concentration profiles for 150 keV, 1×10^{14} cm^{-2} S implants in GaAs after capless rapid thermal annealing at 1000°C for 2 sec (O) and after Si$_3$N$_4$ capped furnace annealing at 850°C for 15 min (●). [From Kuzuhara et al. (1983).]

capped, 850°C, 15 min furnace annealing. An appreciable S diffusion is observed for both annealing methods but is less significant for rapid thermal annealing, resulting in a steeper carrier concentration profile. The enhanced activation observed near the surface was ascribed to the damage-enhanced outdiffusion of implanted S ions. A peak electron concentration of 5–6 × 10^{18} cm^{-3} and a lowest sheet resistivity of 54 Ω/□ were obtained on the rapid thermal annealed sample.

Bensalem et al. (1983) reported rapid thermal annealing of room temperature, 300 keV, 5×10^{14} cm^{-2} Sn implants in GaAs. Samples were encapsulated with AlN and subsequently annealed at 1000°C using a graphite strip heater. Figure 34 shows the resultant carrier concentration profile for Sn implants, where a peak electron concentration of 5×10^{18} cm^{-3} is obtained despite the profile broadening due to Sn diffusion during annealing.

A potential advantage of heavily doped layers formed by rapid thermal annealing is the thermal stability of doping levels against any subsequent heat treatment. This is in contrast with the lack of thermal stability observed on heavily doped layers activated by pulse annealing (Amano et al., 1980; Pianetta et al., 1980b, 1981). Figure 35 illustrates the changes in sheet electron concentration with subsequent furnace heat treatment for 10 min

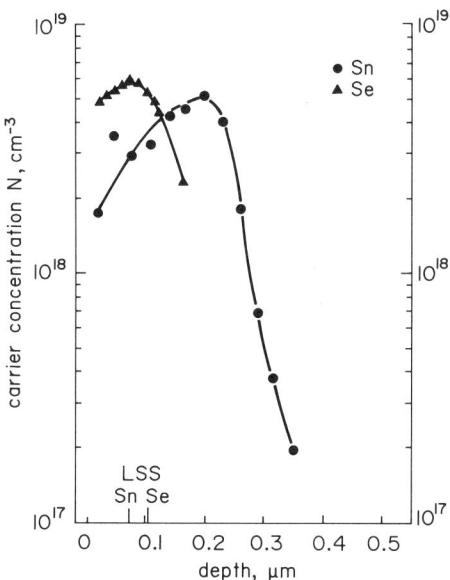

FIG. 34. Carrier concentration profile for 300 keV, 5×10^{14} cm^{-2} Sn implants in GaAs after AlN capped, 1000°C rapid thermal annealing (●). Results for Se implants (300 keV, 5×10^{14} cm^{-2}) are also shown (▲). [From Bensalem et al. (1983).]

(Davies et al., 1983). Rapid thermal annealed samples (1.5×10^{14} Si$^+$ cm^{-2} and 4×10^{14} Si$^+$ cm^{-2}) preserve their initial doping levels at least up to ~600°C. On the other hand, conduction loss is evident for the heavily doped layer formed by diffusing Se with a pulsed electron beam, as indicated by the broken-line curve (Davies et al., 1981). This thermal stability is of great importance in the device application, since most devices are exposed to heat treatments, such as ohmic alloying or CVD film deposition, after implant annealing.

An attractive application of heavily doped layers is nonalloyed ohmic contact formation. Nonalloyed ohmic contacts have been formed on heavily doped n-type layers prepared by rapid thermal annealing (Kuzuhara et al., 1985). In order to increase the electron concentration near the surface, samples were subjected to a dual energy Si implantation (150 keV, 4×10^{14} cm^{-2} + 50 keV, 1.2×10^{14} cm^{-2}) and then annealed at 1150°C for 5 sec with an SiO$_x$N$_y$ encapsulant. Ohmic contacts were formed by depositing AuGe–Ni followed by Ti–Au as overlayers. During metallization, sample temperatures were kept at less than 85°C to avoid a self-alloying effect. Figure 36 shows the I–V characteristics for the nonalloyed ohmic contacts measured between 10-μm-spaced, 200-μm-wide ohmic pads. Almost

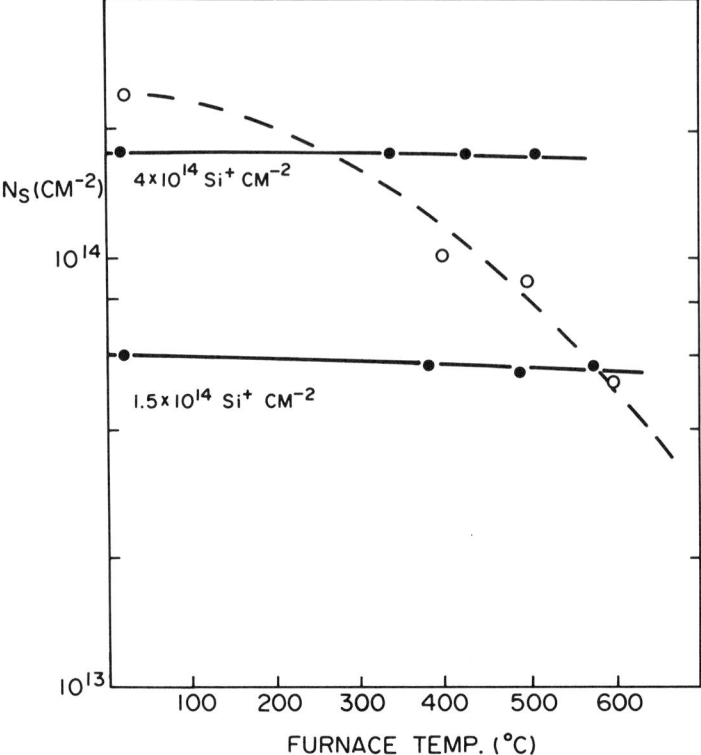

FIG. 35. Changes in sheet electron concentration for rapid thermal annealed n^+ layers as a function of subsequent furnace heat treatment for 10 min, indicating thermal stability at least up to ~600°C. Conduction loss typical of pulse annealed layers is shown by a broken curve. [From Davies et al. (1983).]

linear characteristics were obtained with a specific contact resistance of $9 \times 10^{-5}\ \Omega\ cm^{-2}$. This value corresponds to a surface electron concentration of $1 \times 10^{19}\ cm^{-3}$ when a 0.7 eV barrier height is assumed, in the theoretical calculation by Chang et al. (1971). The specific contact resistance was significantly improved to a value as low as $6 \times 10^{-6}\ \Omega\ cm^2$, when the sample was alloyed at 300°C. The low-temperature (300°C) alloying resulted in no difference in surface morphology compared with that of a nonalloyed sample, while the 420°C alloyed contact exhibited a nonuniform morphology due to the regrowth of heterogeneous structures (see Fig. 37). Thus, the low-temperature alloying technique is considered, from a practical viewpoint, as a promising technique to achieve a low contact resistance without degrading the smooth surface morphology of the contact metal.

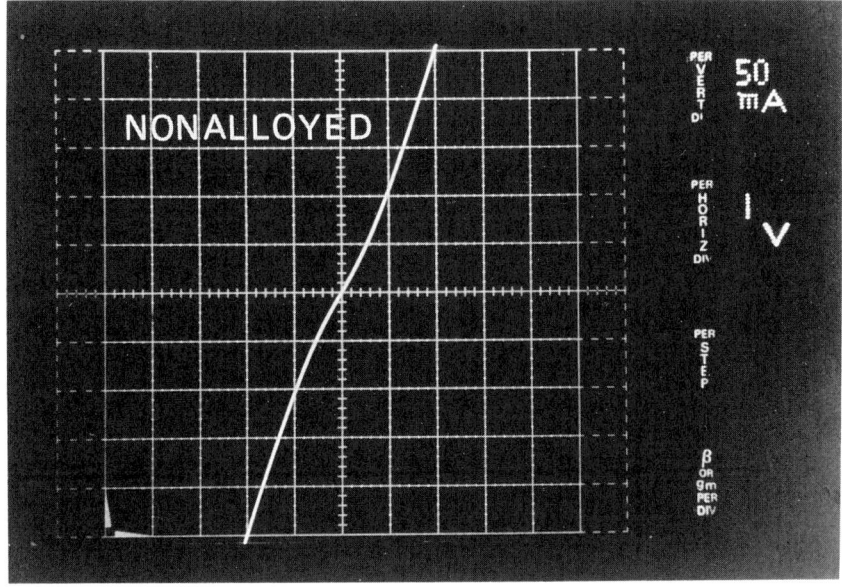

FIG. 36. I-V characteristics for nonalloyed ohmic contact formed on a dual energy Si-implanted (150 keV, 4×10^{14} cm^{-2} + 50 keV, 1.2×10^{14} cm^{-2}) GaAs annealed at 1150°C for 5 sec with an SiO$_x$N$_y$ (RI = 1.75) encapsulant. Measurements were made between 10-μm-spaced, 200-μm-wide AuGe-Ni contact pads. [From Kuzuhara et al. (1985).]

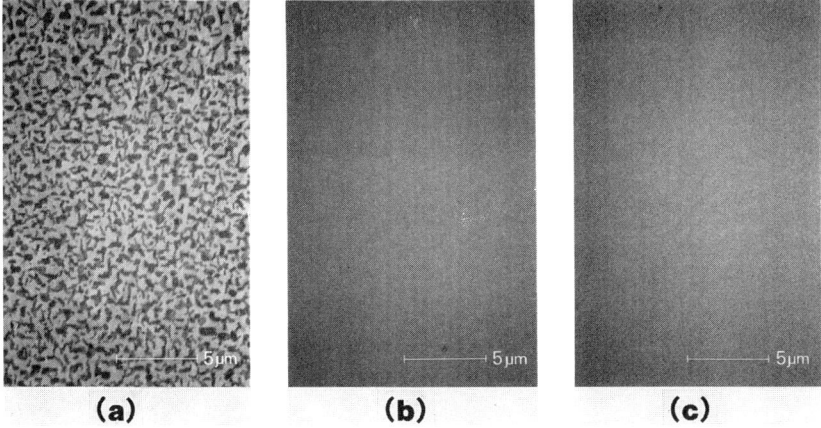

FIG. 37. Scanning electron micrographs of AuGe-Ni surface morphologies. 420°C alloyed contact (a) shows a nonuniform morphology due to regrown heterogeneous structures, while 300°C alloyed contact (b) shows as smooth a morphology as that for the nonalloyed contact (c). [From Kuzuhara et al. (1985).]

9. p-Type Implants

In the p-type layer formation by ion implantation, Be, Mg, Zn, and Cd usually have been used as an acceptor impurity. These p-type dopants usually redistribute during post-implantation annealing, especially at relatively higher doses. In the device application, the designed profile should be obtained with higher electrical activation in a controlled manner. Rapid thermal annealing is an effective method to suppress dopant redistribution, because of the short duration of annealing time. Since many factors are involved in dopant redistribution and electrical activation, optimization of these factors is necessary on the basis of mechanisms involved in dopant redistribution and electrical activation. In this section, the state of the art in the formation of ion-implanted p-type layers using rapid thermal annealing is presented, including several results obtained by furnace annealing.

a. Zn Implants

Tiwari et al. (1985) examined implanted Zn redistribution using SIMS profile measurement. Following Zn implantation at 200 keV with doses of 1×10^{13}, 1×10^{14}, and 1×10^{15} cm^{-2}, rapid thermal annealing was carried out at temperatures ranging from 760 to 1000°C for 3–6 sec. Two banks of infrared quartz lamps were used on either side of the wafer. The samples were placed with the implanted side on a Si wafer. For a 1×10^{13} cm^{-2} dose, the measured SIMS profile showed no measurable diffusion up to an 1060°C annealing temperature. For a 1×10^{14} cm^{-2} dose, a slight impurity profile broadening was observed at higher annealing temperatures (>950°C). However, in the case of a 1×10^{15} cm^{-2} dose, significant diffusion was observed both at maximum concentrations and at the tail end. The most pronounced effect is redistribution at the peak, which accumulates at the surface and moves rapidly as a function of temperature into the bulk. It was concluded that Zn accumulation at the surface is due either to point defects generated during implantation or to amorphous–single crystal material transition during annealing. Zn diffusion at the tail end was postulated to be caused by an interaction with point defects.

In the case of conventional furnace annealing for high-dose implants, deep penetration of Zn is well known (Zolch et al., 1977; Inada et al., 1979). It was reported that carrier concentration profiles extended to as deep as 1 μm beyond the anticipated 0.2 μm depth.

Even in the case of high-dose Zn implants, Davies and McNally (1983) reported minimized redistribution with annealing time reduced to ~1 sec. The samples were implanted at 200 keV with 1×10^{15} Zn$^+$ cm^{-2}. Annealing was performed with a system having two 1000 W quartz halogen lamps using a 1000-Å-thick Si$_3$N$_4$ encapsulant. The results are shown in Fig. 38, where

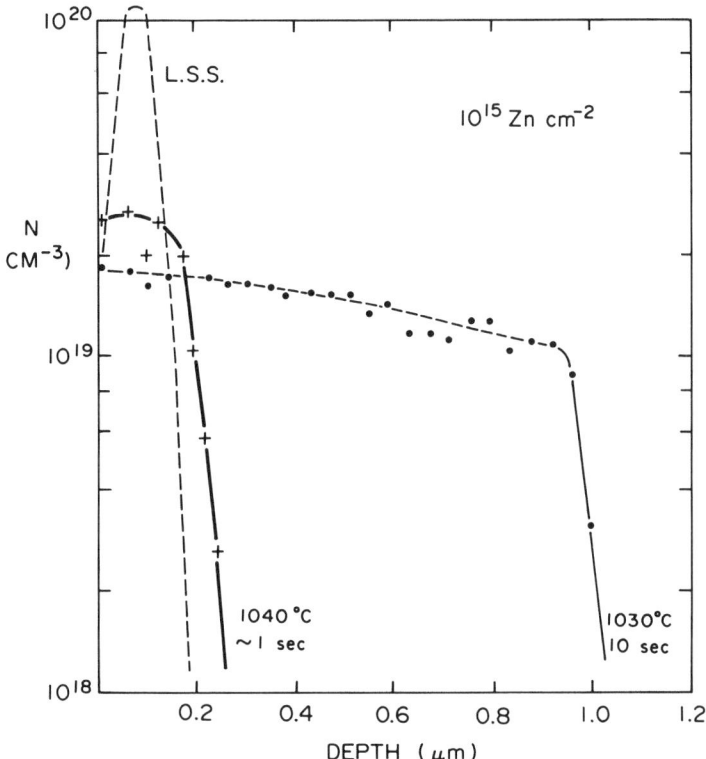

FIG. 38. Distribution of 200 keV, 1×10^{15} cm^{-2} Zn implants in GaAs, showing (●) a pronounced redistribution after 10 sec (1030°C) and (+) its prevention on annealing for only ~1 sec (1030°C). [From Davies and McNally (1983).]

marked improvement in carrier concentration profile was achieved for ~1 sec annealing time. Electron beam annealing was also experimentally confirmed to be effective to suppress enhanced diffusion (Barrett et al., 1985). A high percentage electrical activation with a hole concentration of 7–8 × 10^{19} cm^{-3} was obtained without significant indiffusion for 60 keV, 1 × 10^{15} cm^{-2} Zn implants. This was achieved by irradiating for a total time of 3 sec at a power density of 50 and 60 W cm^{-2}, which corresponds to a calculated peak temperature (Shahid et al., 1983) of about 1040 and 1120°C, respectively.

An interesting phenomenon was reported by Kasahara et al. (1982) about the redistribution of implanted Zn. Zn piled up at around the depth of $R_p + 1.5\Delta R_p$. Figure 39 shows atomic concentration profiles obtained by SIMS as a function of implantation energies of 50, 100, and 150 keV

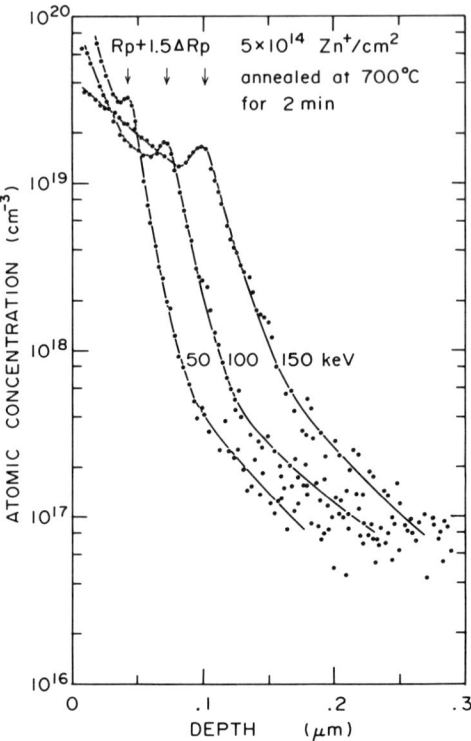

FIG. 39. Redistributed Zn atom profiles by SIMS for 5×10^{14} cm^{-2}, 50–150 keV Zn implants in GaAs after annealing at 700°C for 2 min. Arrows indicate the depth at $R_p + 1.5\Delta R_p$, corresponding to ion energies. [From Kasahara et al. (1982).]

after annealing at 700°C for 2 min. The profiles were not Gaussian and had a hump. Depths at the peak of the pileup were 420, 710, and 990 Å for 50, 100, and 150 keV implantations, respectively. Similar results for 5×10^{14} Zn cm^{-2} implants with 70 keV were also reported in a paper by Liu and Narayan (1984). The depth at the peak of the pileup, which was close to $R_p + 1.5\Delta R_p$, was found to be almost the same as that for the amorphous layer thickness, which was obtained by Rutherford backscattering (RBS) analysis. Thus, it was concluded that the pileup of implanted Zn atoms occurs near the interface between the single crystal and the amorphous layer induced by high-dose implantation. Hole concentration profiles agreed well with atomic concentration profiles for 700°C annealed samples. However, at higher annealing temperatures of 800 and 850°C, discrepancies between the atomic and electrical profiles were observed. The discrepancy increased with increasing annealing temperature, suggesting the formation of complexes of Zn with host atoms and/or vacancies.

It is well-known that Zn redistribution and electrical activation during annealing are greatly affected by the difference in annealing method, such as capped or capless annealing, kinds of encapsulants, encapsulant thickness and arsenic overpressure existence in capless annealing. Kasahara and Watanabe (1981b) examined the differences in carrier concentration profiles using capped and capless annealing, both of which were carried out with and without AsH_3/H_2 ambient. The samples were implanted at 150 keV with a dose of 1×10^{15} cm^{-2}. Conventional furnace annealing at 850°C for 2 min was carried out using a 3000-Å-thick Si_3N_4 film as an encapsulant. From the carrier concentration profiles shown in Fig. 40, two characteristic results were pointed out: (1) Zn diffusion during post-implantation annealing is enhanced by the capping, and (2) the arsenic pressure in the annealing ambient suppresses the enhanced diffusion of Zn in the implanted GaAs. It

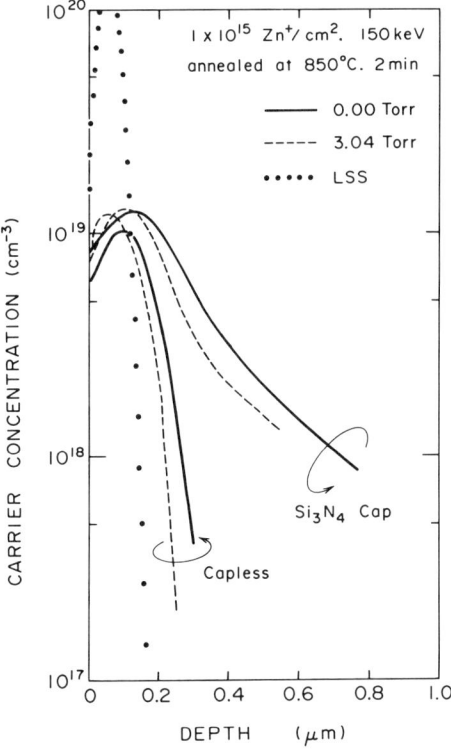

FIG. 40. Comparison between carrier concentration profiles obtained in capped and capless annealing at 850°C for 2 min for samples implanted at 150 keV with 1×10^{15} Zn cm^{-2}. Solid lines: profiles after annealing with 0.00 torr arsenic pressure; dashed lines: profiles after annealing with arsenic pressure (3.04 torr); dotted lines: LSS. [From Kasahara *et al*. (1981).]

was considered that the thermal stress between the capping film and the GaAs leads to enhancement of diffusion in implanted GaAs. In the case of capless annealing in an arsenic ambient, the decrease in Zn diffusion coefficient due to the decrease in As-vacancy concentrations resulting from arsenic overpressure (Fujimoto et al., 1967), as well as the stress-free effect in capless annealing, was considered to be responsible for the suppression of Zn enhanced diffusion.

Similar results regarding capping and As overpressure effect on Zn redistribution in graphite strip heating were reported by Barrett et al. (1985). The origin of carrier concentration broadening in the Si_3N_4 capped annealing was pointed out as being the release of Zn from damage complexes, which are usually formed in high-dose Zn-implanted GaAs, due to the strain at the encapsulant–GaAs interface. It is generally believed that Zn diffuses in GaAs via the substitutional–interstitial mechanism. A Zn atom diffuses interstitially until it reacts with a Ga vacancy:

$$Zn_i + V_{Ga} \rightarrow Zn_{Ga}$$

The Zn atom on the Ga lattice site acts as a single acceptor and is essentially immobile. Zn incorporation on Ga lattice sites is greatly influenced by the defect formation, which is affected in the following reactions (Hurle, 1979):

$$As_{As} + V_i \leftrightarrows As_i + V_{As}; \quad [As_i][V_{As}] = K_{fa}$$

$$\tfrac{1}{2}As_2(g) + V_i \leftrightarrows As_i; \quad [As_i]P_{As_2}^{-1/2} = K_{As_2^i}$$

$$Ga_{Ga} + V_i \leftrightarrows Ga_i + V_{Ga}; \quad [Ga_i][V_{Ga}] = K_{fg}$$

$$V_{Ga} + V_{As} \leftrightarrows 0; \quad [V_{Ga}][V_{As}] = K_s$$

where As_{As}, V_i, As_i, and V_{As} represent an arsenic atom on an arsenic site, a vacant interstitial site, an arsenic atom occupying an interstitial site and a vacant arsenic site, respectively. A similar representation is also applied for Ga atoms. Square brackets denote concentrations of the species and K represents a mass-action constant. From these equations, the following considerations were made. If the As pressure increases, $[As_i]$ increases and hence $[V_{As}]$ decreases. This leads to an increase in $[V_{Ga}]$, which retards the diffusion of Zn_i, giving rise to a sharper, peaked profile.

In the case of annealing with a SiO_2 encapsulant, increased activation with higher peak carrier concentration, as compared to the Si_3N_4 encapsulation, was reported for 1×10^{15} Zn cm^{-2} implants annealed at 1060°C for ~1 sec (Davies and McNally, 1984). This originates from Ga vacancy formation, which is favorable for Zn activation, caused by loss of Ga into SiO_2.

Dual implants of Zn and As are also effective to suppress impurity profile broadening and to achieve higher peak carrier concentration (Davies and

1. ACTIVE LAYER FORMATION BY ION IMPLANTATION

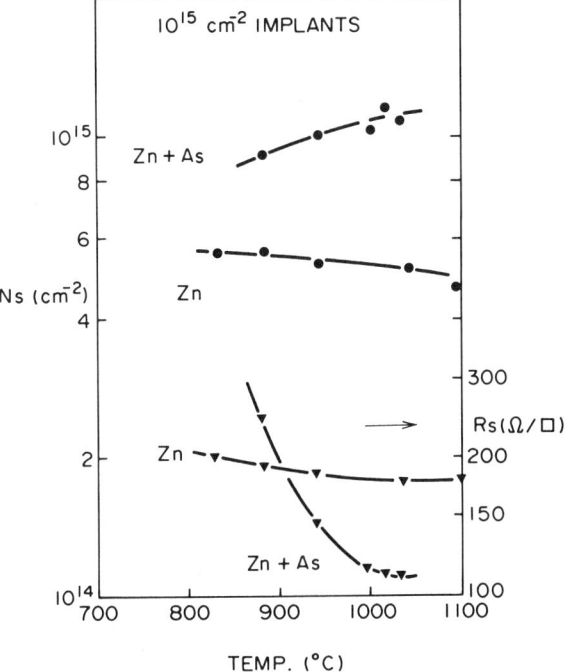

FIG. 41. Activation of 200 keV, 1×10^{15} cm^{-2} Zn implants in GaAs, when implanted alone and with an equal dose implantation of 220 keV As. [From Davies and McNally (1984).]

McNally, 1984). After an initial 1×10^{15} Zn cm^{-2} implant, part of the wafer was reimplanted with the same dose of As. Implant energies were 200 keV for Zn and 220 keV for As. Chemical vapor deposited Si$_3$N$_4$ films were used as encapsulants. Elliptical mirrors focus the output onto the GaAs, and illumination times of 1.5–2.5 sec typically provide annealing temperatures of >1000°C. As is clearly shown in Fig. 41, increased activation is found for all of the annealing temperatures in the dual implant samples. Figure 42 shows carrier concentration profiles for single and dual implants, where a 7.7×10^{19} cm^{-2} peak carrier concentration, which is an increase by a factor of 3 compared to that of a single implant, was obtained for the dual implant. It was also noted that the increased activation for dual implants is not brought about by extending the doping depth. Similar results were reported by Barrett *et al.* (1985). It is considered that the implantation of As produces Ga vacancies, which is favorable for Zn activation without profile broadening.

The dependence of electrical activation on heating rate was studied by Suzuki *et al.* (1983). Ion energy and dose were 150 keV and 5×10^{14} cm^{-2},

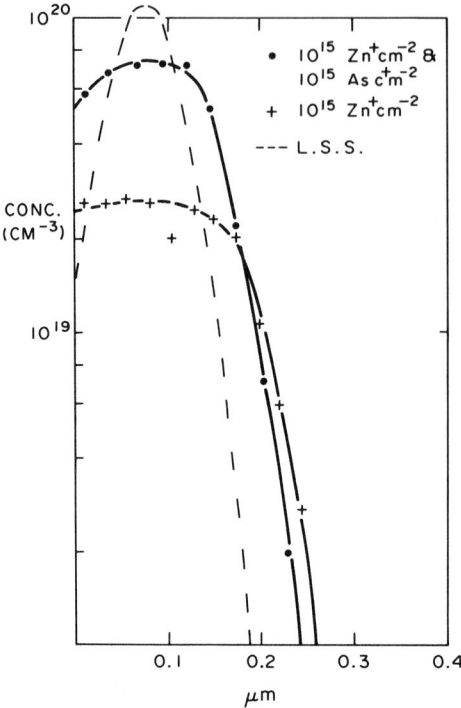

FIG. 42. Carrier concentration profiles for 200 keV, 1×10^{15} cm^{-2} Zn implants in GaAs showing an increase in peak concentrations with dual implants. With (●) and without (+) reimplantation with 1×10^{15} cm^{-2} As. [From Davies and McNally (1984).]

respectively. Annealing was carried out setting the samples face-to-face on a Si wafer in a N_2 gas ambient. Figure 43 shows activation efficiency dependence on annealing temperature for two heating rates with a hold time of 0 sec. As compared to the 20°C/sec heating rate, higher activation efficiency was obtained for the 100°C/sec heating rate over the entire annealing temperature range; the highest activation efficiency of over 90% was exhibited at 800°C. Atomic Zn profiles obtained by SIMS analysis and carrier profiles for samples annealed at 800°C with 100°C/sec heating rate exhibited double peaks, similar to the results shown in Fig. 39, indicating the redistribution of Zn during annealing. However, profile broadening at the tail region was not observed. In the samples annealed with the 20°C/sec heating rate, an inferior degree of implantation-induced damage annihilation at annealing temperatures from 600 to 800°C was observed from RBS measurements, as compared to the samples annealed with the 100°C/sec heating rate.

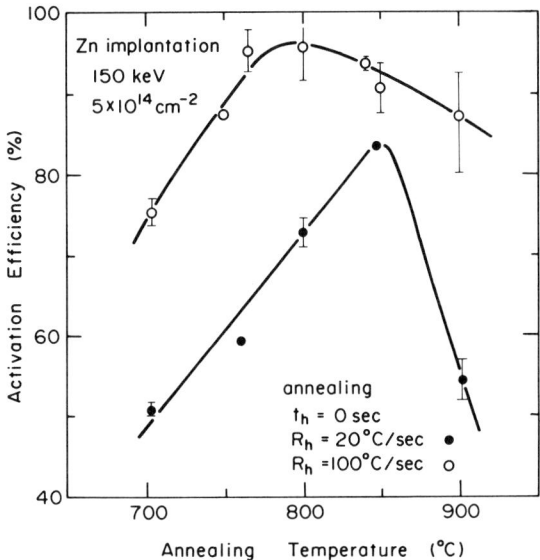

FIG. 43. Activation efficiency dependence on annealing temperature for 150 keV, 5×10^{14} cm^{-2} Zn implants in GaAs after rapid thermal annealing with different heating rates: (●) 20°C/sec; (○) 100°C/sec. Hold time: 0 sec. [From Suzuki et al. (1983).]

As described above, many factors are involved in Zn redistribution and electrical activation during post-implantation annealing. Thus, optimization of these factors should progress with each device application.

b. Mg *Implants*

Noticeable carrier concentration profile broadening was reported for furnace-annealed Mg-implanted GaAs (Choe et al., 1980). As is the case for Zn implants, rapid thermal annealing is effective to suppress carrier concentration profile broadening.

Kuzuhara et al. (1984) reported the electrical characteristics for samples implanted at 80 keV with a dose of 2×10^{14} cm^{-2}. Rapid thermal annealing at 700, 800, and 900°C for 2 sec was carried out in a N_2 gas ambient without encapsulation. Figure 44 shows the carrier concentration profiles. Even in the case of 900°C annealing, only a slight Mg diffusion was observed at the tail region. Peak carrier concentrations increased with increasing annealing temperature. Sheet carrier concentrations also increased with increasing annealing temperature, and 66% activation efficiency was achieved at 900°C annealing. In the sample annealed at 900°C, a double peak, which is similar to the Zn-implant case shown in Fig. 39, was observed in the carrier

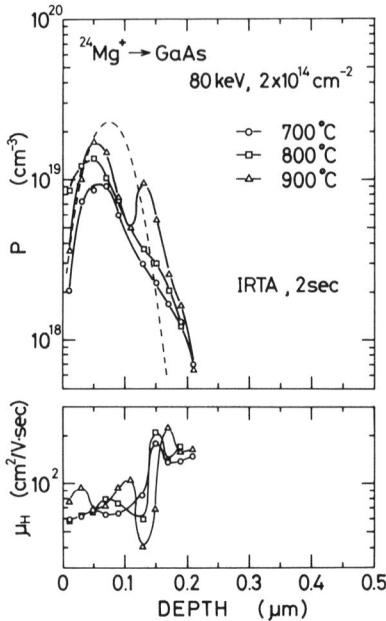

FIG. 44. Carrier concentration profiles for 80 keV, 2×10^{14} cm^{-2} Mg implants in GaAs after rapid thermal annealing at (○) 700, (□) 800, and (△) 900°C for 2 sec. [From Kuzuhara et al. (1984).]

concentration profile. Using SIMS profile measurements, Tiwari et al. (1985) examined profile broadening for samples implanted at 200 keV with doses of 1×10^{14} and 1×10^{15} cm^{-2}. Rapid thermal annealing without encapsulation was carried out at 770–1060°C for 3–6 sec. For a 1×10^{14} cm^{-2} dose, Mg showed no noticeable diffusion up to 1060°C. However, for a 1×10^{15} cm^{-2} dose, considerable anomalous diffusion is observed, as in the case of Zn implants. The highest electrical activation was obtained at around 900°C, exhibiting ~70% and ~30% for doses of 1×10^{14} and 1×10^{15} cm^{-2}, respectively.

Blunt et al. (1984) reported electrical characteristics on samples into which Mg was implanted through an Si$_3$N$_4$ film with a dose of 1×10^{14} cm^{-2} at 150 keV and which were annealed at temperatures ranging from 700 to 900°C for 5 sec by rapid thermal annealing. The highest activation efficiency of 70% was obtained at 800°C with a sharp carrier-concentration profile. In contrast, the 900°C annealing resulted in considerable profile-broadening due to diffusion. Patel and Sealy (1985) examined a Si$_3$N$_4$ capped sample in rapid thermal annealing. They agreed with Blunt's results that the optimum annealing condition for 100 keV, 1×10^{14} cm^{-2} Mg implants occurs at

800°C. However, for doses greater than 1×10^{14} cm^{-2}, profile broadening accompanied with peak hole-concentration decrease was observed with increasing annealing temperature even though the sheet resistivity decreased.

c. *Be Implants*

Tabatabai-Alavi *et al.* (1983b) examined Be implant activation using an argon arc lamp. Double implants of 4.4×10^{14} cm^{-2} at 40 keV plus 5.1×10^{14} cm^{-2} at 150 keV were annealed from 950 to 1160°C using an SiO$_2$ cap. Significant damage-enhanced diffusion was observed for the sample annealed at 800–900°C for 10–15 min using a conventional furnace. On the other hand, no Be diffusion took place upon rapid thermal annealing, and a high peak concentration was maintained. The best results were obtained for a 950°C/10 sec anneal, exhibiting 72% electrical activation with a peak carrier concentration of about 2×10^{19} cm^{-3}.

Liu and Narayan (1984) also reported a carrier concentration profile free from Be diffusion. Pearton *et al.* (1985) reported very little Be redistribution for 1×10^{14} and 1×10^{15} cm^{-2} Be implants annealed at 900–950°C for 1–3 sec.

Barrett *et al.* (1984) compared electrical characteristics for samples annealed with and without an Si$_3$N$_4$ encapsulant. Be was implanted at 40 keV with a dose of 5×10^{14} cm^{-2}. Annealing was carried out at 900°C for 0–5 sec. Almost the same electrical profiles were obtained for both samples, indicating that both activation and diffusion of Be are not affected by surface strain generated at the Si$_3$N$_4$/GaAs interface.

Yuen *et al.* (1985) examined the dependence of sheet carrier concentration on annealing temperature for Be-implanted GaAs with a dose of 1×10^{14} cm^{-2}. At annealing temperatures higher than 850°C, activation efficiency exceeded 85%.

Activation of implanted Be after low-temperature annealing was reported by Vaidyanathan and Dunlap (1984). Be was implanted at 200 keV with doses of 1×10^{13} and 1×10^{14} cm^{-2}. Rapid thermal annealing was carried out at 550°C with an SiO$_2$ encapsulant. For both doses, activation efficiency increased with increasing annealing time and reached ~60% after 20–30 sec.

10. MINIMIZATION OF ANNEAL-INDUCED DEFECTS

A number of interesting experimental results for rapid thermal annealing (RTA) have been reported with regard to activating impurities, which are introduced into GaAs using an ion implantation technique. In addition to impurity activation, minimization of RTA-induced defect generation is essential in the application of RTA to device fabrication.

Because of the temperature nonuniformity over an entire wafer, which originates from rapid heating or cooling in the RTA process, slip line formation at the periphery of the wafer is well known. Blunt et al. (1985) examined the stage of the slip line formation in 2-in. GaAs substrates, annealed in an incoherent light furnace at 900°C for 5 sec. It was observed that the slip lines in Si_3N_4-encapsulated GaAs are rarely decorated with blisters or thermal etch pits, despite the fact that the slip must cause discontinuities in the Si_3N_4 film. If the slip occured during an earlier stage of the annealing, more evidence of blisters along the slip lines would have been expected. Therefore, it was concluded that most of the slip is generated during the cool-down stage of the annealing. Blunt et al. also reported that noticeable suppression of slip line formation is possible with the use of a stack of two annular guard rings of Si coated with Si_3N_4 film, since they are effective to reduce temperature gradient at the peripheral region of the annealed wafer. The use of a similar guard ring has already been examined in suppressing temperature gradients near the wafer periphery in Si device processing (Komatsu and Kajiyama, 1984).

Generation and/or annihilation of deep levels during the annealing process should be clarified and controlled in order to establish controllability and reproducibility in an active-layer formation. Since the rapid thermal annealing process includes rapid heating and cooling in the process cycle, there is a possibility that some deep levels are introduced in GaAs as a result of lattice defect formation or stoichiometry change during annealing. Using deep level transient spectroscopy (DLTS) measurements, Kuzuhara and Nozaki (1986) examined the behavior of deep levels in n-type bulk GaAs wafers, which were processed in rapid thermal annealing. It was found that a new electron trap (termed EN1), which has an activation energy of 0.20 eV from the conduction band edge, is introduced by capless rapid thermal annealing at temperatures greater than 800°C and disappears at 950°C annealing. The results are shown in Fig. 45, where EL2 concentration changes are also presented. The EN1 trap formation was closely related to the rapid heating stage in an RTA process and was never observed after conventional furnace annealing. EN1 showed a similar concentration change behavior to that of EL2 in various RTA conditions: (1) A sudden decrease in concentrations of both traps was observed for 950°C capless annealing (Fig. 45); (2) concentrations of both traps decreased in a similar manner with increasing annealing time for 850°C capless annealing; and (3) both traps exhibited similar concentration changes with changes in the refractive index of SiO_xN_y encapsulating films. From these results, it was proposed that the RTA-induced EN1 trap be ascribed to compositional defect complexes, including As_{Ga} antisite defects such as $V_{As}As_{Ga}$.

One of the potential advantages of rapid thermal annealing over

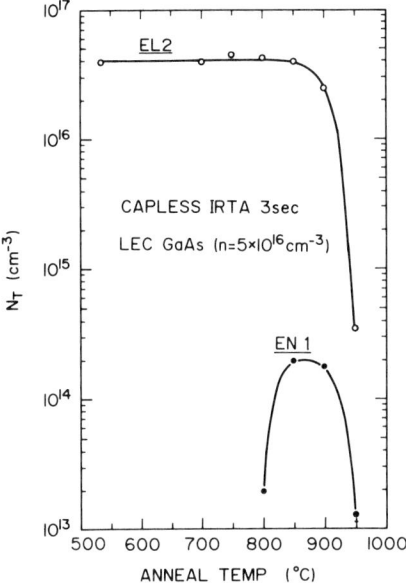

FIG. 45. Electron trap concentrations for (●) EN1 and (○) EL2 in n-GaAs as a function of annealing temperature. Annealing was carried out for 3 sec without encapsulation. [From Kuzuhara and Nozaki (1986).]

conventional furnace annealing is the capability for capless annealing. This is based on the concept that surface dissociation will be negligible due to the short annealing time. However, in order to obtain controllability and reproducibility in the RTA process, the extent of surface dissociation has to be quantitatively clarified and the corresponding device characteristics should be evaluated. Using neutron activation analysis (NAA), Rose et al. (1983) measured As and Ga losses during rapid thermal annealing. Three kinds of heating sources were investigated: pulsed ruby laser, incoherent light source, and vitreous carbon strip heater. In each case, As and Ga lost during the annealing process were deposited on quartz catcher slides located above and close to, but not touching, the GaAs samples. The NAA is based on the following neutron capture and decay equations:

$$^{75}\text{As} + n \rightarrow {}^{76}\text{As} \xrightarrow{26.3\,\text{h}} {}^{76}\text{Se}, 559\,\text{keV}\,\gamma\text{-ray}$$

$$^{71}\text{Ga} + n \rightarrow {}^{72}\text{Ga} \xrightarrow{14.1\,\text{h}} {}^{72}\text{Ge}, 834\,\text{keV}\,\gamma\text{-ray}$$

In addition to NAA, Rutherford backscattering analysis (RBS) was also employed. It was confirmed that the NAA results are consistent with RBS data. Good annealing of 100 keV, 3×10^{15} Te cm^{-2} implanted samples was obtained with laser pulse energies of ~0.9 J cm^{-2}, where NAA indicated Ga and As loss of 2 and 4 × 10^{15} cm^{-2}, respectively. Equivalent loss measured with solid phase annealed samples was an order of magnitude higher. The correlation between the degree of surface dissociation and impurity profile variation is not clear at this stage. However, in establishing controllable and reliable rapid thermal annealing process, suppression of surface dissociation will be necessary. In this respect, rapid thermal annealing under As overpressure (Liu and Narayan, 1984; Hiramoto et al., 1985) will be promising.

Since investigations of RTA-induced defects have just been started, enough data for minimization of RTA-induced defects are not yet available. In the future, the increase in device packing density will be advanced by the use of large-diameter wafers. In these situations, the restriction of defect generation by the RTA process will be strengthened. Therefore, detailed investigations, including theoretical considerations, should be continued for minimization of RTA-induced defects.

11. Device Application and Future Subjects

Rapid thermal annealing techniques have been used to produce high-quality, abrupt channel regions and low-resistivity, shallow source and drain regions for the fabrication of high-performance GaAs FETs. The controllability of the doping profile and the high doping levels that can be achieved are principal advantages in applying rapid thermal annealing to device fabrication.

The feasibility of rapid thermal annealing for use in FET channel formation was first demonstrated by Kuzuhara et al. (1982); the improved device performance was then reported by the same group (Kohzu et al., 1983). Channel regions were formed by implanting 100 keV Si ions into semi-insulating LEC GaAs substrates to a dose of 5 × 10^{12} cm^{-2}, followed by capless rapid thermal annealing at 950°C for 2 sec. Planar GaAs MESFETs with 1 µm gate length and 300 µm gate width were fabricated using the Al side-etching technique (Katano et al., 1981; Furutsuka et al., 1981). Figure 46a shows drain characteristics for an FET fabricated on a channel layer activated by rapid thermal annealing. For comparison, characteristics for an FET fabricated by Si$_3$N$_4$-encapsulated furnace annealing (850°C, 15 min) are shown in Fig. 46b. Both FET characteristics are summarized in Table III. Anomalous characteristics, such as low gate breakdown voltage and/or high leakage current, were not observed for the FET made by rapid thermal annealing. These facts indicate that capless rapid thermal annealing in a N$_2$ atmosphere does not affect GaAs surface conditions and is actually acceptable for GaAs device fabrication. By the use of rapid thermal annealing, transconductance was improved by 30–40% compared to that for a conventional FET. This higher transconductance is due to higher peak carrier

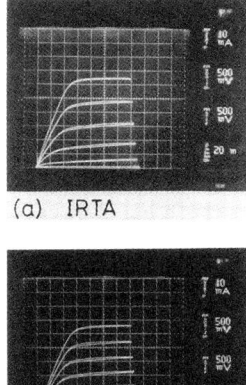

FIG. 46. Comparison between drain characteristics for GaAs MESFETs fabricated on channel regions (100 keV, 5×10^{12} Si$^+$ cm^{-2}) activated by (a) capless rapid thermal annealing (950°C, 2 sec) and by (b) Si$_3$N$_4$ capped furnace annealing (850°C, 15 min). [From Kohzu et al. (1983).]

concentration with a steeper carrier concentration profile, which is the principal feature of rapid thermal annealing (see Fig. 27).

Badawi and Mun (1984) reported a similar enhancement in transconductance of MESFETs, which were fabricated by Si-implantation at 80 keV with a 5×10^{12} cm^{-2} dose and capless-annealed at 900°C for 2 sec. They measured the threshold voltage uniformity over a 2-in. GaAs wafer. Despite the higher transconductance, the spread in threshold voltage was rather large ($\sim 20\%$) compared to the spread for furnace-annealed FETs ($\sim 10\%$).

TABLE III

COMPARISON OF GaAs MESFET CHARACTERISTICS FABRICATED BY RAPID THERMAL ANNEALING AND FURNACE ANNEALING

	MESFET		
Item	RTA	Furnace anneal	Conditions
Drain saturation current (mA)	64.0	65.5	$V_{ds} = 3$ V, $V_{gs} = 0$ V
Transconductance (mS)	34	24	$V_{ds} = 3$ V, $V_{gs} = 0$ V
Threshold voltage (V)	2.7	3.6	$V_{ds} = 3$ V, $I_{d}s = 100\,\mu$A
Gate leakage current (nA)	27	26	$V_{ds} = 0$ V, $V_{gs} = -1$ V
Gate breakdown voltage (V)	10–12	14–18	$V_{ds} = 0$ V, $I_{gs} = 10\,\mu$A

Characteristics for GaAs power MESFETs fabricated by rapid thermal annealing were reported by Kanber et al. (1985b). Channel regions were formed by implanting 300 keV Si ions to doses of 6, 7, and 8×10^{12} cm^{-2} into undoped LEC GaAs substrates. Power GaAs MESFETs, with a nominal 1 μm × 2400 μm gate structure, were fabricated on the channel regions, which were activated by capless rapid thermal annealing at 890°C for 10 sec. The fabricated device showed an output power of 1.73 W with 4.9 dB associated gain, 30% power-added efficiency and 8.1 dB linear gain at 10 GHz. These values are comparable with the best data for current ion-implanted and furnace-annealed power FETs.

The application of rapid thermal annealing to the activation of n^+ source and drain regions in the refractory metal gate self-aligned MESFETs has been reported by Ohnishi et al. (1984) and Matsumoto et al. (1984). The use of rapid thermal annealing resulted in suppression of lateral diffusion in the source and drain regions, thus significantly alleviating a threshold voltage shift to the negative side for gate lengths shorter than 1.5 μm (see Chapter 3 for more details).

Unfortunately, no reports have so far been published with regard to the application of rapid thermal annealing to GaAs integrated circuit fabrication. This is in most part due to the lack of threshold voltage uniformity over the wafer, which is a matter of paramount importance in IC fabrication. One of the factors that affect the threshold voltage uniformity is undoubtedly slip line formation, as discussed in the previous section. More work is needed to eliminate the slip lines and eventually to realize uniform activation over the entire wafer.

As described in Section 8, currently available n-type doping levels are limited to below 2×10^{18} cm^{-3} under typical furnace annealing conditions. However, much higher doping levels with shallower dimensions will be required in the advanced GaAs device technology as a result of scaling-down of device dimensions. Thus, rapid thermal annealing is expected to become a key technology in future device fabrication. To achieve still higher n-type doping, the doping mechanism should be studied with regard to stoichiometry and implantation-induced defects. For stoichiometry control during annealing, rapid thermal annealing under an arsenic overpressure (Liu and Narayan, 1984; Hiramoto et al., 1985) will be promising. An example of such an annealing system is shown in Fig. 47, where the arsenic pressure is controlled by an AsH$_3$ flow. The effect of implantation-induced defects can be quantitatively analyzed by the Boltzmann transport approach developed by Christel and Gibbons (1981). Figure 48 shows calculated defect profiles in GaAs implanted with 150 keV Si ions. Because of the recoil processes during implantation, excess concentrations of Ga and As vacancies are generated near the surface region, while excess concentrations of Ga and

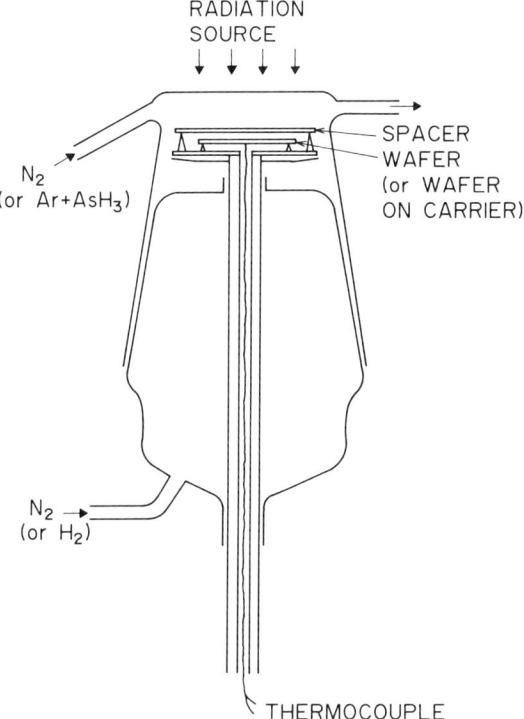

FIG. 47. Vertical quartz tube rapid thermal annealing system, where AsH$_3$ can be introduced at controlled flow rates. [From Liu and Narayan (1984).]

As atoms are generated at deeper regions. Since these defect concentrations increase with increasing ion doses, this effect will play an important role in the activation of high-dose shallow implants.

Another important subject is to understand and control the GaAs surface properties. To achieve ultrashallow doping by rapid thermal annealing, more rigorous characterization of GaAs surfaces will be indispensable. A question arises as to how shallow a doping region is obtainable under the constraints of Fermi level pinning (Mead and Spitzer, 1964). Another question arises as to whether the Fermi-level pinning position, i.e., the barrier height, remains unchanged under high-dose shallow implantation conditions. These surface factors, together with the achievable doping levels, will determine the parasitic source resistances including Ohmic contact resistance of the devices, which will become available in the future GaAs integrated circuit fabrication technology.

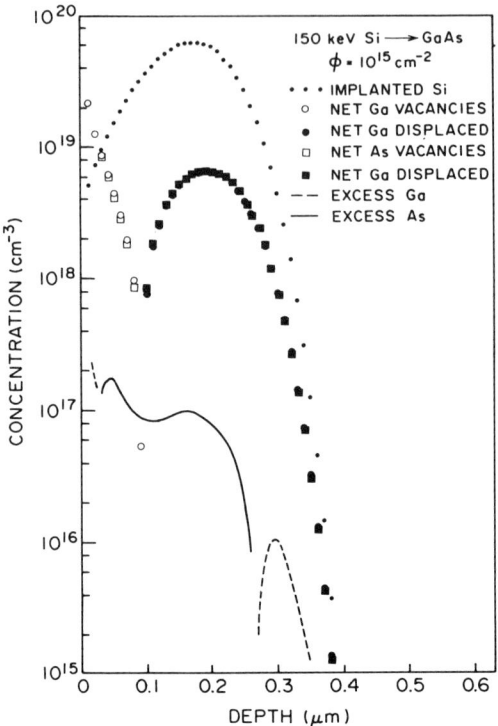

FIG. 48. Calculated stoichiometric distribution in GaAs implanted with 150 keV Si to a dose of 1×10^{15} cm^{-2}. Calculations were made using a Boltzmann transport equation approach. [From Christel and Gibbons [1981].]

References

Amano, J., Pianetta, P. A., and Stolte, C. A. (1980). *Appl. Phys. Lett.* **37**, 948.
Antell, G. R. (1977). *Appl. Phys. Lett.* **30**, 432.
Arai, M., Nishiyama, K., and Watanabe, N. (1981). *Jpn. J. Appl. Phys.* **20**, L124.
Asbeck, P. M., Miller, D. L., Petersen, W. C., and Kirkpatrick, C. G. (1982). *IEEE Electron. Devices Lett.* **EDL-3**, 366.
Badawi, M. H., and Mun, J. (1984). *Electron. Lett.* **20**, 125.
Banwell, T. C., Maenpaa, M., Nicolet, M-A., and Tandon, J. L. (1983). *J. Phys. Chem. Solids* **44**, 507.
Barrett, N. J., Grange, J. D., Sealy, B. J., and Stephens, K. G. (1984). *J. Appl. Phys.* **56**, 3503.
Barrett, N. J., Grange, J. D., Sealy, B. J., and Stephens, K. G. (1985). *J. Appl. Phys.* **57**, 5470.
Bensalem, R., Barrett, N. J., and Sealy, B. J. (1983). *Electron. Lett.* **19**, 112.
Blaauw, C. (1983). *J. Appl. Phys.* **54**, 5064.
Blunt, R. T., Szweda, R., Lamb, M. S. M., and Cullis, A. G. (1984). *Electron. Lett.* **20**, 444.
Blunt, R. T., Lamb, M. S. M., and Szweda, R. (1985). *Appl. Phys. Lett.* **47**, 304.
Chang, C. Y., Fang, Y. K., and Sze, S. M. (1971). *Solid State Electron.* **14**, 541.

Chapman, R. L., Fan, J. C. C., Donnelly, J. P., and Tsaur, B. Y. (1982). *Appl. Phys. Lett.* **40,** 805.
Chetterjee, P. K., Vaidyanathan, K. V., Durschlag, M. S., and Streetman, B. G. (1975). *Solid State Commun.* **17,** 1421.
Choe, B. D., Yeo, Y. K., and Park, Y. S. (1980). *J. Appl. Phys.* **51,** 4742.
Christel, L. A., and Gibbons, J. F. (1981). *J. Appl. Phys.* **52,** 5050.
Davis, L. E., MacDonald, N. C., Palmberg, P. W., Riach, G. E., and Weber, R. E. (1976). *In* "Handbook of Auger Electron Spectroscopy." Physical Electronics, Eden Prairie Minnesota.
Davies, D. E., and McNally, P. J. (1983). *IEEE Electron Device Lett.* **EDL-4,** 356.
Davies, D. E., and McNally, P. J. (1984). *Appl. Phys. Lett.* **44,** 304.
Davies, D. E., Kennedy, J. K., and Ludington, C. E. (1975). *J. Electrochem. Soc.* **122,** 1374.
Davies, D. E., Lorenzo, J. P., and Ryan, T. G. (1980). *Appl. Phys. Lett.* **37,** 612.
Davies, D. E., Ryan, T. G., Lorenzo, J. P., and Kennedy, E. F. (1981). *In* "Laser and Electron Beam Solid Interactions and Material Processing" (J. F. Gibbons, L. D. Hess, and T. W. Sigmon, eds.), p. 247. Elsevier/North-Holland, New York.
Davies, D. E., McNally, P. J., Lorenzo, J. P., and Julian, M. (1982). *IEEE Electron Device Lett.* **EDL-3,** 102.
Davies, D. E., McNally, P. J., Ryan, T. G., Soda, K. J., and Comer, J. J. (1983). *Conf. Ser.— Inst. Phys. No. 65,* p. 619.
Donnelly, J. P. (1981). *Nucl. Instrum. Methods* **182/183,** 553.
Donnelly, J. P., Lindley, W. T., and Hurwitz, C. E. (1975). *Appl. Phys. Lett.* **27,** 41.
Eisen, F. H. (1980). *Radiant. Eff.* **47,** 99.
Eisen, F. H., Welch, B. M., Muller, H., Gamo, K., Inada, T., and Mayer, J. W. (1977). *Solid State Electron.* **20,** 219.
Ezis, A., Yeo, Y. K., and Parks, Y. S. (1984). *Mat. Res. Soc. Symp. Proc.* **23,** 681.
Foxon, C. T., Harvey, J. A., and Joyce, B. A. (1973). *J. Phys. Chem. Solids* **34,** 1693.
Foyt, A. G., Donnelly, J. P., and Lindley, W. T. (1969). *Appl. Phys. Lett.* **14,** 372.
Fujimoto, M., Sato, Y., and Kudo, K. (1967). *Jpn. J. Appl. Phys.* **6,** 848.
Furutsuka, T., Tsuji, T., Katano, F., Higashisaka, A., and Kurumada, K. (1981). *Electron. Lett.* **17,** 944.
Gamo, K., Inada, T., Krekeler, S., Mayer, J. W., Eisen, F. H., and Welch, B. M. (1977). *Solid State Electron.* **20,** 213.
Gibbons, J. F., Johnson, W. S., and Mylroie, S. W. (1975). "Projected Range Statistics." Dowden, Hutchinson & Ross, Stroudsburg, Pennsylvania.
Gyulai, J., Mayer, J. W., Mitchell, I. V., and Rodriguez, V. (1970). *Appl. Phys. Lett.* **17,** 332.
Hallais, J., Micrea-Roussel, A., Farges, J. P., and Poibleud, G. (1976). *Conf. Ser.—Inst. Phys. No. 336,* p. 220.
Harris, J. S., Eisen, F. H., Welch, B., Haskell, J. D., Pashley, R. D., and Mayer, J. W. (1972). *Appl. Phys. Lett.* **21,** 601.
Hiramoto, T., Saito, T., and Ikoma, T. (1985). *Jpn. J. Appl. Phys.* **24,** L193.
Hunsperger, R. G., and Marsh, O. J. (1969). *J. Electrochem. Soc.* **116,** 488.
Hurle, D. T. J. (1979). *J. Phys. Chem. Solids* **40,** 613.
Immorlica, A. A., Jr., and Eisen, F. H. (1976). *Appl. Phys. Lett.* **29,** 94.
Inada, T., Miwa, H., Kato, S., Kobayashi, E., Hara, T., and Mihara, M. (1978a). *J. Appl. Phys.* **49,** 4571.
Inada, T., Ohkubo, T., Sawada, S., Hara, T., and Nakajima, M. (1978b). *J. Electrochem. Soc.* **125,** 1525.
Inada, T., Kato, S., Maeda, Y., and Tokunaga, K. (1979). *J. Appl. Phys.* **50,** 6000.
Ito, K., Yoshida, M., Otsubo, M., and Murotani, T. (1983). *Jpn. J. Appl. Phys.* **22,** L299.

Jaccodine, R. J., and Schlegel, W. A. (1966). *J. Appl. Phys.* **37**, 2429.
Kanber, H., Cipolli, R. J., Henderson, W. B., and Whelan, J. M. (1985a), *J. Appl. Phys.* **57**, 4732.
Kanber, H., Henderson, W. B., Rush, R. C., Siracusa, M., and Whelan, J. M. (1985b). *Appl. Phys. Lett.* **47**, 120.
Kasahara, J., and Watanabe, N. (1981). *Appl. Phys. Lett.* **38**, 798.
Kasahara, J., Arai, M., and Watanabe, N. (1979a). *J. Appl. Phys.* **50**, 541.
Kasahara, J., Arai, M., and Watanabe, N. (1979b). *J. Electrochem. Soc.* **126**, 1997.
Kasahara, J., Arai, K., Kato, Y., Dohsen, M., and Watanabe, N. (1981). *Electron. Lett.* **17**, 621.
Kasahara, J., Sakurai, H., Kato, Y., and Watanabe, N. (1982). *Jpn. J. Appl. Phys.* **21**, L103.
Katano, F., Furutsuka, T., and Higashisaka, A. (1981). *Electron. Lett.* **17**, 236.
Kirkby, P. A., Seuray, P. R., and Westbrook, L. D. (1979). *J. Appl. Phys.* **50**, 4567.
Klein, P. B., Nordquist, P. E. R., and Siebenmann, P. G. (1980). *J. Appl. Phys.* **51**, 4861.
Kohzu, H., Kuzuhara, M., and Takayama, Y. (1983). *J. Appl. Phys.* **54**, 4998.
Komatsu, R., and Kajiyama, K. (1984). *J. Appl. Phys.* **56**, 486.
König, U., and Sasse, E. (1983). *J. Electrochem. Soc.* **130**, 950.
Kuiper, A. E. T., Koo, S. W., Habraken, F. H. P. M., and Tamminga, Y. (1983). *J. Vac. Sci. Technol.* **B1**, 62.
Kuzuhara, M., and Kohzu, H. (1984). *Appl. Phys. Lett.* **44**, 527.
Kuzuhara, M., and Nozaki, T. (1986). *J. Appl. Phys.*, **59**, 3131.
Kuzuhara, M., Kohzu, H., and Takayama, Y. (1982). *Appl. Phys. Lett.* **41**, 755.
Kuzuhara, M., Kohzu, H., and Takayama, Y. (1983). *J. Appl. Phys.* **54**, 3121.
Kuzuhara, M., Kohzu, H., and Takayama, Y. (1984). *Mat. Res. Soc. Symp. Proc.* **23**, 651.
Kuzuhara, M., Nozaki, T., and Kohzu, H. (1985). *J. Appl. Phys.* **58**, 1204.
Leigh, P. A. (1982). *Int. J. Electronics* **52**, 23.
Lidow, A., Gibbons, J. F., and Magee, T. (1977). *Appl. Phys. Lett.* **31**, 158.
Lidow, A., Gibbons, J. F., Magee, T., and Peng, J. (1978). *J. Appl. Phys.* **49**, 5213.
Lindhard, J., Scharff, M., and Schiott, H. E. (1963). *Mat. Fys. Medd. Dan Vid. Selsk.* **33**, no. 14.
Liu, S. G., and Narayan, S. Y. (1984). *J. Electron. Materials* **13**, 897.
Liu, S. G., Wu, C. P., and Magee, C. W. (1980). In "Laser and Electron Beam Processing of Materials" (C. W. White and P. S. Peercy, eds.), p. 341. Academic Press, New York.
Malbon, R. M., Lee, D. H., and Whelan, J. M. (1976). *J. Electrochem. Soc.* **123**, 1413.
Masuyama, A., Nicolet, M-A., Golecki, I., Tandon, J. L., Sadana, D. K., and Washburn, J. (1980). *Appl. Phys. Lett.* **36**, 749.
Matsumoto, K., Hashizume, N., and Atoda, N. (1984). *Electron Lett.* **20**, 940.
Mayer, J. W., Marsh, O. J., Mankarious, R., and Bower, R. (1967). *J. Appl. Phys.* **38**, 1975.
Mead, C. A., and Spitzer, W. G. (1964). *Phys. Rev.* **134**, A713.
Morgan, D. V., Eisen, F. H., and Ezis, A. (1981). *IEEE Proc.* **128**, 109.
Mozzi, R. L., Fabian, W., and Piekarski, F. J. (1979). *Appl. Phys. Lett.* **35**, 337.
Nishiyama, K., Arai, M., and Watanabe, N. (1980). *Jpn. J. Appl. Phys.* **19**, L563.
Oberstar, J. D., and Streetman, B. G. (1983). *Thin Solid Films* **103**, 17.
Ohdomari, I., Mizutani, S., Kume, H., Mori, M., Kimura, I., and Yoneda, K. (1978). *Appl. Phys. Lett.* **32**, 218.
Ohnishi, T., Yamaguchi, Y., Inada, T., Yokoyama, N., and Nishi, H. (1984). *IEEE Electron Device Lett.* **EDL-5**, 391.
Okamura, S., Nishi, H., Inada, T., and Hashimoto, H. (1982). *Appl. Phys. Lett.* **40**, 689.
Oraby, A. H., Yuba, Y., Takai, M., Gamo, K., and Namba, S. (1984). *Jpn. J. Appl. Phys.* **23**, 326.
Pashley, R. D., and Welch, B. M. (1975). *Solid State Electron.* **18**, 977.

Patel, K. K., and Sealy, B. J. (1985). *Radiant. Eff.* **91**, 53.
Pearton, S. J., Cummings, K. D., and Vella-Coleiro, G. P. (1985). *J. Appl. Phys.* **58**, 3252.
Pianetta, P. A., Stolte, C. A., and Hansen, J. L. (1980a). *Appl. Phys. Lett.* **36**, 597.
Pianetta, P. A., Stolte, C. A., and Hansen, J. L. (1980b). *In* "Laser and Electron Beam Processing of Materials" (C. W. White and P. S. Peercy, eds), p. 328. Academic Press, New York.
Pianetta, P. A., Amano, J., Woodhouse, G., and Stolte, C. A. (1981). *In* "Laser and Electron Beam Solid Interactions and Material Processing" (J. F. Gibbons, L. D. Hess, and T. W. Sigmon, eds.). p. 239. Elsevier/North-Holland, New York.
Rose, A., Pollock, J. T. A., Scott, M. D., Adams, F. M., Williams, J. S., and Lawson, E. M. (1983). *Mat. Res. Soc. Symp. Proc.* **13**, 633.
Rosenblatt, D. H., Hitchens, W. R., Shatas, S., Gat, A., and Betts, D. A. (1984). *Mat. Res. Soc. Symp. Proc.* **23**, 669.
Roughan, P. E., and Manchester, K. E. (1969). *J. Electrochem. Soc.* **116**, 278.
Sanders, I. R., Peake, A. H., and Surridge, R. K. (1980). *In* "Semi-Insulating III–V Materials" (G. J. Rees, ed.), p. 349. Shiva Publishing, Nantwick, UK.
Schroeder, J. B., and Dieselman, H. D. (1967). *Proc. IEEE* **55**, 125.
Sealy, B. J., Surridge, R. K., Kular, S. S., and Stephens, K. G. (1979). *Conf. Ser.—Inst. Phys. No. 46*, p. 476.
Shah, N. J., Ahmed, H., Sanders, I. R., and Singleton, J. F. (1980). *Appl. Phys. Lett.* **16**, 433.
Shah, N. J., Ahmed, H., and Leigh, P. A. (1981). *Appl. Phys. Lett.* **39**, 322.
Shahid, M. A., Moffatt, S., Barrett, N. J., Sealy, B. J., and Puttick, K. E. (1983). *Radiant. Eff.* **70**, 291.
Stolte, C. A. (1984). *In* "Semiconductors and Semimetals" (R. K. Willardson and A. C. Beer, eds.), Vol. 20, Chap. 2. Academic Press, New York.
Surridge, R. K., Sealy, B. J., D'Cruz, A. D. E., and Stephens, K. G. (1977). *Conf. Ser.—Inst. Phys. No. 33a*, p. 161.
Suzuki, T., Sakurai, H., and Arai, M. (1983). *Appl. Phys. Lett.* **43**, 951.
Tabatabaie-Alavi, K., Masum Choudhury, A. N. M., Fonstad, C. G., and Gelpey, J. C. (1983a). *Appl. Phys. Lett.* **43**, 505.
Tabatabaie-Alavi, K., Masum Choudhury, A. N. M., Kanbe, H., Fonstad, C. G., and Gelpey, J. C. (1983b). *Appl. Phys. Lett.* **43**, 647.
Tandon, J. L., Nicolet, M-A., and Eisen, F. H. (1979). *Appl. Phys. Lett.* **34**, 165.
Tiku, S. K., and Duncan, W. M. (1985). *J. Electrochem. Soc.* **132**, 2237.
Tiwari, S., DeLuca, J. C., and Deline, D. R. (1985). *Conf. Ser.—Inst. Phys. No. 74*, p. 83.
Tokuyama, T., Fujii, Y., Sugita, Y., and Kishino, S. (1967). *Jpn. J. Appl. Phys.* **6**, 1252.
Troeger, G. L., Behle, A. F., Friebertshauser, P. E., Hu, K. L., and Watanabe, S. H. (1979). *IEDM Tech. Dig.*, p. 497.
Vaidyanathan, K. V., and Dunlap, H. L. (1984). *Mat. Res. Soc. Symp. Proc.* **23**, 687.
Vaidyanathan, K. V., Helix, M. J., Wolford, D. J., Streetman, B. G., Blattner, R. J., and Evans, C. A., Jr. (1977). *J. Electrochem. Soc.* **124**, 1781.
Williams, E. W., and Elliott, C. T. (1969). *J. Phys.* **D2**, 1657.
Yuen, A. T., Long, S. I., and Merz, J. L. (1985). *Mat. Res. Soc. Symp. Proc.* **45**, 285.
Zolch, R., Ryssel, H., Krapz, H., Reichl, H., and Ruge, I. (1977). *In* "Ion Implantation of Semiconductors and Other Materials" (F. Chernow, J. A. Borders, and D. K. Brice, eds.), p. 593. Plenum Press, New York.
Zucca, R. (1976). *Conf. Ser.—Inst. Phys. No. 336*, p. 228.
Zuleeg, R., Notthoff, J. K., and Lehovec, K. (1978). *IEEE Trans. Electron Devices* **ED-25**, 628.

CHAPTER 2

Focused Ion Beam Implantation Technology

H. Hashimoto

FUJITSU LABORATORIES LTD.
ATSUGI, JAPAN

I.	INTRODUCTION	63
II.	IMPLANTER	64
	1. *Focusing Column*	64
	2. *LM Ion Source*	65
	3. *Ion Source for* III–V *Compound*	66
	4. *Ion Beam Intensity Profile*	70
III.	FIB IMPLANTS	71
	5. *Fine Pattern Doping*	71
	6. *Implantation-Induced Damage*	74
	7. *Electrical and Optical Characteristics of Implanted Layers*	77
	8. *FIB-Implantation—Semi-Insulating*	79
IV.	COMPOSITIONAL DISORDERING OF GaAs–AlGaAs	82
V.	FIB-DOPING MBE GROWTH	85
	9. *Growth System*	85
	10. *Quality of FIBI–MBE Grown Crystal*	87
VI.	PROSPECT OF FIB TECHNOLOGY	93
	REFERENCES	95

I. Introduction

Recently, maskless ion implantation by scanning a focused ion beam (FIB) has received much attention as a technique for the new impurity doping of semiconductor devices (Kubena *et al.*, 1981; Miyauchi *et al.*, 1983d). This method has various advantages over the conventional implantation with a broad ion beam. Fine pattern doping in crystal can be performed without a mask. The troublesome photolithography for forming a mask on a wafer and subsequent cleaning steps are reduced, making possible the realization of drastically simple device fabrication processes. Using an alloyed liquid metal (LM) ion source, successive interchanging doping with ions for p- and n-type, and for isolation is done by changing the electric field of an ExB mass separator (Miyauchi and Hashimoto, 1986). Arbitrary lateral and depth doping profiles are formed by controlling dwell time and accelerating voltage of ion beams with computer software. These functions of FIB implantation offer a new, powerful tool that cannot be duplicated by the conventional implantation; they also offer the possibility to fabricate novel device structures. The FIB implantation also satisfies requirements for future fabrication processes for semiconductor devices, such as fine structure with submicron size, process simplification, and automation. Thus the maskless process using FIBs has been the final goal since the beginning of development of the ion implantation technology. However, this technique had not appeared to be a reality for lack of a high-brightness ion source that focused ion beams to a small spot.

In 1979, Seliger *et al.* (1979) developed a scanning focused ion probe system with a 1000-Å-diameter spot incorporating a Ga LM ion source and demonstrated the possibility of realizing the dream of maskless ion implantation. Since then, the study of the FIB implantation technology has been stimulated significantly. It is only in the past several years that this technology has been studied in earnest. There are many unresolved scientific issues and engineering problems to be solved before its practical use as fabrication tools for semiconductor devices. Its successful realization will be owed to future studies.

In this chapter, first the maskless ion implanter that brings out the powerful ability in impurity doping is explained. Then, the marked features of FIB implantation technology for *p*- and *n*-type impurity doping in GaAs are described. Its application to attractive processes such as isolation of active layers by forming semi-insulating GaAs, compositional disordering of GaAs/AlGaAs superlattice structure, and pattern-doped crystal growth by a combination of equipment with molecular beam epitaxy are also shown.

II. Implanter

1. FOCUSING COLUMN

The development of an implanter has advanced rapidly over the last few years, and commercial systems with the maximum accelerating voltages of 30 to 200 kV are now available (Wang *et al.*, 1981; Shiokawa *et al.*, 1985). A block diagram of the typical 100 kV maskless implanter is shown in Fig. 1 (Miyauchi and Hashimoto., 1986). The focusing column consists of a field-emission-type LM ion source, a three-electrode-type condenser lens and an Einzel-type objective lens with an ExB mass separator and a postlens deflector. The extracted ion beam from an ion source is accelerated and focused with the condenser lens. The ion species can be selected with the ExB mass separator, focused again with the objective lens and scanned over a wafer by the electrostatic deflector. The minimum spot diameter on the wafer is less than $0.1\,\mu$m. Impurities are implanted precisely in the designated

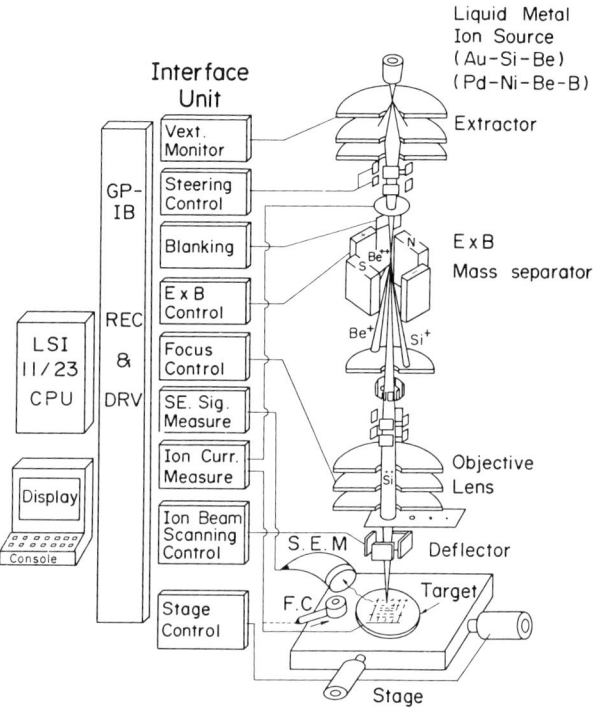

FIG. 1. Schematic diagram of the 100 kV maskless ion implanter. [From Miyauchi and Hashimoto (1986).]

regions by scanning and blanking FIBs without a mask. The implantation procedures such as mass separation, beam scanning, positional alignment and adjustment of optics are controlled by a computer.

To draw a designed pattern by FIB implantation, a positional alignment method similar to that for electron-beam lithography is used (Kubena et al., 1981). A secondary electron signal produced by scanning the FIBs across etched benchmarks on a wafer is used as a guide for sample-stage positioning. Contrast and sharpness of the secondary electron intensity profile at the marker depend on the beam diameter, the incident energy of the FIB and, in particular, the geometry of the benchmark (Morita et al., 1986). The alignment accuracy is better than 0.1 μm under optimized conditions.

2. LM Ion Source

In the FIB implantation technology, the ion source as well as the focusing column is very important in obtaining a fine and stable ion beam. Typical configurations of LM ion sources are shown in Fig. 2. Among them, a

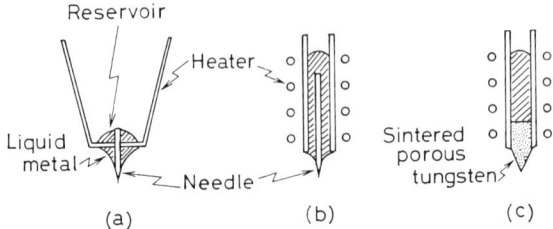

FIG. 2. Typical configuration of LM ion sources: (a) hairpin type, (b) capillary type, (c) impregnated-electrode type.

hairpin-type ion source illustrated in Fig. 2a is the most popular ion source with simple structure. This ion source consists of a sharp emitter needle with an apex radius of several microns wetted with molten source metal; its reservoir; and a heater (Swanson et al., 1979). The molten metal is supplied at the emitter tip along the needle surface from the reservoir. An intense electric field is applied to the apex of the emitter to form a stable protrusion that resembles the so-called Taylor cone. The ions are emitted from the apex of this liquid feature by a process that is thought to be field evaporation. The electric field of ion emission, $\sim 10^8$ V/cm, corresponds to that of the evaporation field (Tsong, 1978).

For liquid metal with high vapor pressure, the ion source shown in Fig. 2a is liable to evaporate source materials from the reservoir and the needle surface. To suppress evaporation, an impregnated-electrode-type ion source shown in Fig. 2c has been proposed (Ishikawa and Takagi, 1984). Molten metal is permeated and supplied through a sintered porous tungsten tip to suppress its evaporation. Thus, this type of ion source is capable of emitting ions from molten metal with high vapor pressure.

In addition to contrivance of the ion-source configuration, eutectic alloy is used to decrease the fusing temperature of the source metal with the high melting point, resulting in lower vapor pressure. Almost all kinds of metal ions can be emitted by this method regardless of the source configuration. As will be described in detail later, the eutectic is effective in not only decreasing the melting point, but also offering skillful technique in conjunction with a mass separator in the column.

3. Ion Source for III–V Compound

As dopants for III–V compound semiconductors, Be, Mg, Zn, Si, and Sn are available. In early studies, ion sources such as Au–Be and Au–Si eutectic alloy (Seliger et al., 1982), which emit a single *p*- or *n*-type ion, were usually loaded in the implanter. However, ion sources that are capable of emitting multiple ions from a single emitter tip are now widely used for convenience.

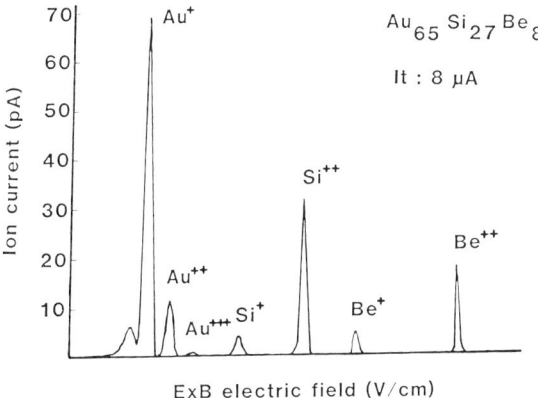

FIG. 3. Mass spectrum of the Au–Si–Be LM ion source. [From Miyauchi et al. (1983a).]

One such ion source is a Au–Si–Be eutectic alloy (Miyauchi et al., 1983a; Gamo et al., 1983). Since both Be and Si have extremely high vapor pressure at the melting point, they are not suitable for a LM ion source in their elemental states. Their alloy with Au has a lower melting point over which the vapor pressure is reduced. The typical mass spectrum emitted from the $Au_{65}Si_{27}Be_8$ ion source operated at 600°C is shown in Fig. 3 (Miyauchi et al., 1983a). In this experiment, a hairpin-type ion source consisting of a tungsten needle and a tantalum capillary reservoir was used. It is observed clearly that both Be ions for p-type and Si ions for n-type are emitted. The double-charged species predominate over the single-charged ions for both Be and Si, while the single-charge peak is much higher than the double-charge one for Au ions. Although the charged state depends upon the current level of the ion beam, these results are expected from the simplified theory of field evaporation (Gomer, 1979). The high emission level of double-charged ions is favorable for an ion implantation process since the ion beam energy is doubled for the same accelerating voltage. With this ion source, p- and/or n-type ions can be implanted arbitrarily by simply changing the electric field of an ExB mass separator. Ion species are interchangeable electrically without readjustment of focusing optics and the cumbersome source-exchanging procedure, which usually requires breaking the vacuum.

The secondary electron images generated by scanning the Si^{2+} and Be^{2+} focused beams of 0.1 μm diameter on a nickel mesh as a target at 160 keV are shown in Fig. 4 (Miyauchi et al., 1983b). These microphotographs were taken by changing the electric field of the ExB mass separator, maintaining a constant magnetic field without readjusting focusing condition. In these photographs, the arrows indicate the same position of a mesh edge. Comparing the relative position of these mesh edges with the bar marks, the

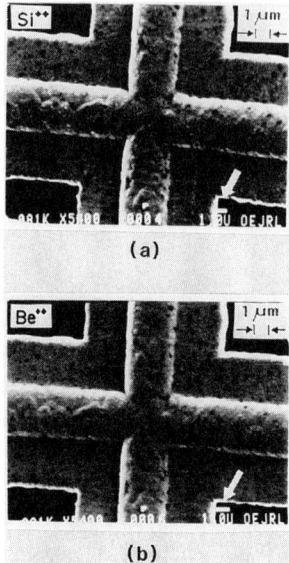

FIG. 4. Microphotographs of SIM images generated by alternating (a) Si and (b) Be ion beams on a nickel mesh at 160 keV. [From Miyauchi et al. (1983b).]

positional shift of Be^{2+} and Si^{2+} ion beams is around 0.5 μm. At present, this value is reduced to less than 0.1 μm by more precise adjustment of the ExB voltage. The SEM image in Fig. 5 shows the cleaved and stained cross-section of p-n junction arrays in GaAs fabricated by forming a wide Be stripe superposed with narrow Si lines (Miyauchi et al., 1983d). The doping geometry similar to a bipolar transistor was formed by alternate implantation of Be and Si at 160 keV, followed by annealing at 800°C for 20 min.

A Pd-Ni-Si-Be-B LM ion source that emits B (boron) ion for electrical isolation between devices as well as Be and Si ions for doping has been

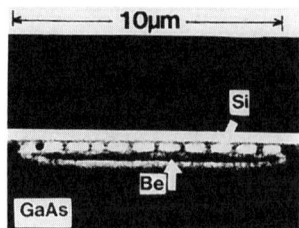

FIG. 5. SEM image of the stained cleavage of successive Be and Si implanted GaAs.

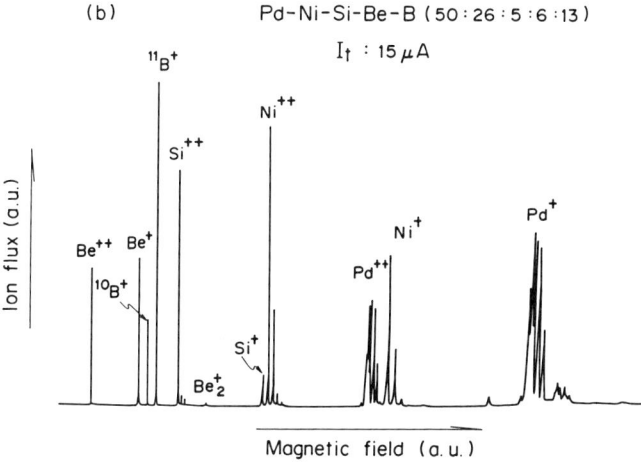

FIG. 6. Mass spectrum of the Pd–Ni–Si–Be–B LM ion source. [From Arimoto et al. (1985).]

developed (Arimoto et al., 1985). This ion source is composed of $Pd_{50}Ni_{26}Si_5Be_6B_{13}$ and has a melting point of approximately 800°C. It operates at 850°C, which is higher than the operating temperature of a Au–Si–Be LM ion source. The mass spectrum from this ion source measured with a 90° magnetic-sector filter is shown in Fig. 6. B ions are emitted together with Be and Si. However, double-charged B^{2+} ion is not observed, while Be^{2+} and Si^{2+} are emitted. This difference is due to the evaporation field (B^+ 6.48 V/Å, B^{2+} 7.65 V/Å) (Müller and Tsong, 1969). Because of the higher operation temperature, Pd–Ni–Si–Be–B alloy is apt to react a little with the tungsten needle and to be contaminated with residual gas such as oxygen and carbon in the source chamber. Although these reactions affect the stability and lifetime of the ion beam, this source operates stably over 100 hours—which is, however, still shorter than the 700 hours of the Au–Si–Be LM ion source. The lifetime can be increased further by optimizing the alloy composition and the needle materials and by reducing residual gas in the ion-source chamber. Using the Pd–Ni–Si–Be–B LM ion source, the required ions for p- and n-type doping and for isolation can be implanted arbitrarily by controlling the ExB voltage at high speed.

The semi-insulating property of GaAs formed by B implantation is thermally stable even after high-temperature treatment at around 800°C (Martin et al., 1982). Therefore, after implantation of B, the wafer can be annealed to activate implanted Be and Si at around 800°C. The semi-insulating characteristics of the focused B-ion implanted GaAs layer will be described later.

4. Ion Beam Intensity Profile

The minimum diameter of ion beams on target is ultimately determined by the source size and the energy spread of the ion. Under ideal conditions without chromatic and spherical aberration, the minimum spot size is almost comparable to the source size because the magnification of the focusing column is approximately unity. The source size of the Au–Si–Be LM ion source has not been measured yet. For Ga of a typical LM ion source, it is around 600 Å (Wagner et al., 1981). The spread of energy distributions of ion beams causes chromatic aberration and affects focusing capability. When ion beam current density increases, coulomb scattering between charged particles occurs in ion beams, causing a broader energy spread. Because of this effect, the energy distributions spread in order of ion mass. In general, the spot size of ion beams becomes smaller as the ion mass decreases. However, some ion species have abnormally long tails in the energy distributions at two orders of magnitude below the peak ion-current intensity. In these cases, the intensity profiles of the FIBs also have larger tail-like spread than the spread estimated from the Gaussian distribution. For FIB implantation, even the lower ion-flux at the tail region takes effect electrically and enlarges doping geometry significantly (Wada et al., 1985; Ward et al., 1987).

Figure 7 shows the measured spreads of Be and Si as a function of ion dose for two ion-emission currents ($10\,\mu A$ and $40\,\mu A$) over the equivalent

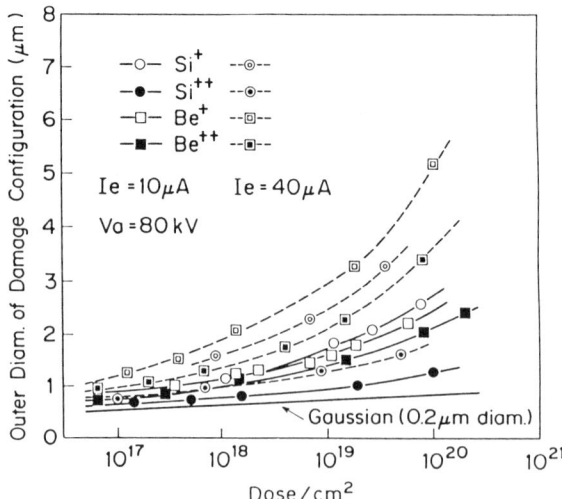

Fig. 7. Measured outer diameter of Be- and Si-implantation-induced damages on GaAs as a function of implantation dose. Solid lines: $I_e = 10\,\mu A$; dashed lines: $I_e = 40\,\mu A$. ○, ◎: Si$^+$; ●, ◉: Si^{2+}; □, ▣: Be$^+$; ■, ▪: Be^{2+}. [From Miyauchi et al. (1987).]

beam-intensity level of the tail region at three to six orders of magnitude below the peak (Miyauchi et al., 1987). These beam spreads were measured using the outer diameter of the visible configuration of implantation-induced damages on GaAs. Single-charged Be^+ and Si^+ have larger spreads than those of double charges (Be^{2+} and Si^{2+}), corresponding to their broad ion-energy distributions with long tails. The magnitude of spreads also increases remarkably with ion dose and ion emission current. The Be^+ and Si^+ ion are probably produced by charge transfer collision between doubly charged ions and neutral atoms and by field ionization of neutral atoms in free space in addition to field evaporation. Therefore, the Be^+ and Si^+ ion beams tends to include many kinds of ion species with various energies, making a large diameter on the target as a consequence of chromatic aberration.

On the other hand, the double-charged Be^{2+} and Si^{2+} ion are mainly generated by field evaporation, and their narrow energy-width forms small-diameter ion beams. In this case, the double-charged Be^{2+} beam has a slightly larger spread than that of Si, especially at higher doses. Considering the effect of coulomb scattering in ion beams, the measured results are contrary to expectation. The reason is not clear at present. In any case, the double-charged Be^{2+} and Si^{2+} FIBs emitted from the Au–Si–Be and Pd–Ni–Si–Be–B LM ion sources are suitable for fine pattern doping of p- and n-type impurities in GaAs.

III. FIB Implants

5. FINE PATTERN DOPING

FIB implantation has capability of fine patern doping and offers a powerful microfabrication technique. For forming a fine doping pattern, it should be taken into account that implanted ions are scattered and spread by collisions with the nuclei in a crystal. The calculated contours of equiconcentration of implanted Si in GaAs are plotted in Fig. 8a (Hashimoto and Miyauchi, 1984). The calculation was carried out by assuming that the 160-keV Si beam of 0.1-μm diameter, which has a Gaussian intensity distribution, was line-scanned over GaAs with a dose of 1×10^{13} cm^{-2}. So, this figure shows a cross-section perpendicular to the implanted line. The estimated line width becomes about 0.5 μm at the concentration contour of 1×10^{15} cm^{-3}, even when the Si-beam diameter is focused down to 0.1 μm.

The photograph in Fig. 8b is a SEM image of Si implanted n-type (100) GaAs (1×10^{15} cm^{-3}) with 7° off the axis under the same conditions as the calculation in Fig. 8a. In this sample, the n-type GaAs layer (1×10^{15} cm^{-3}) of 1.5 μm was overgrown on the implanted layer by molecular beam epitaxy (MBE) to reduce the influence of thermal conversion of the surface region

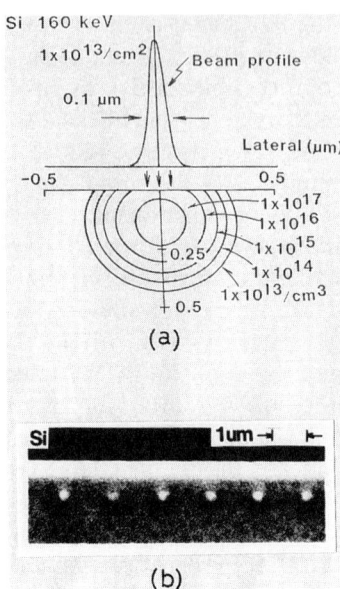

FIG. 8. (a) Calculated Si concentration contour lines in GaAs implanted at 160 keV with a dose of 1×10^{13} cm^{-2}. (b) SEM image of Si-implanted GaAs. [From Hashimoto and Miyauchi (1984).]

during annealing. The Si-implanted lines were buried in GaAs. The sample then was annealed at 800°C for 20 min with a SiO$_2$ encapsulant. The actual doping line width obtained experimentally is about 0.4 μm and agrees approximately with the calculated value.

On the other hand, high-dose FIB implantation brings about lateral spreads of implanted Be and Si in the n-type (100) GaAs epitaxial layers (5×10^{14} cm^{-3}) at 160 keV. Lateral spreads are shown in Fig. 9 as a function of ion dose (Miyauchi *et al.*, 1983c). The figure also shows calculated values under the assumption that an incident ion beam has a step-like profile with 0.1-μm diameter. The solid and dot-and-dash lines show the calculated lateral spreads of Be and Si in a wafer with background concentrations of 1×10^{15} and 1×10^{16} cm^{-3}, respectively. As the dose increases more than 1×10^{13} cm^{-2}, the measured line widths become wider than the calculated ones. This broadened distribution may be associated with superposed phenomena caused by enhanced migration of the implanted atoms—that is, radiation-enhanced migration during implantation (Lidow *et al.*, 1978; Wilson and Jamba, 1981) and thermal diffusion during post-implant annealing (McLevige *et al.*, 1977)—and by the tail effect of incident ion-beam profiles described previously. To make a fine doping pattern, it is necessary to suppress these phenomena.

2. FOCUSED ION BEAM IMPLANTATION TECHNOLOGY

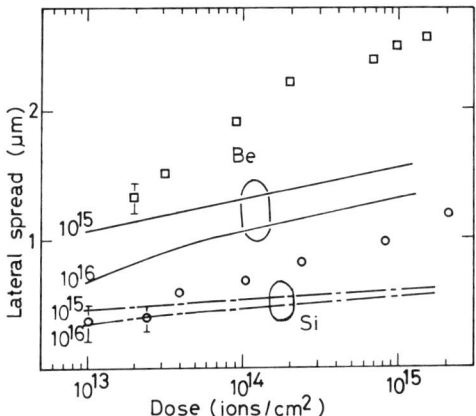

FIG. 9. Dose dependences of lateral spread for 0.1-μm-diameter Be ($\square$) and Si ($\bigcirc$) FIBI at 160 keV. Solid and dot-and-dashed lines indicate calculated impurity concentration levels in atoms/cm^3. [Miyauchi et al. (1983c).]

As a matter of fact, the maskless implantation system can form any doping pattern by computer control. The photograph in Fig. 10, which is a scanning ion microscope (SIM) image of the Be implanted GaAs surface, demonstrates the fine letters that can be written by using 160-keV ion beams of 0.2 μm. Each small letter in Fig. 10 was written in a 3-μm square area. These doping capabilities will also offer a tool to realize automatic device fabrication processes linked with computer-aided design in the near future.

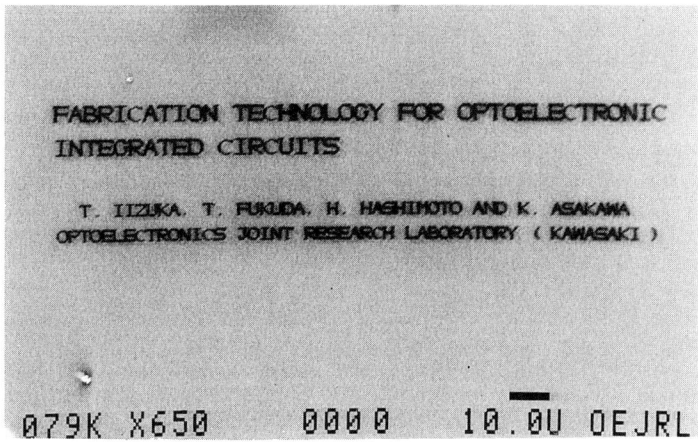

FIG. 10. SIM image of a Be-implanted GaAs surface. The letters were written with 160-keV Be ion beams of 0.2-μm diameter.

6. Implantation-Induced Damage

For FIB implantation, the density of the ion beam is 10^3 to 10^6 times higher than that of the unfocused ion beam (UIB) of a conventional implanter. FIBs usually have a current density of 0.1 to 1 A/cm^2, while UIBs are less than 10^{-3} A/cm^2. Experimental results on dose-rate dependence of implantation-induced damages in GaAs by UIBs have shown that damages increase with increasing current densities (Harris, 1971). Thus it was feared at first that such a high current density would have an undesired influence on damage production. Therefore, lattice damages in the FIB implanted GaAs were investigated by Raman scattering spectroscopy (Bamba et al., 1983; 1984). Figure 11 represents obtained Raman spectra from the Be FIB-implanted layer excited with a 514.5-nm line of Ar laser. The wafer was (100)-oriented, n-type, MBE-grown GaAs (2.7×10^{15} cm^{-3}). The Be FIB had a 0.2-μm diameter and the implantation energy was 160 keV. The doses were varied from 3×10^{12} to 1×10^{15} cm^{-2} along the off-channeling direction. The beam current was about 90 pA, resulting in a current density of approximately 0.2 A/cm^2. The scanning speed of FIB was about 10 cm/sec. With increasing Be-ion dose, the intensity of the longitudinal optical (LO) phonon line decreases, the width broadens, and the frequency-shift reduces. All such features are the same as the Raman spectra from UIB-implanted (100) GaAs (Yaney et al., 1981; Nakamura and Katoda, 1982).

Figure 12 shows the LO phonon frequency shift as a function of the Be dose for both FIB and UIB. The square represents the value for the

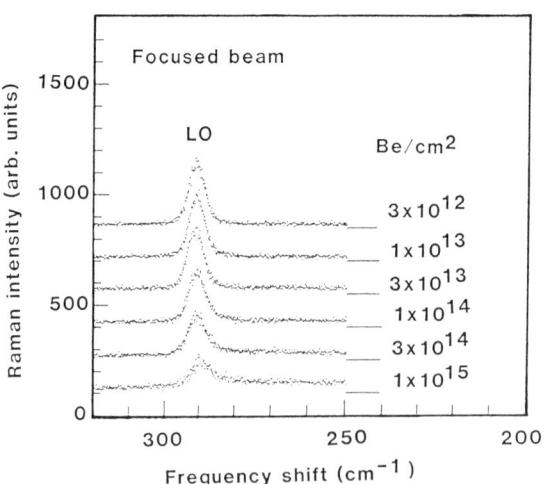

FIG. 11. Raman spectra of 160-keV Be FIB-implanted GaAs with various doses. [From Bamba et al. (1984).]

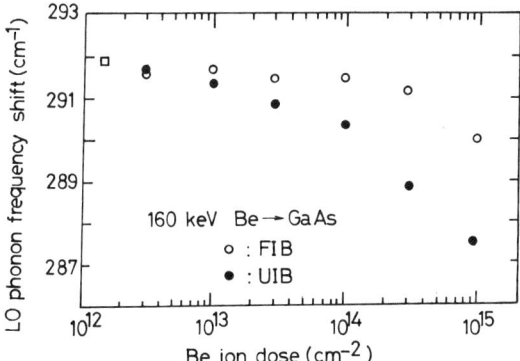

FIG. 12. Dose dependence of LO phonon frequency shift for focused (○) and conventional (●) Be implantations at 160 keV. The LO frequency shift for unimplanted GaAs is indicated by the square symbol. [From Bamba et al. (1984).]

unimplanted MBE GaAs. The lowering of LO phonon frequency is less in this case than in the UIB case. Lowering of LO phonon frequency in (100) GaAs is ascribed to bond weakening caused by strains or lattice-atom displacements. Considering these results, it is deduced that FIB generates damages less than UIB. A similar behavior was also observed for Si FIB implantation in GaAs (Bamba et al., 1983).

To further investigate damages generated by FIB implantation, distributions of implantation-induced damages for both FIB and UIB were measured by Raman spectroscopy in conjunction with successive chemical etching of the implanted layer. Figure 13 shows obtained in-depth distributions of the LO phonon–frequency shift in Be-implanted GaAs layers (Bamba et al., 1984). The ordinate represents the difference in the LO phonon frequency between the implanted (denoted by ω_{impl}) and the unimplanted GaAs (denoted by ω_{unimpl}, 291.9 cm^{-1}). This value is considered to correspond to the degree of implantation-induced damages. The damages for FIB are significantly smaller in the surface region as compared with UIB, whereas a comparable amount of damages exist around the projected range (R_p) for FIB and UIB. Since there are few reports on the reason FIB implantation produces less damage, the mechanism is not fully understood at present.

The degree of implantation-induced damages depends upon the balance between generation and annealing of defects during implantation. With cascade collision along a single particle track, an implanted ion slows down and comes to rest within 10^{-12} sec (Nelson, 1968). Such a time is much shorter than the time in which the track is overlapped with another track of a subsequent implanted ion under the present FIB density of around 1 A/cm^2. The generation of defects may be independent of the implantation

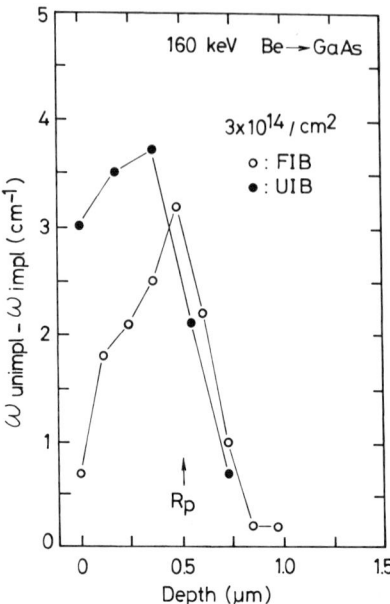

FIG. 13. Depth distribution profiles of the damage in FIB-implanted (○) and UIB-implanted (●) GaAs (Be, 160 keV, 3×10^{14} cm^{-2}). The ordinate represents the difference of the frequency shift between the implanted and the unimplanted GaAs. [From Bamba et al. (1984).]

method, FIB or UIB. For FIB, it is possible that annealing of defects, including their diffusion, annihilation and complex formation, is affected by other defects generated by other ions implanted soon afterward, since the current density of ion beams is very high. However, self-heating may be a minor effect for defect annealing, because the hemispherical heat flow from the small irradiated spot is extremely efficient in extracting the beam power, resulting in a very low temperature rise (Brown and Wagner, 1983). To explain the mechanism of damage reduction in FIB implantation, effects other than self-heating must be considered.

The relaxation time of implantation-induced defects is much longer in GaAs than in Si and becomes comparable to the period of successive impingement of ions in UIB implantation with relatively low current density. Therefore, the annealing stage is influenced by low-current ion beams. Rutherford backscattering measurements of Si and C UIB-implanted GaAs show that induced damage is accumulated and increases with increasing current density (Harris, 1971).

In contrast, FIB implantation causes less damages, as shown in Fig. 12 and Fig. 13. From the fact that damages near the surface are smaller than those for UIB, while they are comparable with those for UIB at around the projection range, R_p (Fig. 13), one possible model for the mechanism is that a high rate of localized electronic energy is transferred to defects, thereby annealing them out (Bamba et al., 1984). In the collision of Be ion with GaAs host atoms, inelastic scattering predominates over elastic scattering because of the light mass of the Be projectile at high accelerating energy. In the near-surface region, the probability of inelastic scattering is significantly higher than that of elastic scattering. Energy transfer via the electronic excitation of target atoms is predominant over nuclear energy loss by elastic scattering in this region. The high rate of energy transfer by forming dense electron–hole pairs with FIB implantation, followed by nonradiative recombination at the defects, probably causes enhanced motion of displaced atoms and annealing. At present, however, knowledge of damages in GaAs with FIB irradiation is insufficient to lead to the concrete mechanism of damage production. Further study is needed to obtain a conclusion.

7. Electrical and Optical Characteristics of Implanted Layers

The electrical and optical characteristics of FIB-implanted GaAs layers after high-tempeature annealing are almost equivalent or somewhat superior to those obtained by UIB implantation (Bamba et al., 1983). Figures 14a and 14b compare the results of the Hall effect measurement of Si FIB- and

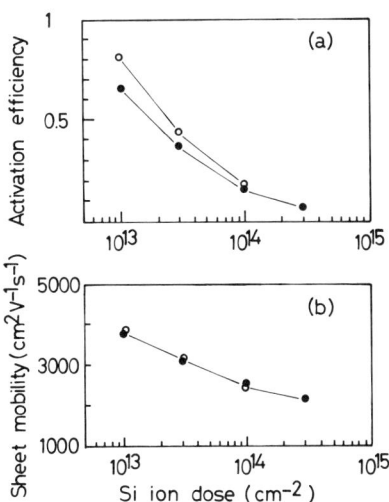

FIG. 14. (a) Electrical activation efficiency and (b) sheet electron mobility of 160-keV Si ion-implanted and 850°C-annealed GaAs with FIB (○) and UIB (●). [From Bamba et al. (1983).]

UIB-implanted GaAs. Si ion beams were implanted in p-type (100)-oriented MBE-grown GaAs (2×10^{15} cm^{-3}) at 160 keV with doses of 1×10^{13} to 3×10^{14} cm^{-2}. Following implantation, the samples were annealed at 850°C for 20 min with SiO$_2$ encapsulation. As shown in Fig. 14a, the electrical activation efficiency of implanted Si decreases with increasing Si dose irrespective of the implantation method. Such behavior is the same as is observed in conventionally implanted GaAs. The activation efficiency for FIB tends to be larger than that for UIB, reflecting less damage.

For sheet electron mobility, the obtained values plotted in Fig. 14b are almost the same in both samples. Since the FIB-implanted layers have the larger activation efficiency, resulting in higher carrier concentration, they are less compensated than those for UIB. Results similar to Si FIB implantation in GaAs were also observed in Be FIB implantation for p-type doping. Hall mobility is comparable to that in a bulk crystal (Kubena et al., 1981). Thus there is no problem in high-density FIB implantation with respect to electrical characteristics.

As for characterizing optical properties, the Si FIB-implanted MBE-grown GaAs layer was investigated by photoluminescence spectroscopy (PL) (Bamba et al., 1983). Figure 15 shows near-band-edge PL spectra for Si

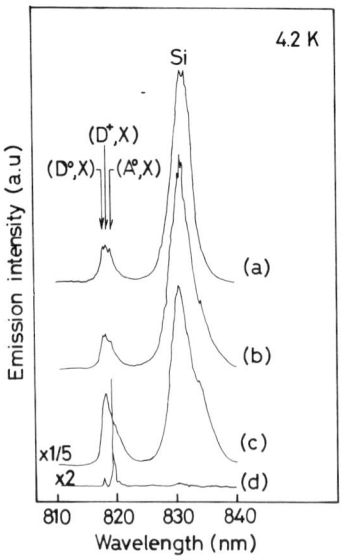

FIG. 15. PL spectra of variously processed GaAs at 4.2 K. Spectra for 160-keV Si (a) FIB- and (b) UIB-implantation at 1×10^{12} cm^{-2} with 850°C annealing; (c) spectrum for intentional Si molecular-beam doping to 3.3×10^{16} cm^{-3}; and (d) spectrum for undoped GaAs. [From Bamba et al. (1983).]

FIB-implanted GaAs and other various GaAs samples. The curves (a), (b), (c), and (d) represent emission spectra from a Si FIB-implanted and 850°C-annealed sample, a UIB-implanted and similarly annealed one, an n-type MBE-grown GaAs (3.3×10^{16} cm^{-3}) doped with Si, and an undoped and unimplanted MBE GaAs, respectively. It is obvious that the spectral shapes and intensities of curves (a) and (b) are almost the same. Emission peaks at 831 nm in curves (a) and (b) are probably related to Si-acceptor, which is also seen as a dominant emission peak at the same wavelength in curve (c). A fine structure is observed in bound exciton peaks in the curves (a), (b), and (d): Emission lines at 819.4, 818.8, and 818. 1 nm may be related to exciton-impurity complexes such as (A^0, x), (D^+, x), and (D^0, x), respectively (Bogardus and Bebb, 1968). Considering these emission spectra, the FIB- and UIB-implanted GaAs have approximately the same optical quality.

8. FIB-Implantation—Semi-Insulating

In addition to p- and n-type doping, FIB has been applied to selectively form semi-insulating and modified crystal structures (Arimoto *et al.*, 1984; Hirayama *et al.*, 1985c). Recently, there has been increasing interest in such techniques as a method for development of new structural devices. To fabricate GaAs ICs and optoelectronic devices composed of GaAs–AlGaAs, it is usually necessary to isolate their active regions. The isolation is often obtained by implantation with ions such as protons (Speight *et al.*, 1977).

However, the semi-insulating property with proton implant damage disappears with a subsequent heat treatment higher than 500°C. After proton implantation, usable fabrication techniques are restricted only to a low temperature process. On the other hand, B (boron)-implanted layers keep high resistivity after a high temperature treatment (Martin *et al.*, 1982). Using B FIB, finely patterned isolation regions are formed successfully in GaAs epitaxial films (Arimoto *et al.*, 1985). 100-keV B FIBs of 0.1-μm diameter from the Pd–Ni–Si–Be–B LM ion source were scanned only once across the mesa-etched n-type epitaxial GaAs films (3×10^{17} cm^{-3}, 0.2-μm thick) grown on the semi-insulating substrates as illustrated in Fig. 16. Estimating the lateral spread of implanted B atoms with scattering in GaAs under this condition, the line width of the isolated region is approximately 0.5 μm.

Figure 17 shows the V–I characteristics of (1) the unimplanted sample, and (2) the B-implanted sample (2×10^{13} ions/cm^2) between both electrodes illustrated in Fig. 16. Even single line-implantation with B FIBs is sufficient to isolate active areas. The resistance of the semi-insulating region formed by B FIB implantation is 10^7 times larger than that of the unimplanted sample, and the breakdown voltage is several tens of volts. Figure 18 shows the V–I characteristics of (1) the unimplanted sample, and (2) a B FIB

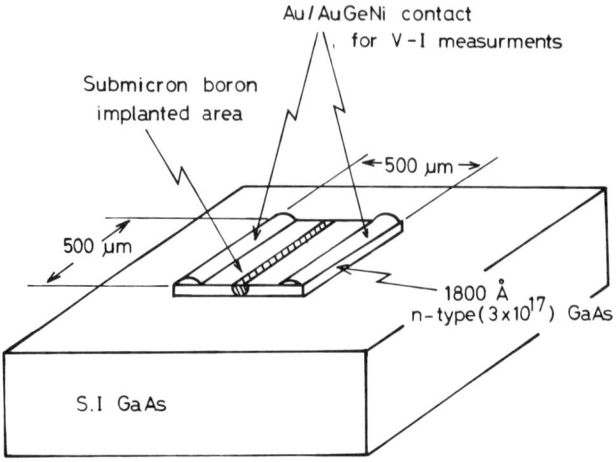

Fig. 16. Schematic structure of the B FIB-implanted sample for resistivity measurements. [From Arimoto et al. (1985).]

implanted sample at 2×10^{13} ions/cm^2, (3) at 2×10^{14} ions/cm^2, and (4) at 2×10^{15} ions/cm^2, after heat treatment at 800°C for 20 min. The semi-insulating regions resulting from implantation doses greater than 2×10^{14} ions/cm^2 retain high resistivity with breakdown voltage of 4 to 10 V, whereas those resulting from a 2×10^{13} ions/cm^2 dose cannot be maintained through a high-temperature device fabrication process. The resistivity of the

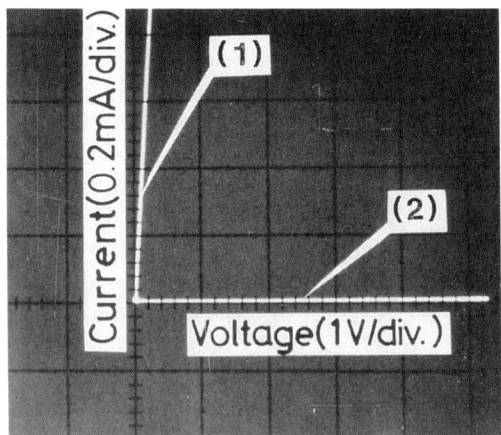

FIG. 17. Current–voltage characteristics of (1) the unimplanted sample, and (2) the B FIB-implanted sample at 100 keV and a dose of 2×10^{13} cm^{-2}. [From Arimoto et al. (1985).]

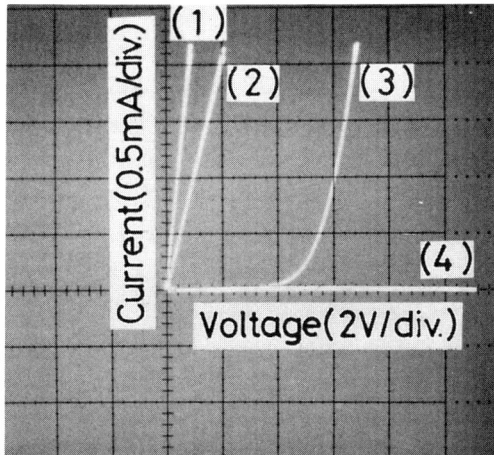

FIG. 18. Current–voltage characteristics of (1) the unimplanted sample and B FIB-implanted samples at (2) 2×10^{13} cm^{-2}, (3) 2×10^{14} cm^{-2}, and (4) 2×10^{15} cm^{-2} after heat treatment at 800°C for 20 min. [From Arimoto et al. (1985).]

isolated regions implanted with doses of 2×10^{14} to 2×10^{15} ions/cm^2 is 10^4 to 10^5 Ω-cm. Implanted Be and Si can be successfully activated at high temperature after successive implantation including B for isolation using the Pd–Ni–Si–Be–B LM ion source.

Similar semi-insulating GaAs layers are formed with Ga FIB implantation (Nakamura et al., 1985; Hirayama et al., 1985c). Figure 19 represents the resistance change of Ga FIB implanted GaAs regions as a function of heat-treatment temperature (Hirayama et al., 1985c). The resistance is normalized in the ratio of Ga-implanted to unimplanted samples. Ga FIBs of 0.1-μm diameter were implanted in the mesa-etched p^+-type (Be doped, 2×10^{18} cm^{-3}) and n^+-type (Si doped, 2–6 $\times 10^{18}$ cm^{-3}) epitaxial films (thickness 0.08–0.1 μm) on the semi-insulating GaAs substrates at 100 keV with the doses of 5×10^{13} and 4×10^{14} cm^{-2}. The sample structure is fundamentally the same as shown in Fig. 16. Both as-implanted layers in p^+- and n^+-type GaAs have high resistivity. The samples of Ga FIB-implanted n^+-type GaAs retain high resistivity after heat treatment as high as 800°C, while for p^+-type GaAs, resistivity enhancement disappears at temperatures above 650°C.

The scientific reason why B and Ga implantation form such thermally stable semi-insulating layers is now under study. If simple implant damage causes high resistivity, it is reordered by annealing around 500°C, resulting in a decrease in resistivity enhancement, as seen in proton-implanted GaAs. Thus such resistivity enhancement probably originates with another

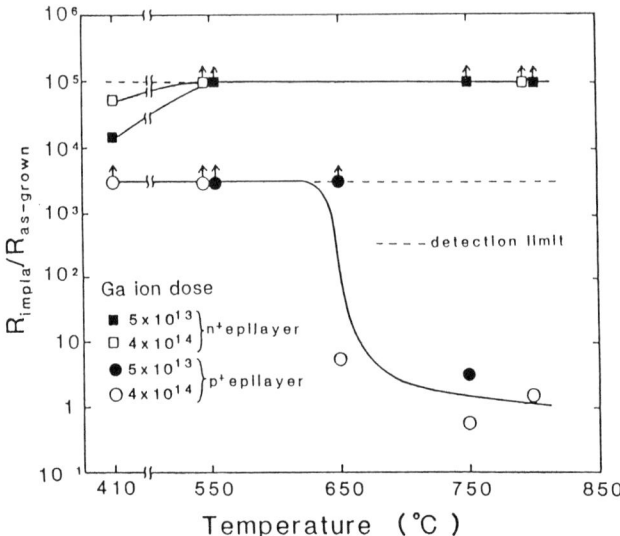

FIG. 19. Resistance enhancement due to Ga FIB implantation into n^+ and p^+ epilayers as a function of annealing temperature. n^+ epilayer: (■) 5×10^{13} cm^{-2} Ga, (□) 4×10^{14} cm^{-2} Ga; p^+ epilayer: (●) 5×10^{13} cm^{-2} Ga, (○) 4×10^{14} cm^{-2} Ga. [From Hirayama and Okamoto (1985c).]

mechanism. Photoluminescence studies have shown the presence of deep acceptor levels in B- and Ga-implanted GaAs layers followed by annealing at 850°C (Dansas, 1985). These levels are tentatively ascribed to Ga_{As} and B_{As} antisite defects that are produced with the stoichiometric imbalance induced by implantation and the influence of B implantation and presumably relate to resistivity enhancement. Whatever the mechanism, fine pattern isolated regions can be formed by a simple process using FIB implantation.

IV. Compositional Disordering of GaAs–AlGaAs

Material modification by FIB implantation offers new possibilities for fabrication of novel device structures. It is well known that Zn and Si diffusion into GaAs–$Al_xGa_{1-x}As$ superlattice (SL) or quantum-well (QW) heterostructures enhance the Al–Ga interdiffusion at the heterointerfaces and create uniform, compositionally disordered $Al_{x'}Ga_{1-x'}As$ (Laidig et al., 1981; Meehan et al., 1984). Heterointerfaces free of impurity are ordinarily stable to high-temperature heat-treatment (Chang and Koma, 1976), whereas with high concentration doping of impurities they become unstable and cause disordering even at much lower temperatures. Ion implantation, followed by thermal annealing, similarly brings about this phenomenon (Coleman *et al.*, 1982). The heterostructures also can be converted to compositionally disordered crystal with various impurities in addition to Zn and Si (Hirayama *et al.*, 1985a). Multiple quantum-well (MQW) lasers with index-guided buried heterostructure that exhibits both optical and electrical confinement have been made successfully by this simple process of selective Zn- or Si-diffusion-induced disordering (Fukuzawa *et al.*, 1984; Gavrilovic *et al.*, 1985).

Recently, selective-area disordering in SL was demonstrated using implantation by line-and-space-scanning Ga or Si FIBs (Hirayama *et al.*, 1985b; Fukunaga *et al.*, 1985). Pattern-drawing with FIBs offers the possibility of fabricating finely disordered structures such as the gratings incorporated in integrated optical devices. This FIB technique is also attractive for high-speed devices consisting of heterostructure crystal as well as optical devices. If high-quality, ultrafine GaAs-quantum-well wires of GaAs–AlGaAs heterocrystal structures are fabricated by compositional disordering, quasi-one-dimensional electron gas with extremely high mobility will be induced in a GaAs region (Sakaki, 1980), leading to a new high-speed device.

The photograph shown in Fig. 20 is a SEM image of a cleaved and stained cross section of $Al_{0.5}Ga_{0.5}As$(30 nm)–GaAs (30 nm) MQW disordered with Si FIB doping (Fukunaga *et al.*, 1985). Si FIBs of 0.2-μm diameter were implanted selectively in MQW at 80 keV with a dose of 3.3×10^{14} cm^{-2} by line-scan. Following implantation, the sample was heat-treated at 850°C for

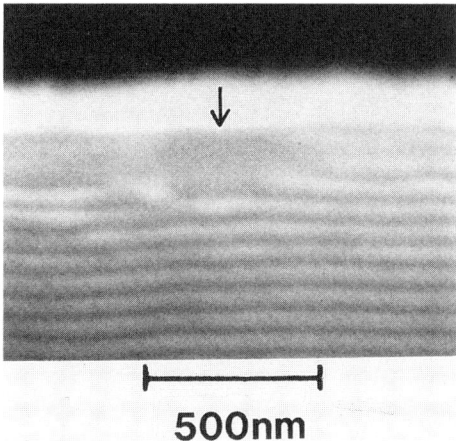

FIG. 20. SEM image of cross section of $Al_{0.5}Ga_{0.5}As$(30 nm)–GaAs(30 nm) MQW line-implanted with 80 keV, 0.2-μm-diameter Si FIB and annealed at 850°C for 1 h. The dose is 3.3×10^{14} cm^{-2}. [From Fukunaga *et al.* (1985).]

1 h in an evacuated quartz ampoule with an As overpressure. The dark and white in the photograph indicate the AlGaAs and GaAs layers, respectively. The top two surface layers of MQW were removed by stain etching. Si FIB implantation–induced disordering is observed clearly in the arrowed region in Fig. 20.

Disordering is also confirmed from the depth profiles of Ga in the crystal by sputtering Auger electron spectroscopy (AES) as shown in Fig. 21 (Fukunaga *et al.*, 1985). In this case, the samples were doped with Si by overlapping 80-keV Si FIB scans on areas large enough for AES measurements and annealed under the identical conditions as in previous experiment in Fig. 20. The implant doses in MQW were varied from 6.7×10^{13} to 7.0×10^{14} cm^{-2}. The disordered regions expand with increasing dose and are distributed beyond the as-implanted Si profiles. Their deeper distributions relate to abnormally enhanced Si diffusion at high concentration, which is accompanied by compositional disordering of GaAs–AlGaAs MQW. The measurement results by secondary ion mass spectroscopy (SIMS)

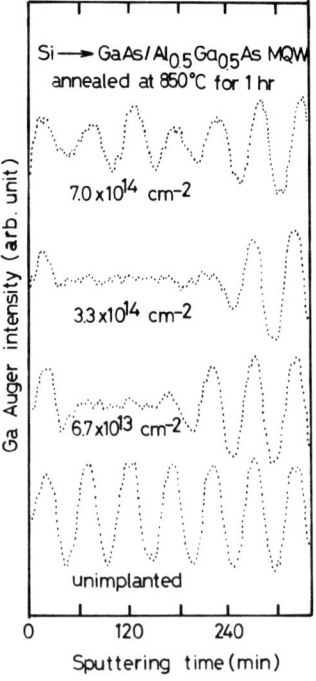

FIG. 21. The Ga Auger signal depth profile of $Al_{0.5}Ga_{0.5}As$(30 nm)–GaAs(30 nm) MQW implanted with 80-keV Si ions at various doses, followed by annealing at 850°C for 1 h. [From Fukunaga *et al.* (1985).]

show that the extension of disordered regions coincides with the Si distribution after diffusion (Kobayashi et al., 1986). With a further increase in the Si dose to 7.0×10^{14} cm^{-2}, however, the complete disordering is not observed, as shown in Fig. 21. Also, the surface top layers are not disordered irrespective of the high Si doses.

It is well known that Si diffuses rapidly in GaAs at high concentrations. For low concentrations, Si ordinarily diffuses through vacancies as isolated Si atoms, producing very low diffusivity. However, at high concentrations, Si tends to form a neutral donor–acceptor pair on the nearest neighboring Ga–As sites (Greiner and Gibbons, 1984). The measurement result of amphoteric Si location in GaAs by channeling particle-induced X-ray emission (PIXE) confirms that the Si atoms in the high-doped region occupy both Ga and As sites with about equal probabilities (Narusawa et al., 1985). These paired Si atoms probably can move substitutionally by exchanging sites with either a Ga or an As vacancy, and they have large diffusion coefficients. Such Ga or As vacancies generated with Si diffusion seem to cause Al–Ga intermixing at the heterointerface, leading to disordering of MQW. The precise mechanism of disordering is now under study. This disordering phenomenon gives us interesting information for study of the physics of atom migration in III–V compound semiconductors, along with its practical application to device fabrication processes.

V. FIB-Doping MBE Growth

9. Growth System

Owing to its maskless patterning capability, FIB implantation is able to combine with other vacuum processes and develop new technology. The combination of MBE growth system with an FIB implanter, which is termed FIBI–MBE, has been studied (Takamori et al., 1984; Miyauchi and Hashimoto, 1986).

Because it offers precise controllability of composition, thickness, and doping profile in the direction of growth on an atomic scale, MBE is used to grow crystal for high-speed devices such as high electron mobility transistors (HEMTs) (Mimura et al., 1980). However, in the case of GaAs and AlGaAs MBE, it is difficult to form a high-quality interface between an epitaxial layer and a substrate that has once been exposed to air or dry nitrogen. The metamorphic layer which is formed at the growth-interrupted interface gives rise to depletion of carriers and decreases in photoluminescence intensity (Kawai et al., 1982; Takamori et al., 1984), causing difficulty in fabricating devices. Contamination on the crystal surface with exposure to these atmospheres probably generates defects acting as electron

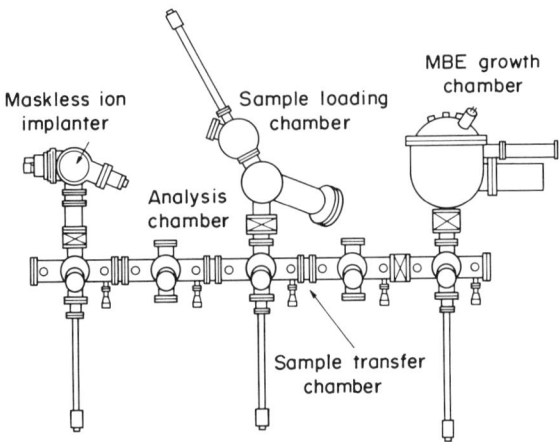

FIG. 22. Schematic structure of the FIBI–MBE system. [From Takamori et al. (1984).]

traps and nonradiative centers. Because of these phenomena, it is difficult to fabricate devices by using conventional ion implantation and photolithography followed by crystal growth.

In order to prevent such difficulties and to grow a crystal with selective area doping, the FIBI–MBE growth system has been developed. As shown in Fig. 22, this system consists of an MBE growth chamber, a 100-kV FIB implanter, and an analysis chamber equipped with Auger electron spectroscopy, each of which is connected with a sample transfer chamber in an ultrahigh vacuum below 5×10^{-10} torr (Takamori et al., 1984). The multilayered crystal with three-dimensional pattern doping structures is grown by transferring the substrate back and forth between the growth chamber and the implanter.

Figure 23 demonstrates the patterned Be- and Si-doped MBE-grown GaAs multilayer with positional alignment. The photograph is an SEM image of a cleaved and stained cross section of the wafer. In this sample, first Si was

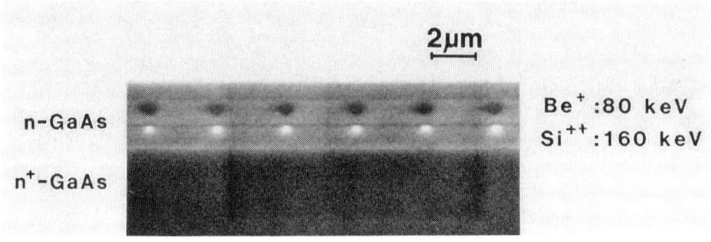

FIG. 23. SEM image of cross section of Be- and Si-doped GaAs grown by FIBI–MBE.

FIB-implanted with a diameter of 0.2 μm into a p-type GaAs epitaxial layer (1×10^{16} cm^{-3}, 1 μm) grown on a semi-insulating substrate with a dose of 1×10^{14} cm^{-2} and an acceleration energy of 160 keV. This implantation formed impurity-doping lines with an interval of 3 μm. Then, a p-type GaAs epitaxial layer 1 μm thick was grown and Be was FIB-implanted with the identical pattern with reference to the benchmarks made on the wafer by etching for pattern alignment. After Be implantation, a 0.5-μm-thick GaAs layer was overgrown by MBE, and a CVD SiO$_2$ film of 0.1-μm thickness was deposited for encapsulation. Finally, the sample was annealed at 850°C for 20 min to activate the implanted atoms. The upper and lower doping regions are well aligned within an accuracy of 0.2 μm. Without overgrowth on the benchmarks, FIBs can be implanted in the designated regions within an alignment accuracy of 0.1 μm. After the overgrowth, the benchmark features are deformed because of the anisotropic growth of crystallographic orientation, causing misalignment. However, it is enough to align and implant FIBs into the designated regions within 0.3 μm even when 2-μm-thick GaAs is overgrown.

10. Quality of FIBI–MBE Grown Crystal

FIBI–MBE-grown GaAs and GaAs/AlGaAs have good electrical and optical characteristics in consequence of a through process in ultrahigh vacuum (UHV). As already mentioned, once a growing GaAs surface is exposed to the atmosphere or the degraded vacuum, a crystal regrown on it has deteriorative interface properties. Carrier concentration profiles in selectively Si-doped GaAs that was grown with interruption for FIB implantation under various vacuum grades were investigated (Takamori et al., 1985a).

The sample structure is shown in Fig. 24a. First, an n-type GaAs layer (1×10^{16} cm^{-3}) doped with Si was MBE-grown on an n-type GaAs substrate to a thickness of 2 μm at 600°C. Then, 160-keV Si FIBs of 0.2-μm diameter were raster-scanned over a 500 μm × 500 μm square in this epitaxial layer with a dose of 1×10^{13} cm^{-2}. After implantation, the sample was not annealed at high temperature, but was thermally cleaned at 630°C for 3 min prior to the second growth, as is usually done for GaAs MBE. Then, the 1.0-μm layer of n-type GaAs (1×10^{16} cm^{-3}) and 0.5 μm of p-type GaAs (1×10^{19} cm^{-3}) were grown at 600°C without interruption. After growth, the sample was taken out of the FIBI–MBE vacuum system and annealed at 850°C for 20 min with SiO$_2$ encapsulation. Finally, Au/Zn ohmic contacts were formed, and discrete p–n junction diodes were fabricated by mesa etching.

The carrier concentration profiles in the Si-implanted and unimplanted

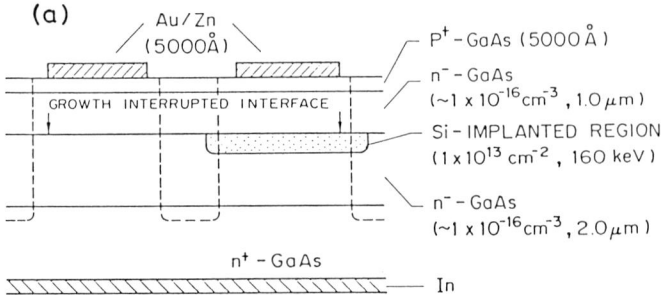

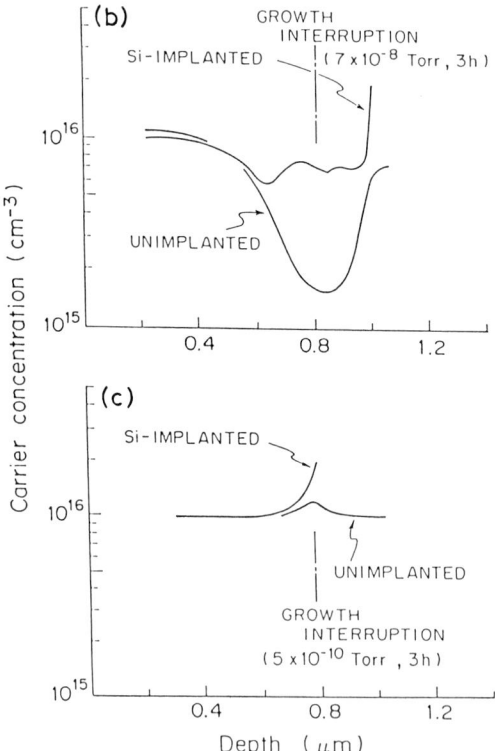

FIG. 24. (a) Schematic structure used for C–V measurements around the growth-interrupted interface. (b) Carrier concentration depth profile of selectively Si-implanted n-GaAs with interruption of 3 h at 7×10^{-8} torr, (c) 3 h at 5×10^{-10} torr. [From Takamori et al. (1985a).]

epitaxial layers, with growth interruption for 3 h in 7×10^{-8} torr, were determined from the C–V measurement and are shown in Fig. 24b. In the unimplanted region, a carrier depletion layer about $0.4\,\mu$m thick is formed around the growth-interrupted interface. The carrier depletion layer is also formed at the Si-implanted interface, but the dip is much shallower. The carrier profile is not uniform around the interface on the implanted layer. In the Si-implanted region, the tail of the carrier profile appears to recede about $0.2\,\mu$m into the substrate. This suggests that implanted Si donors may be compensated by acceptors generated near the interface by growth interruption. The interface state density estimated from the carrier depletion is about 2×10^{11} cm^{-2}, which is of the same order of magnitude as the data reported for interfaces exposed to the atmosphere (Kawai *et al.*, 1982; Chang and Kroemer, 1984).

In contrast, the carrier profiles in the sample growth with interruption for 3 h in 5×10^{-10}-torr vacuum are shown in Fig. 24c. No carrier depletion is observed at the interface, regardless of Si implantation. It is speculated that the carrier depletion is caused by interface states related to adsorbed contaminants during interruption of MBE growth.

Figure 25 shows the depth distribution of carbon, oxygen, and implanted Si in GaAs epitaxial layers grown with interruption for implantation. The profiles were measured by secondary ion mass spectroscopy (SIMS) (Takamori *et al.*, 1985a). The sample structure is almost the same as that illustrated in Fig. 24a, but with no top *p*-layer. In the sample whose interface was exposed to a 5×10^{-10}-torr atmosphere for 3 h during growth interruption, no contamination of carbon and oxygen was observed at the interface, as demonstrated in Fig. 25a. It should be noted that the contamination was not detected even on the Si-implanted region. For the sample exposed to the atmosphere for a few minutes during growth interruption, carbon was significantly increased at the interface, but oxygen was not, as shown in Fig. 25b. Adsorbed oxygen is readily removed by thermal cleaning prior to regrowth at 630°C, but carbon remains.

Carbon ordinarily acts as a shallow acceptor in GaAs, so that carrier compensation with it is expected in the *n*-type sample. However, photoluminescence intensity also decreases remarkably at the interface in a *p*-type epitaxial layer grown on a *p*-GaAs substrate after air exposure. This means that a high density of deep levels acting as nonradiative centers exists at the interface. Considering these experimental results, it is speculated that the carrier depletion at the interface as observed in Fig. 24b is probably caused by interface deep states related to carbon contamination.

The FIBI–MBE system enables us to grow high-quality GaAs/AlGaAs multilayers even though AlGaAs is more chemically active. The interfacial quality of $Al_{0.3}Ga_{0.7}As$ grown with interruption in 5×10^{-10} torr for ion

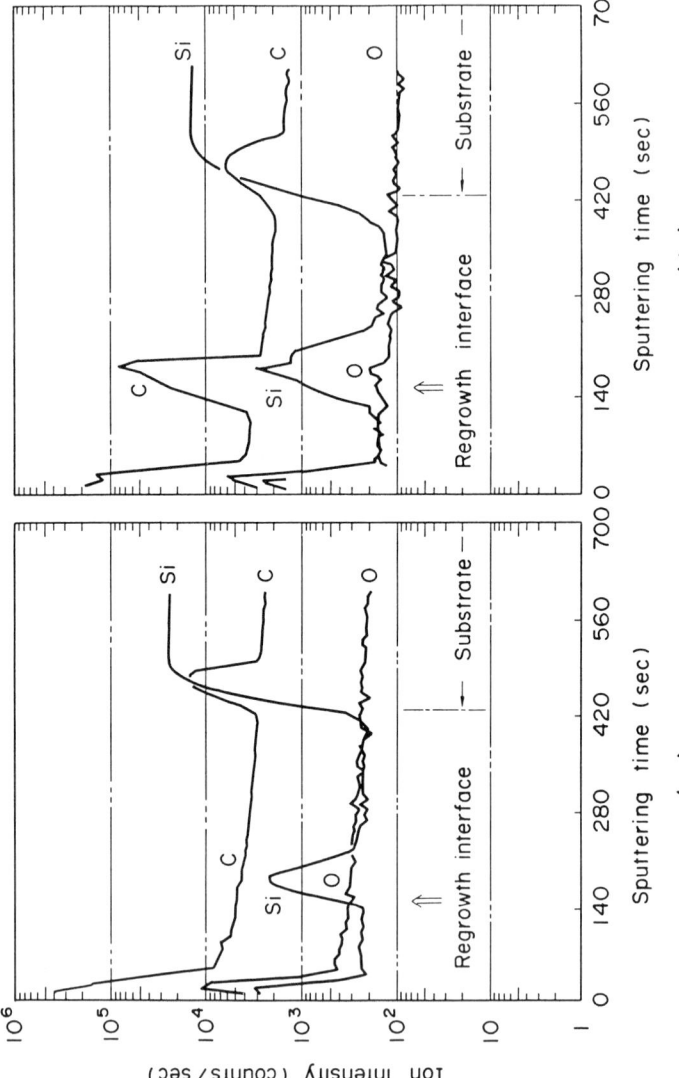

FIG. 25. Depth profiles of carbon, oxygen, and implanted Si in growth-interrupted samples (a) kept at a background vacuum of 5×10^{-10} torr during interruption for 3 h, (b) exposed to air. [From Takamori et al. (1985a).]

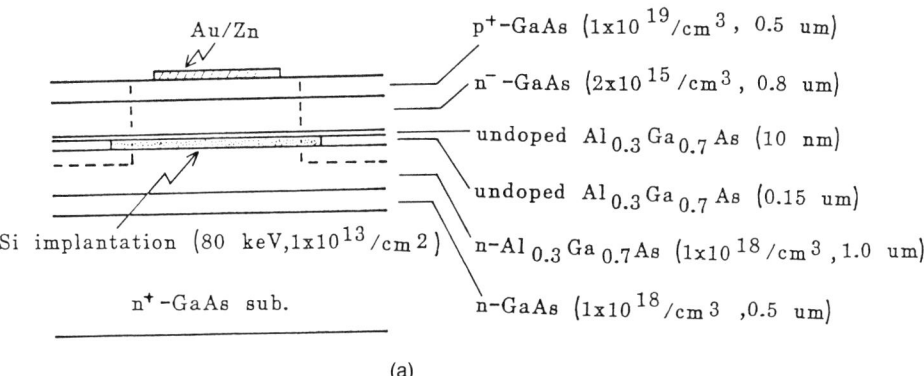

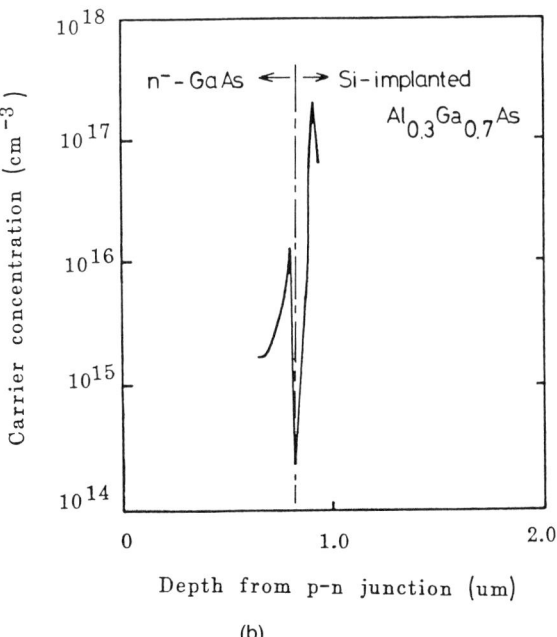

FIG. 26. (a) Schematic structure of the sample and (b) the carrier concentration profile in GaAs/FIB-implantation-doped AlGaAs.

implantation doping is almost comparable to the standard MBE-grown crystal without growth interruption. Excellent interfacial properties obtained with the FIBI–MBE system offer the possibility of developing a new type of device such as a complementary HEMT that has a two-dimensional electron and hole gas induced at a GaAs/AlGaAs interface by selective FIB doping in the AlGaAs layer.

To investigate the possibility of this type of HEMT, a preliminary experiment was carried out. Figure 26a shows a sample structure for measuring a carrier profile in GaAs/n-AlGaAs doped by Si FIB implantation. Si was FIB-implanted in the undoped AlGaAs layer at 80 keV with a dose of 1×10^{13} cm^{-2}, followed by growth of the 10-nm undoped $Al_{0.3}Ga_{0.7}As$ spacer layer, n^--GaAs (2×10^{15} cm^{-3}, 0.8 μm) and p^+-GaAs (1×10^{19} cm^{-3}, 0.5 μm) at 670°C. After the growth, the sample was annealed at 850°C for 20 min to activate implanted Si. As shown in Fig. 26b, selective Si FIB implantation doping in $Al_{0.3}Ga_{0.7}As$ induces a two-dimensional electron gas in the GaAs side of the hetero-interface. This modulation doping technique using FIBI–MBE may make possible the realization of a novel HEMT in which the concentration of the two-dimensional electron and hole gas is controlled by ion implantation.

Crystal quality of GaAs and AlGaAs with growth interruption, of course, depends upon the growth conditions. Conditions of growth interruption to keep the interface of high quality are represented in Fig. 27 as a function of vacuum pressure and growth-interruption time (Takamori et al., 1985b). Open circles indicate samples having no carrier depletion, while closed circles show the samples that have interface-state density higher than 2×10^{11} cm^{-2}. The hatched areas indicate critical conditions. Referring to the results of

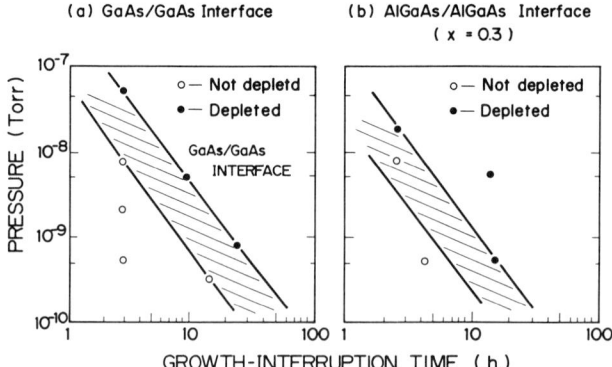

FIG. 27. Growth condition with and without carrier depletion around the growth-interrupted (a) GaAs/GaAs and (b) $Al_{0.3}Ga_{0.7}As/Al_{0.3}Ga_{0.7}As$ interface as a function of growth-interruption time and pressure. ○, not depleted; ●, depleted. [From Takamori et al. (1985b).]

Fig. 27, the conditions of vacuum pressure and interrupt time allowable for growth of FIB-doping GaAs and AlGaAs with a high-quality interface can be estimated. The photoluminescence data measured at the interface of GaAs/GaAs and AlGaAs/AlGaAs grown with various interruption conditions also give the same result, as shown in Fig. 27.

VI. Prospect of FIB Technology

As already described, the FIB implantation certainly has many potential advantages over conventional ion implantation. This technology, however, has just been introduced, and many engineering issues must be improved for its implementation in the actual production line. One such problem is a poor throughput. Figure 28 shows an area throughput given by Brown and Wagner (1983). The implanted area throughput is represented as a function of beam diameter. For this calculation, ion fluence is estimated by assuming the largest value of a 1 A/cm^2 current density for scanning ion beams. The quantity of implanted ions decreases as beam diameter becomes smaller. Even if the reduction in number of process steps by the maskless method of FIB implantation is considered, the practical throughput is very small as compared with the conventional UIB-implantation at present. Considering the emission mechanism of a LM ion source, it is difficult to increase the ion current density drastically. Therefore, it is necessary to study other methods to improve the throughput. Actual devices do not always necessitate the implantation of impurities over the whole area of a wafer using fine-focused

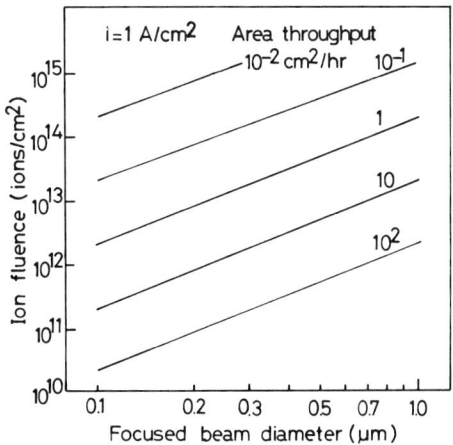

FIG. 28. Area throughput of FIB implantation as a function of beam diameter and ion dose for a 1 A/cm^2 beam. [From Brown and Wagner (1983).]

ion beams. As a matter of course, the largest throughput can be achieved by using the largest beam size that is consistent with the required pattern size. Development of a variable-shaped ion beam implanter like a high-speed electron beam exposure system (Moor et al., 1981) is the most promising way to increase the throughput.

Smooth operation of the FIB implanter is also an inevitable requirement for its practical use. The major factor governing successful operation of the FIB implanter is stable emission of ion beams from a LM ion source. The Au–Si–Be LM ion source usually has a long lifetime (more than 700 hours) and can operate stably for a long time. However, abnormal ion-emission and fluctuations of ion beam current that happen to move the beam position on a target are sometimes observed (Arimoto et al., 1986a). Disturbance of the LM flow on the tip of the ion source under an intense applied electric field probably causes this phenomenon. Recently it was found that contamination of the liquid metal from an ambient atmosphere gave rise to irregular flow. To avoid contamination, it is effective to operate ion sources in an ultrahigh vacuum. For the Au–Si–Be LM ion source, stable emission is observed in a vacuum better than 2×10^{-7} torr (Arimoto et al., 1986b).

Another problem with the ion source is erosion of the emitter tip during high-temperature operation. For more reactive metals and alloys, their corrosive reactions with the materials of the needle vary its apex shape and subsequently move the Taylor cone position in addition to the flow disturbance. In order to realize smooth operation of the FIB implanter, further studies on the ion source are needed.

The ability of the FIB implanter operated by a computer depends greatly upon its software. The software must be improved still more to implant dopants arbitrarily in specified patterns with various conditions of ion species, accelerating energy, dose and beam size in accordance with device design. When the above issues are overcome, FIB implantation will truly emerge as a useful and powerful technology.

The maskless method using FIBs is easy to combine with other fabrication processes in a vacuum, making it possible to offer a novel technique. The combination of a FIB implanter with an MBE growth chamber is one example, as explained in this chapter. Arbitrary pattern-doping is performed without air exposure of growing crystal surfaces during growth. High-quality crystals can be grown because of the freedom from contamination. Such a new process technique fabricates a novel structure that cannot be grown by a conventional method, leading to development of a novel device with high performance. Furthermore, an automated, single-enclosed semiconductor process facility in UHV composed of maskless implantation, etching and so on using FIB is expected in the final stage. These promising techniques will be realized in the near future.

REFERENCES

Arimoto, H., Takamori, A., Miyauchi, E., and Hashimoto, H. (1984). *Jpn. J. Appl. Phys.* **23**, L165.
Arimoto, H., Takamori, A., Miyauchi, E., and Hashimoto, H. (1985). *J. Vac. Sci. Technol.* **B3(1)**, 54.
Arimoto, H., Morita, T., Miyauchi, E., and Hashimoto, H. (1986a). *Jpn. J. Appl. Phys.* **25**, L507.
Arimoto, H., Miyauchi, E., and Hashimoto, H. (1986b). *Jpn. J. Appl. Phys.* **25**, L567.
Bamba, Y., Miyauchi, E., Arimoto, A., Kuramoto, K., Takamori, A., and Hashimoto, H. (1983). *Jpn. J. Appl. Phys.* **22**, L650.
Bamba, Y., Miyauchi, E., Arimoto, H., Takamori, A., and Hashimoto, H. (1984). *Jpn. J. Appl. Phys.* **23**, L515.
Bogardus, E. H., and Bebb, H. B. (1968). *Phys. Rev.* **176**, 993.
Brown, W. L., and Wagner, A. (1983). *Proc. Int. Ion Engineering Congress, Kyoto*, p. 1738 A.
Chang, L. L., and Koma, A. (1976). *Appl. Phys. Lett.* **29**, 138.
Chang, Y., and Kroemer, H. (1984). *Appl. Phys. Lett.* **45**, 449.
Coleman, J. J., Dapkus, P. D., Kirkpatrick, C. G., Camras, M. D., and Holonyak, N., Jr. (1982). *Appl. Phys. Lett.* **40**, 904.
Dansas, P. (1985). *J. Appl. Phys.* **58**, 2212.
Fukunaga, T., Ishida, K., Kuroda, T., Matsui, K., Narusawa, T., Morita, T., Miyauchi, E., Hashimoto, H., and Nakashima, H. (1985). *Inst. Phys. Conf. Ser.* **79**, 439.
Fukuzawa, T., Semura, S., Saito, H., Ohta, T., Uchida, Y., and Nakashima, H. (1984). *Appl. Phys. Lett.* **45**, 1.
Gamo, K., Matsui, T., and Namba, S. (1983). *Jpn. J. Appl. Phys.* **22**, L692.
Gavrilovic, P., Meehan, K., Guido, L. J., Holonyak, N., Jr., Eu, V., Feng, M., and Burnham, R. D. (1985). *Appl. Phys. Lett.* **47**, 903.
Gomer, R. (1979). *Appl. Phys.* **19**, 365.
Greiner, M. E., and Gibbons, J. F. (1984). *Appl. Phys. Lett.* **44**, 750.
Harris, J. S. (1971). In "Proceedings of the 2nd International Conference on Ion Implantation in Semiconductors" (Ruge, I., and Graul, J., eds.), p. 157. Springer, Berlin.
Hashimoto, H., and Miyauchi, E. (1984). *Ext. Abst. 16th (1984 Int.) Conf. Solid State Devices and Materials, Kobe*, p. 121.
Hirayama, Y., and Okamoto, H. (1985c). *Jpn. J. Appl. Phys.* **24**, L965.
Hirayama, Y., Suzuki, Y., and Okamoto, H. (1985a). *Jpn. J. Appl. Phys.* **24**, 1498.
Hirayama, Y., Suzuki, Y., Tarucha, S., and Okamoto, H. (1985b). *Jpn. J. Appl. Phys.* **24**, L516.
Ishikawa, J., and Takagi, T. (1984). *J. Appl. Phys.* **56**, 3050.
Kawai, N. J., Wood, C. E. C., and Eastman, L. F. (1982). *J. Appl. Phys.* **53**, 6208.
Kobayashi, J., Nakajima, M., Bamba, Y., Fukunaga, T., Matsui, K., Ishida, K., Nakashima, H., and Ishida, K. (1986). *Jpn. J. Appl. Phys.* **25**, L385.
Kubena, R. L., Anderson, C. L., Seliger, R. L., Jullens, R. A., Stevens, E. H., and Lagnado, I. (1981). *J. Vac. Sci. Technol.* **19(4)**, 916.
Laidig, W. D., Holonyak, N., Jr., Camras, M. D., Hess, K., Coleman, J. J., Dapkus, P. D., and Bardeen, J. (1981). *Appl. Phys. Lett.* **38**, 776.
Lidow, A., Gibbons, J. F., Deline, V. R., and Evans, C. A., Jr. (1978). *Appl. Phys. Lett.* **32**, 149.
Martin, G. M., Secordel, P., and Venger, C. (1982). *J. Appl. Phys.* **53**, 8706.
McLevige, W. V., Helix, M. J., Vaidyanathan, K. V., and Streetman, B. G. (1977). *J. Appl. Phys.* **48**, 3342.
Meehan, K., Holonyak, N., Jr., Brown, J. M., Nixon, M. A., Gavrilovic, P., and Burnham, R. D. (1984). *Appl. Phys. Lett.* **45**, 549.

Mimura, T., Hiyamizu, S., Fujii, T., and Nanbu, K. (1980). *Jpn. J. Appl. Phys.* **19**, L225.
Miyauchi, E., and Hashimoto, H. (1986). *J. Vac. Sci. Technol.* **A4**(3), 933.
Miyauchi, E., Hashimoto, H., and Utsumi, T. (1983a). *Jpn. J. Appl. Phys.* **22**, L225.
Miyauchi, E., Arimoto, H., Hashimoto, H., Furuya, T., and Utsumi, T. (1983b). *Jpn. J. Appl. Phys.* **22**, L287.
Miyauchi, E., Arimoto, H., Bamba, Y., Takamori, A., Hashimoto, H., and Utsumi, T. (1983c). *Jpn. J. Appl.* **22**, L423.
Miyauchi, E., Arimoto, H., Hashimoto, H., and Utsumi, T. (1983d). *J. Vac. Sci. Technol.* **B1**(4), 1113.
Miyauchi, E., Arimoto, H., Morita, T., and Hashimoto, H. (1987). *Jpn. J. Appl. Phys.* **26**, L145.
Moor, R. D., Caccoma, G. A., Pfeiffer, H. C., Weber, E. V., and Woodard, O. C. (1981). *J. Vac. Sci. Technol.* **19**, 950.
Morita, T., Miyauchi, E., Arimoto, H., Takamori, A., Bamba, Y., and Hashimoto, H. (1986). *J. Vac. Sci. Technol.* **B4**(4), 829.
Müller, E. W., and Tsong, T. T. (1969). "Field Ion Microscopy Principles and Applications." Elsevier, New York.
Nakamura, T., and Katoda, T. (1982). *J. Appl. Phys.* **53**, 5870.
Nakamura, K., Nozaki, T., Shiokawa, T., Toyoda, K., and Namba, S. (1985). *Jpn. J. Appl. Phys.* **24**, L903.
Narusawa, T., Uchida, Y., Kobayashi, K. L., Ohta, T., Nakajima, M., and Nakashima, H. (1985). *Inst. Phys. Conf. Ser.* **74**, 127.
Nelson, R. S. (1968). "The Observation of Atomic Collisions in Crystalline Solids," p. 270. North Holland, Amsterdam.
Sakaki, H. (1980). *Jpn. J. Appl. Phys.* **19**, L735.
Seliger, R. L., Ward, J. W., Wang, V., and Kubena, R. L. (1979). **34**, 310.
Seliger, R. L., Kubena, R. L., and Wang, V. (1982). *Jpn. Appl. Phys. Supple. 21-1*, 3.
Shiokawa, T., Kim, P. H., Toyoda, K., Namba, S., Gamo, K., Aihara, R., and Anazawa, N. (1985). *Jpn. J. Appl. Phys.* **24**, L566.
Speight, J. D., O'Sullivan, P., Leigh, P. A., McInlyre, N., Cooper, K., and O'Hara, S., (1977). *Inst. Phys. Conf. Ser.* **33A**, 275.
Swanson, L. W., Schwind, G. A., Bell, A. E., and Brady, J. E. (1979). *J. Vac. Sci. Technol.* **16**, 1864.
Takamori, A., Miyauchi, E., Arimoto, H., Bamba, Y., and Hashimoto, H. (1984). *Jpn. J. Appl. Phys.* **23**, L599.
Takamori, A., Miyauchi, E., Arimoto, H., Bamba, Y., Morita, T., and Hashimoto, H. (1985a). *Jpn. J. Appl. Phys.* **24**, L414.
Takamori, A., Miyauchi, E., Arimoto, H., Morita, T., Bamba, Y., and Hashimoto, H. (1985b). *Inst. Phys. Conf. Ser.* **79**, 247.
Tsong, T. T. (1978). *Surface Sci.* **70**, 211.
Wada, Y., Shukuri, S., Tamura, M., Masada, H., and Ishitani, T. (1985). *Proc. Electrochemical Soc. Meeting, Las Vegas, Oct. 1985,* p. 433.
Wagner, A., Venkatesan, T., Petroff, P. M., and Barr, D. (1981). *J. Vac. Sci. Technol.* **19**(4), 1186.
Wang, V., Ward, J. W., and Seliger, R. L. (1981). *J. Vac. Sci. Technol.* **19**(4), 1158.
Ward, J. W., Utraut, M. W., and Kubena, R. L. (1987). *J. Vac. Sci. Technol.* **B5**(1), 169.
Wilson, R. G., and Jamba, D. M. (1981). *Appl. Phys. Lett.* **39**, 715.
Yaney, P. P., Baird, W. E., Jr., and Park, Y. S. (1981). *In* "Proceedings of SPIE—The International Society for Optical Engineering (Aspnes, D. E., So, S., and Potter, R. F., eds.), Vol. 276, p. 84. SPIE, Washington.

CHAPTER 3

Device Fabrication Process Technology

T. Nozaki

OPTOELECTRONICS RESEARCH LABORATORIES
NEC CORPORATION
KAWASAKI, JAPAN

A. Higashisaka

COMPOUND SEMICONDUCTOR DEVICE DIVISION
NEC CORPORATION
KAWASAKI, JAPAN

I.	INTRODUCTION	97
II.	FUNDAMENTAL PROCESS TECHNOLOGIES	99
	1. Ohmic Contact Formation	99
	2. High-Stability Schottky Barrier Gate Technology	104
	3. Dry Etching Process Technology	111
III.	SELF-ALIGNED MESFET TECHNOLOGY	114
	4. Refractory Metal Gate Technology	115
	5. SAINT (Self-Aligned Implantation for n^+-Layer Technology)	119
	6. SWAT (Side-Wall Assisted Self-Alignment Technology) . .	123
	7. LDD (Lightly Doped Drain) Technology	127
IV.	DEVICE TECHNOLOGY EFFECTIVE FOR IMPROVING CHARACTERISTICS IN SUBMICRON GATE MESFETs	131
	8. Alleviation of Short-Channel Effects	131
	9. Suppression of the Orientation Effect	146
V.	SUMMARY	153
	REFERENCES	155

I. Introduction

Since the first GaAs ICs were developed in 1974 (Van Tuyl and Liechti, 1974) progress in device fabrication process technology has continued to realize high-speed and high packing density GaAs ICs. To meet the demand in the planar process, which is necessary for attaining high-yield large-scale integration (LSI)-compatible processes, active layer formation technology by selective ion implantation into semi-insulating GaAs was investigated. In this technology, donor impurities, such as Si and Se, are implanted and electrically activated by post-implantation annealing. This technology plays an important role in LSI fabrication. However, substrate quality improvements are vital for achieving good controllability of threshold voltage values. Present status and future subject in active-layer formation are described in Chapter 1.

Among the several kinds of logic gates, the DCFL (direct coupled field effect transistor [FET] logic) circuit has been attracting much attention in the approach to LSIs, owing to its circuit simplicity and low power dissipation. The main problems with DCFL technology are the poor reproducibility of device characteristics for enhancement-mode GaAs metal-semiconductor FETs (MESFETs), and their high parasitic resistance. To overcome these problems, several device structures have been investigated, such as closely spaced electrode structure (Furutsuka et al., 1981). Pt-buried gate structures (Hojo et al., 1981) and n^+ self-aligned gate structures. Two kinds of n^+ self-aligned gate structures have been investigated. One is the refractory silicide gate (Yokoyama et al., 1981). Another one is SAINT (self-aligned implantations for n^+-layer technology; Yamasaki et al., 1982a). In both structures, parasitic resistance reduction is possible, since n^+ layers are embedded adjacent to the gate electrode. Using these n^+ self-aligned gate structures, high performance GaAs LSIs, such as static RAM and gate array, have been fabricated, as described in Chapter 5.

Generally, gate shortening is effective to achieve further progress in performance and packing density in LSIs. However, in submicron-gate MESFETs, so-called short-channel effects emerge, which result in an undesired decrease in transconductance value and degradation in threshold voltage controllability due to negative shifts in threshold voltage value. To reduce short-channel effects, several technologies have been investigated. In addition to short-channel effects due to the piezoelectric property of GaAs crystals, the orientation effect has been recognized—variations in device characteristics, such as threshold voltage and saturation drain current, according to the gate-finger orientation with respect to the $\langle 100 \rangle$ GaAs substrates. At this stage, to avoid these differences in device characteristics, a gate direction is arranged along one crystal orientation in GaAs LSIs. From these situations, this gate alignment is of great importance in developing process technologies that are effective in reducing short-channel effects as well as in controlling the orientation effect.

In this chapter, the state of the art and future subjects in device fabrication process technology are presented and discussed. In Part II, fundamental process technology is presented, including high-stability Schottky barrier gate technology, ohmic contact formation and dry etching process technology. Self-aligned MESFET technology, which is described in Part III, is vital for achieving high performance and high packing density GaAs LSIs. Several process technologies for attaining high device performance through parasitic resistance reduction are presented. In Part IV, undesired characteristics that emerged in submicron-gate MESFETs—that is, short-channel effects and the orientation effect—are presented, together with a description of how to reduce and how to control these effects. Finally, in Part V, the state of the art is summarized, and necessary improvements in device fabrication process technology are discussed.

II. Fundamental Process Technologies

In comparison with Si device technologies, GaAs device fabrication technologies have some serious limitations, which arise from stoichiometric problems, the high vapor pressure of the group V elements and mechanical fragility. The stoichiometric problems make it difficult to understand the mechanisms of ohmic contact formation and the activation process of the implanted impurities, resulting in poor controllability of these technologies. The high vapor pressure of the constituent elements limits high-temperature treatments, such as thermal oxide formation and thermal diffusion of impurities. Mechanical fragility is disadvantageous for introducing automatic transportation and loading of wafers using the cassette-to-cassette system.

These limitations make GaAs device fabrication technologies difficult and complex. However, to overcome these problems, many fabrication technologies have hitherto been developed and applied to practical fabrications. In this section, some fundamental process technologies, such as ohmic contact formation, high-stability Schottky barrier gate technology and dry etching process technology are described.

1. Ohmic Contact Formation

The term "ohmic" refers in principle to an interface to the external metal that is noninjecting and has linear current–voltage characteristics. The quality of the ohmic contact is one of the most significant factors affecting the performance of GaAs high-speed devices.

According to the simple Schottky model, when a metal is in close contact with a semiconductor, a depletion layer is formed with a depth of $d = \{q\Phi_B/2\varepsilon N_D\}^{1/2}$, where Φ_B is defined as the barrier height of the Schottky contact and is given by Φ_m (metal work function) $- \chi_s$ (semiconductor electron affinity). When the semiconductor is doped so heavily that the depletion layer thickness is sufficiently thin to give a tunneling carrier transport, a good ohmic contact is achieved at the metal–semiconductor interface. On the other hand, when Φ_B is sufficiently low, the number of thermally excited electrons over the Schottky barrier is large enough to give an ohmic carrier transport at the interface. However, a high concentration of the surface states is present on a GaAs surface, which pins the Fermi level near mid-gap so that Φ_B is essentially independent of the metal used and is governed mainly by the interface state. Therefore, a Φ_B-lowering approach is not realistic in a practical stage. One interesting attempt toward lowering barrier height is to grow a heavily doped Ge on top of GaAs by molecular-beam epitaxy (MBE) (Stall *et al.*, 1979). The barrier height at the Ge/GaAs interface is as little as about 50 mV and the thermally excited current is extremely large to give the ohmic characteristics. Metal-n^+Ge-GaAs heterojunction is applied to high-speed GaAs MESFET fabrication (where the AuGe-n^+Ge-nGaAs metal system is adopted to form self-aligned contact regions). Evaporated Au-Ge eutectic (88% and 12%, respectively) with a Ni overlayer is most widely used in the production of GaAs devices (Braslau, 1981). In this system, Ge is shown to be a donor in GaAs when on a Ga site, and Ga vacancies that are produced in an alloying process are populated by indiffusing Ge to a density on the order of 10^{19} cm^{-3}. The heavily doped thin GaAs layer gives an ohmic conduction because of tunneling carrier transport. Metallurgical behavior of alloyed Ni-AuGe/GaAs contact has been investigated through the use of microprobe Auger spectroscopy, X-ray diffraction analysis and SEM observation. One of the models proposed to illustrate the metallurgical interaction is shown in Fig. 1 (Ogawa, 1980). At the early stage of the reaction, Ge outdiffuses rapidly toward the contact surface as the Ni sink for Ge. At the same time, GaAs is broken down at the interface mainly through the reaction between molten Au and GaAs

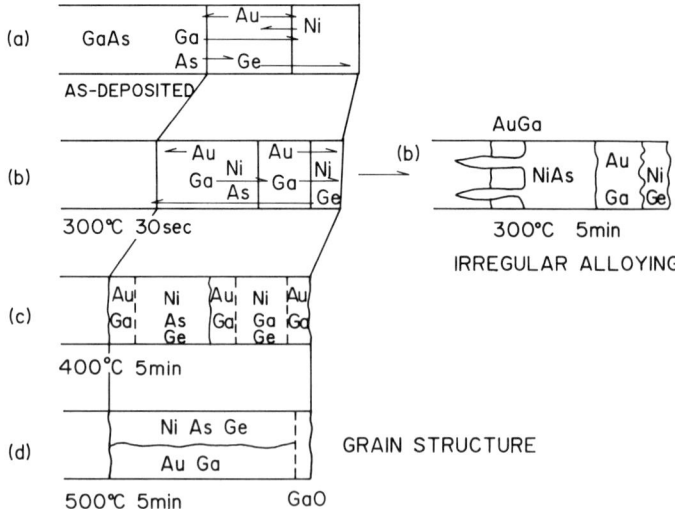

FIG. 1. Model of Ni-AuGe/GaAs ohmic contact alloying. [From Ogawa (1980).]

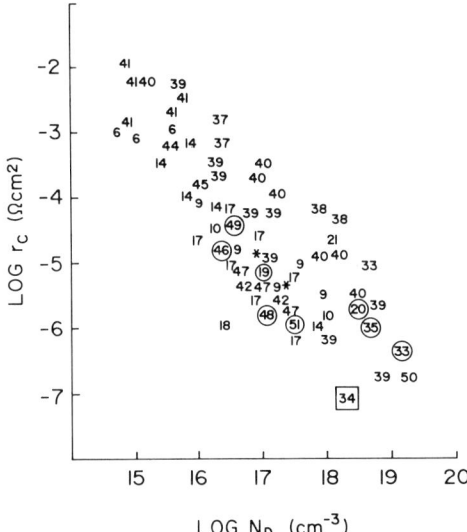

FIG. 2. Observed contact resistance as a function of n-GaAs doping. Numbers identify the reference number used in the paper written by Braslau (1981). Circled points are laser- or electron-beam-annealed. The point in a square is for an MBE-grown heterojunction. Starred points were results reported by Braslau (1981). [From Braslau (1981).]

(the eutectic temperature of the GaAs/Au system is 356°C), and between Ni and GaAs: Au + GaAs → AuGa + As, Ni + As → NiAs. At 400°C, Ge diffuses inward and make the Ni–As–Ge phase. In the cooling process, epitaxial growth takes place, leaving Ge on Ga vacancies, and as a result, the heavily doped layer is created at the metal–semiconductor interface. Ni plays the role of a protector against "balling up" during alloying, as well as that of the driving force for Ge diffusion. The amount of Ni influences the contact reliability. In the case of a thick Ni layer, unreacted Ni is left and further growth of NiAs compounds is the result, giving a Ni–As–Ge phase by trapping Ge. On the other hand, too large an amount of Au results in too much Au–Ga eutectic, producing a large amount of Ga vacancies, which act as acceptors. Macksey (1976) has found the critical value for the Ni layer is 50% of the Au–Ge (12% Ge) layer thickness. Ogawa (1980) defined the optimum Ni thickness as less than that of AuGe.

Contact resistance also depends on the underlying semiconductor doping concentration. Figure 2 shows the observed specific contact resistance r_c as a function of n-GaAs doping level, exhibiting an apparent n_D^{-1} dependence of r_c (Braslau, 1981). For ohmic contact to GaAs, InP, and III–V mixed semiconductors, Au, Ag, and In can be used as base metals in alloyed contacts, but Ag and In are rarely used in practical devices because these

TABLE I

PULSE-ANNEALED OHMIC CONTACTS TO GaAs[a]

Metallization		Annealing conditions			Semiconductor doping (cm^{-3})	Specific contact resistance ($\Omega\ cm^2$)
Contact system	Thickness (Å)	Irradiation	Energy density (J/cm^2)	Pulse duration		
AuGe/Ni/Au			0.64	0.43 cm/s		4.8×10^{-6}
AuGe/Pt/Au			0.61	0.43 cm/s		1.5×10^{-5}
AuGe/Ag/Au	1200–1700/400/500–600	cw argon laser	0.66	0.43 cm/s	10^{17}	2.0×10^{-4}
AuGe/Ti/Au			1.35	0.20 cm/s		1.8×10^{-5}
AuGe/In			0.56	0.43 cm/s		1.3×10^{-6}
AuGe	200/100/Ge top layer	"Free-running" pulsed ruby laser	~10	1 ms	10^{17}	2×10^{-6}
Au/Ge/Ni	200/100/20	$\lambda = 0.6943\ \mu m$				2×10^{-6}
Au/Ge/Ni/Au	1500/400/500	Pulsed ruby laser $\lambda = 0.6943\ \mu m$	0.8	30 ns	2×10^{17}	10^{-5}
AuGe	3000	Q-switched ruby laser $\lambda = 0.6943\ \mu m$	1.02	15 ns	3×10^{16}	$7.04\text{–}5.14 \times 10^{-5}$
AuGe	300–1200	Backside irradiation Q-switched Nd: glass laser $\lambda = 1.06\ \mu m$	0.3–0.5	20 ns	$>10^{17}$	2×10^{-5}
AuGe/Ni/Au			0.28			3.5×10^{-5}
AuGe/Pt/Au	1200–1700/400/500–600	Pulse electron $E = 10\text{–}12\ keV$	0.28	100 ns	10^{17}	8.6×10^{-5}
AuGe/Ag/Au			0.32			2.3×10^{-5}
AuGe/Pt	1300/300	Pulse electron $E = 20\ keV$	0.3–0.5	10^{-7} s	7×10^{17}	4×10^{-7}

[a] From Piotrowska et al. (1983).

metals react readily with the atmosphere. As doping species, Si, Sn, Se, or Te are used for n-type, and Zn, Cd, Be, or Mg are used for p-type GaAs compounds and mixed III–V semiconductors.

In a fabrication process, alloying is usually performed by classical furnace alloying, in which the sample temperature is raised rapidly to 420–500°C, held for from 15 sec to 5 min, and then cooled rapidly in an atmosphere of hydrogen or forming gas. Pulsed annealing using a laser, electron beams or an infrared lamp is promising for high-speed device fabrications, because it gives a superior surface morphology by suppressing the loss of the volatile component of the III–V semiconductor, and it also forms a shallow and uniform contact by suppressing the interdiffusion between metal and semiconductor. A summary of results on pulsed annealed contact is presented in Table I (Piotrowska *et al.*, 1983).

Nonalloyed ohmic contact is also important for the same reason as pulsed annealing. A key point of this technique is to obtain a sufficiently heavy impurity doping ($>10^{19}$ cm^{-3}) for underlying semiconductors to form a tunnelling contact. The laser- or electron-beam-annealed Se$^+$-implanted GaAs exhibited a high impurity concentration of over 10^{19} cm^{-3}, giving a low contact resistance of 10^{-6}–$10^{-7}\,\Omega$ cm^2 (Mozzi *et al.*, 1979). Kuzuhara achieved a specific contact resistance of $6 \times 10^{-6}\,\Omega$ cm^2 by sintering at 300°C (Kuzuhara *et al.*, 1985). In the experiment, heavily doped n-type layers have been prepared by high-temperature infrared rapid thermal annealing using SiO$_x$N$_y$ encapsulants. An electron concentration as high as 9×10^{18} cm^{-3} has been obtained for high-dose Si$^+$ implantation. Low contact resistance of $6 \times 10^{-6}\,\Omega$ cm^2 (Fig. 3) is accomplished by 300°C alloying without melting the AuGe eutectic, resulting in a smooth surface

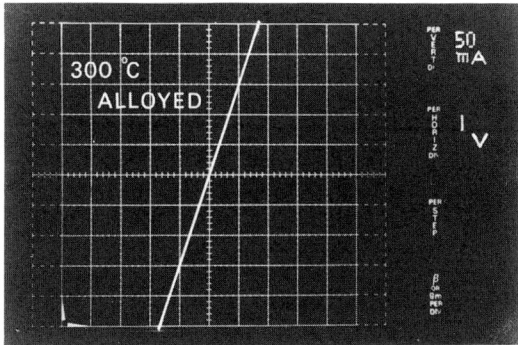

FIG. 3. I–V characteristics for the 300°C alloyed ohmic contact formed on a dual energy Si$^+$-implanted (150 keV, 4×10^{14} cm^{-2} + 50 keV, 1.2×10^{14} cm^{-2}) sample. Measurements were made between 10-μm-spaced, 200-μm-wide contact pads. [From Kuzuhara *et al.* (1985).]

morphology. By using MBE, a highly doped semiconductor can be obtained with sufficient reproducibility and controllability, and so, an ideal non-alloyed ohmic contact can be realized. Because the nonalloyed ohmic contact technique is indispensable to realize fine-pattern, high-speed devices, much effort should be focused on developing this promising technique.

2. High-Stability Schottky Barrier Gate Technology

A good Schottky barrier gate metal for high-speed GaAs device applications should possess the following properties:

(1) Good adhesion and thermal expansion coefficient matching that of GaAs.
(2) Stability against interdiffusion between metal and GaAs.
(3) Stability against humid and oxidizing atmospheres.
(4) High-temperature stability or refractoriness in high-temperature treatment, enabling high-temperature annealing of selectivity ion-implanted layers (Kohn, 1979).
(5) Large Schottky barrier height, preventing reverse biased leakage current.

With these points in view, many materials have hitherto been investigated as possible candidates for a good, reliable Schottky barrier metal. Al has been most widely used in practical GaAs MESFETs (microwave devices) because of its low resistivity, easy manufacturability, and evaporation ease. The Al/GaAs Schottky barrier is stable up to 345°C, above which interdiffusion between GaAs and Al becomes prominent. An Al/GaAs Schottky barrier contact with good adhesion and good electrical properties cannot be formed by the lifting-off technique unless the surface (interfacial layer) is free from contamination. The most recommendable delineation technique is wet etching using H_3PO_4 or HCl, by which means a submicron Schottky barrier gate can be achieved at a practical level (Furutsuka et al., 1981).

Cr-overlaid Pt–Au exhibits good adhesion to GaAs and good electrical properties. However, interdiffusion of Pt to Cr and interaction between Cr and Ga at temperatures above 300°C prevent application of this metal system to practical devices. Ti–Pt–Au multilayer metallization also gives a good adhesion, but at about 450°C, Ti–Ga and Ti–As compounds are produced, giving a soft breakdown to Schottky barrier contact.

In the Pt/GaAs system, a solid-phase reaction takes place above 250°C and produces a Pt–PtGa–PtAs–GaAs multistructure layer. Figure 4 shows the reactions of Pt with GaAs after each stage of annealing for 1 hour at temperatures ranging from 250°C to 550°C (Mukherjee et al., 1979). Secondary ion mass spectrometric (SIMS) analysis after the final anneal

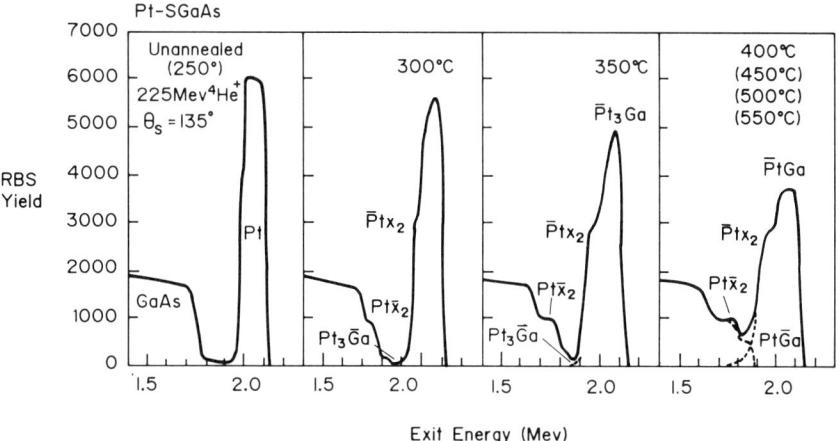

FIG. 4. Rutherford backscattering spectra of Pt–GaAs (for 2.25-MeV ^{4}He$^+$ incident perpendicular to the sample and detected at $\theta_s = 135°$) after various stages of anneal (1 h each in flowing undried argon) showing compounds formed. X indicates Ga or As. A bar over one of the constituents indicates RBS yield from that element present in the proposed compound. Bracketed temperatures imply that the spectrum shown here is essentially identical with that at the bracketed temperature. For example, spectra recorded after the 450°, 500°, and 550°C anneals differ from that after the 400°C anneal only in their progressively longer Pt indiffusion tails. [From Mukherjee et al. (1979).]

indicates an arsenic-rich layer (PtAs$_2$) adjacent to GaAs and a gallium-rich layer (PtGa$_2$, PtGa) on the outside. These reactions suggest that the Schottky barrier interface (PtAs$_2$ front) moves inward in high-temperature annealing (Ogawa et al., 1970). This property of the Pt/GaAs system is successfully utilized to adjust the threshold voltage of a MESFET by adjusting the effective active-layer thickness beneath the gate electrode (Toyoda et al., 1981).

In contrast to the low-resistance metals described above, the refractory metals such as W and Mo, their silicides and nitrides, WAl and TiW are also very important not only because of their high-temperature reliability, but also because of self-aligned gate GaAs MESFET technologies, including the high-temperature (700°C) annealing of ion-implanted GaAs.

W and Mo films are formed by sputter deposition or chemical vapor deposition (CVD). Sputter deposition induces interfacial defects, which act as recombination centers to increase the leakage current and produce soft breakdown. These defects can be partially annealed out at temperatures between 450°C and 650°C. In the CVD method, W and Mo films are deposited on GaAs through the reactions of WF$_6$ and H$_2$, and MoCl$_5$ and H$_2$, respectively. W/GaAs Schottky barrier contact exhibits no degradation at high temperatures (450–500°C). However, when film thickness is 0.2 μm or greater, peeling-off and blistering occur because of the thermal expansion coefficient mismatch (W: $3.5 \times 10^{-6}/°C$, GaAs: $6.5 \times 10^{-6}/°C$). The difference in thermal expansion coefficient between Mo and GaAs is smaller than that between W and GaAs. Therefore, Mo is more advantageous for use as a Schottky barrier metal than W.

Nitrides and silicides of W, Mo, and Ta, as well as some other compounds (WAl, TiW, TiWSi), have been used for refractory gate-electrode metallization in self-aligned GaAs MESFET technology (Yokoyama et al., 1981). Among these refractory metals, a binary alloy, WSi_x, is now widely used. WSi_x film is formed by using a multitarget magnetron cosputtering apparatus. Si content of WSi_x is controlled by varying the sputtering rate ratio for W and Si. Etching of WSi_x film is performed with a parallel-plate reactive ion etching technique using a CF_4 and O_2 gas mixture or NF_3 gas. Figure 5 shows the barrier height and ideality factor for WSi_x Schottky barrier contact on n-GaAs substrates as a function of annealing temperature and compares them with those of TiW and TiWSi films. WSi film was deposited to a thickness of 45 nm by the RF cosputtering technique. Annealing was carried out in a $N_2 + H_2$ gas mixture, with a 100-nm-thick SiO_2 encapsulation film, for 15 minutes. Electron microanalysis and X-ray diffraction analysis show that the WSi_x film is amorphous in the as-deposited state and is crystallized to phase W_5Si_3 through annealing at temperatures

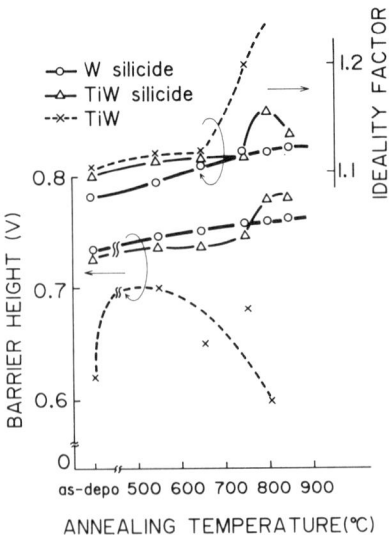

Fig. 5. Barrier heights and ideality factors of (×) TiW, (△) TiW silicide, and (○) W silicide Schottky contacts as functions of annealing temperature. [From Yokoyama et al. (1983a).]

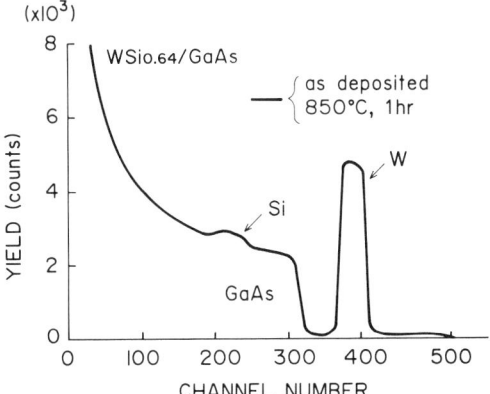

FIG. 6. RBS spectrum of $WSi_{0.4}$/GaAs contact before and after annealing at 850°C for 60 min. [From Ohnishi *et al.* (1983).]

higher than 750°C. It is shown in Fig. 5 that both the barrier height and the ideality factor for WSi_x remain comparatively constants at temperatures up to 850°C (Yokoyama *et al.*, 1983a). Figure 6 shows the RBS spectrum for the W_5Si_3/GaAs system in the as-deposited condition and after annealing at 850°C for 1 hour. Clearly there is no evidence of metallurgical interactions between WSi and GaAs (Ohnishi *et al.*, 1983). Dependences of the barrier height and the ideality factor on Si content evaluated after annealing at 800°C for 20 minutes are shown in Fig. 7. These parameters depend on Si

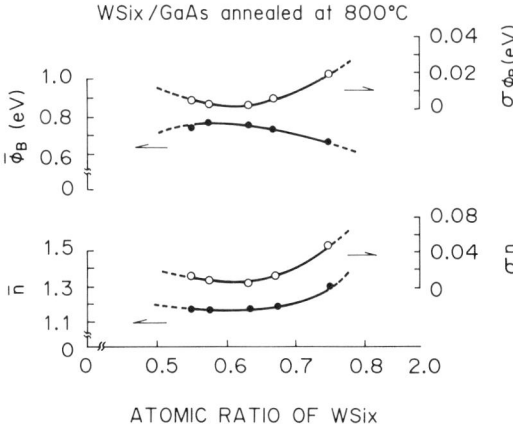

FIG. 7. Average values and standard deviation values of barrier height (Φ_B, σ_{Φ_B}) and ideality factor (n, σ_n) of samples annealed at 800°C for 20 min as a function of Si content. [From Ohnishi *et al.* (1983).]

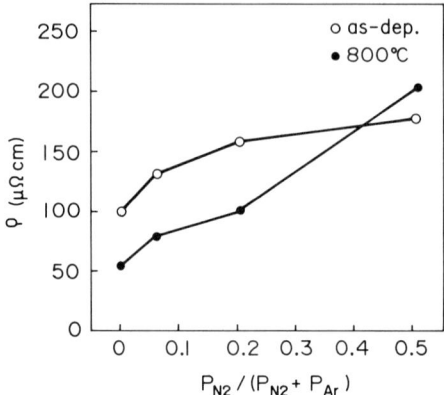

FIG. 8. Resistivity dependence of WN films on N_2 content in N_2-Ar mixed gas. [From Uchitomi et al. (1984).]

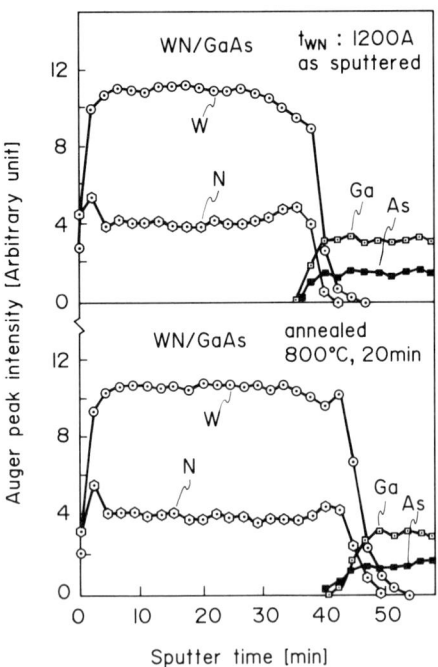

FIG. 9. AES depth profile spectra of a WN/GaAs system, after deposition and after annealing at 800°C for 20 min. [From Yamagishi (1984).]

content and, at around $x = 0.6$, the barrier height and ideality factor take the maximum and the minimum value, respectively. Standard deviation values in barrier height and ideality factor take the minimum value.

WN is more advantageous than WSi with respect to high barrier height (0.84 eV) and low resistivity (less than 100 $\mu\Omega$-cm; Uchitomi et al., 1984). WN_x is deposited with RF reactive sputtering in N_2 and Ar mixed atmosphere. Figure 8 shows the electrical resistivity as a function of N_2 content in mixed gas. The resistivity increases for higher N_2 content and is about 70 $\mu\Omega$-cm at 6% N_2 content. Figure 9 indicates AES depth profiles for as-sputtered and annealed WN film deposited on semi-insulating GaAs substrate for 20% N_2 content. The nitrogen concentration in WN film is almost constant throughout the film, except for a slight increase at the surface and interface regions, indicating that the films are composed of uniformly distributed nitrogen atoms. After annealing at 800°C, the whole spectrum is seen to have changed only slightly (Yamagishi, 1984). Anisotropic fine-pattern etching of WN on GaAs is easily performed by RIE with $CF_4 + O_2$ mixed gas. The etching rate for the WN film and the etching selectivity with the GaAs substrate show maxima when the O_2 content in a mixed gas is 50%.

Sputter-deposited W–Al alloy and W/Al double layer metal are also suitable as thermally stable refractory gate metals for GaAs MESFETs (Matsuura et al., 1984). In Table II the barrier height, ideality factor and resistivity among W, W–Al alloy, and W/Al multilayer metals on GaAs substrate are compared. For the n-value, these metals are quite thermally stable up to 800°C. The resistivity of W/Al film can be as low as 18 $\mu\Omega$-cm, which is only 50% higher than that for W film, Because of the thermal

TABLE II

SCHOTTKY PROPERTIES AND RESISTIVITIES OF W, W–Al ALLOYS, AND W/Al FILM WITH ~0.1 μm THICKNESS[a]

	Schottky property[b]		Resistivity[c] ($\mu\Omega$-cm)
	Φ_B (V)	n-value	
W	0.72 ± 0.02	1.16 ± 0.03	12 ± 2
W–Al (1.7%)	0.76 ± 0.02	1.08 ± 0.03	42 ± 3
W/Al (Al 20 Å)	0.78 ± 0.03	1.09 ± 0.03	18 ± 3

[a] From Matsuura et al. (1984).
[b] From 800°C annealing.
[c] As deposited.

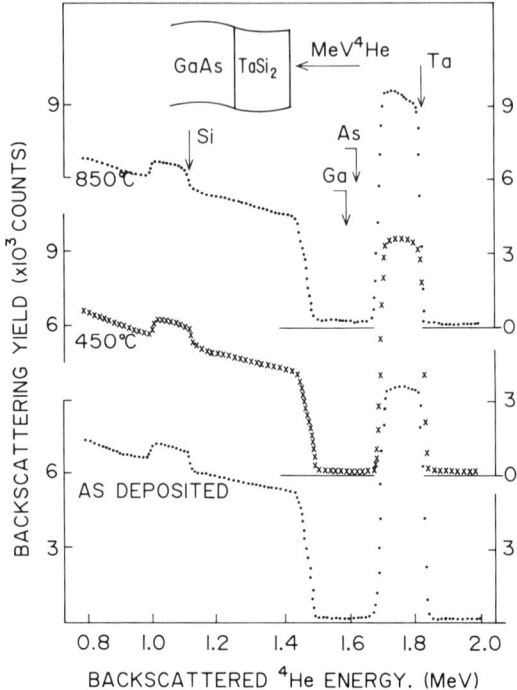

FIG. 10. Backscattering spectra for three samples after annealing at different temperatures. The shape and the height of the Ta and Si signal remain relatively unchanged after annealing up to 850°C. [From Tseng et al. (1983).]

expansion coefficient mismatch, peeling off or blistering occurs for W–Al films thicker than 0.2 μm.

RF-sputtered $TaSi_2$/GaAs gives no observable Ta and/or Si migration into the GaAs substrate at temperatures up to 850°C, although some migration of Ga and/or As is observed (Fig. 10; Tseng et al., 1983). The initial resistivity of the as-deposited film is 480 μΩ-cm. However, by annealing at 850°C (30 minutes), the resistivity decreases to 120 μΩ-cm because of the crystallization of the as-sputtered amorphous TaSi layer at 500°C.

Amorphous Si–Ge–B is a unique Schottky barrier material that has a barrier height as great as 0.9–1.0 V. Amorphous Si–Ge–B film can be deposited by low-pressure CVD using a $SiH_4 + GeH_4 + B_2H_6$ system (Suzuki et al., 1983). The resistivity is as great as 9 Ω-cm, and when this material is used as the gate metal of GaAs MESFETs, low resistivity metal such as Al must be overlaid.

A series of Schottky-barrier metals for the GaAs substrate have been described. Besides the selection of Schottky-barrier metal, another important technical point in obtaining a good Schottky-barrier contact is how to form (deposit) the Schottky-barrier metal. There are three typically methods: a combination of metal deposition and lifting-off; a combination of deposition and etching; and electroplating. The first method is, as mentioned above, not free from surface contamination because the existence of the photoresist prevents sufficient surface treatments such as heat cleaning just prior to metal deposition, as well as the problem of the outgas from the photoresist. Metal/GaAs systems involving an intentional interdiffusion, such as Pt/GaAs and Ni/GaAs (Ida *et al.*, 1981) systems, remove the interfacial layer in the annealing process, resulting in a good Schottky contact. A combination of metal deposition and etching, which is widely used in Al/GaAs and the refractory metal/GaAs systems, is suitable because it is easy to keep a clean surface before metal deposition. However, GaAs surface damage and/or contamination induced by the etching process, especially the dry etching process, must be removed before proceeding to the following fabrication process. The surface irregularities are partially removed by heat treatment and some acid treatments (Uetake *et al.*, 1985).

The Schottky barrier height is governed mainly by the Fermi level pinning effect as the joint action of the surface states and the interfacial oxide layer on the GaAs surface. Therefore, ideal Schottky barrier contact (ideality factor $n = 1$) is hardly realized in a practical production line. The metal/GaAs system was investigated physically using MBE apparatus, indicating an ideality factor of nearly unity ($n = 1.02$–1.10) and low barrier height of 0.67 eV (Sakaki *et al.*, 1981).

3. Dry Etching Process Technology

In order to realize high-performance and higher-packing-density LSIs, device size is becoming smaller and smaller. To achieve device-size shrinkage, progress in dry etching technology is vital for providing a fine pattern definition in a controllable and reproducible manner. In addition, it is essential to establish a reliable process for the dry etching method, which ultimately means a contamination-free and damage-free process. In Si very large-scale integration (VLSI) fabrication, the dry etching process plays an important role in microstructure formation. Some technologies developed in Si device fabrication processes can be applied for GaAs device fabrication processes. However, owing to the characteristics of GaAs material, a suitable technology for GaAs devices should be developed. In GaAs LSI fabrication, the dry etching process includes gate-electrode fine pattern definition, contact hole opening and fine pattern formation of the second-level interconnect.

For gate-electrode fine-pattern definition, there have been several reports published on the characteristics of reactive ion etching (RIE) for refractory

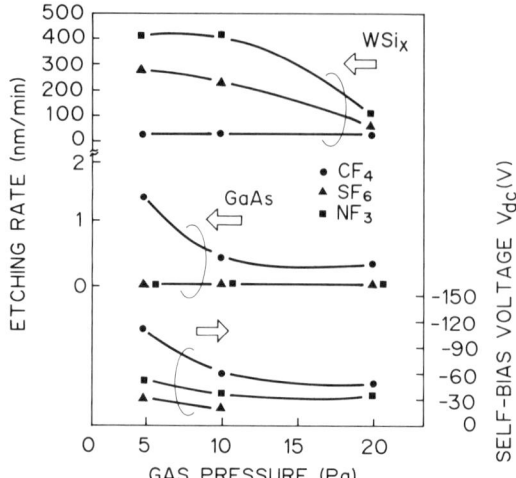

FIG. 11. WSi$_x$ and GaAs etching rate and self-bias voltage as a function of gas pressure for (●) CF$_4$, (▲) SF$_6$, and (■) NF$_3$ gases. 0.1 W/cm^2 power density, 20 ml/min gas flow. [From Miyazaki et al. (1984).]

metal silicides. Since GaAs itself is etched by chlorine-containing plasmas, fluorine-containing etching species are employed. Yokoyama et al. (1983c) reported the formation of a WSi$_x$ fine pattern, having a good rectangular cross-section, by a parallel-plate RIE using a CF$_4$ and O$_2$ gas mixture.

Miyazaki et al. (1984) reported on dry etching characteristics for WSi$_x$ gate metal using three kinds of etching species, CF$_4$, SF$_6$, and NF$_3$, with a parallel-plate RIE system. Figure 11 shows WSi$_x$ and GaAs etching rates and self-bias voltages as a function of etching species gas pressure. The WSi$_x$ etching rate is higher for NF$_3$ gas and lower for SF$_6$ and CF$_4$ gases in that order. Negligible etching was observed for SF$_6$ and NF$_3$ gas, while for CF$_4$ gas, a slight amount of etching in the GaAs surface was observed. These results originate from the magnitude of the self-bias voltage V_{dc}, which is also shown in Fig. 11. At 0.5 Pa gas pressure, the selectivity values between WSi$_x$ and photoresist (OFPR-800) for CF$_4$, SF$_6$, and NF$_3$ were 0.8, 3.9, and 4.1, respectively. Poor selectivity for CF$_4$ gas, which is caused by higher V_{dc}, indicates the difficulty of applying CF$_4$ gas for device fabrication processes. Almost isotropic etching characteristics were observed for SF$_6$ gas, resulting in a large amount of side etching. On the other hand, a good pattern shape, which is desirable for device application, was obtained for NF$_3$ gas. C, N, O, and F were detected by Auger electron spectroscopy (AES) measurements on the GaAs surface exposed after WSi$_x$ film removal by NF$_3$ dry etching. However, these AES signals vanished after removal of a 7-nm-thick GaAs surface layer. No change in sheet resistance was observed for Si^+-implanted samples, fabricated with and without reactive ion etching (RIE) using NF_3 gas. These results indicate that degradation in the GaAs surface, induced by the contamination and/or damage, is restored by high-temperature annealing at 800°C for 20 min, which is necessary for implanted atoms to be electrically activated.

Modification of surface characteristics in GaAs, induced by a hydrogen plasma treatment, was examined by Chung *et al.* (1985). Chamber pressure, power density and exposure time were 0.5 torr, ~0.3 W/cm², and 15 min, respectively. After a hydrogen plasma treatment to *n*-type GaAs, carrier removal in the surface region was observed. The depth of the carrier-depleted region increased with decreasing carrier concentration in *n*-type GaAs. For example, 0.12 and 0.23 μm carrier-depleted regions were observed for 2×10^{18} and 6×10^{17} cm^{-3} carrier concentrations, respectively. However, after 350–400°C annealing, the carrier concentration almost recovered to its initial value, as shown in Fig. 12. From Raman spectra observations, donor–defect complex formation is believed to be the principal reason for carrier removal after plasma exposure. Pang (1984) also reported the degradation in GaAs surface exposed by CF_4 gas RIE. He observed a lowering of doping concentrations at the interface between the etched surface and the VPE-grown epitaxial layers.

There have been several reports published on the GaAs surface cleaning, which is vital for obtaining reliable dry processing. Muraguchi *et al.* (1984) reported that hydrogen plasma treatment with power density as low as 10 mW/cm² for 20 sec is effective for formation of a good and reliable Schottky barrier on a GaAs surface that was exposed by a SiN film removal

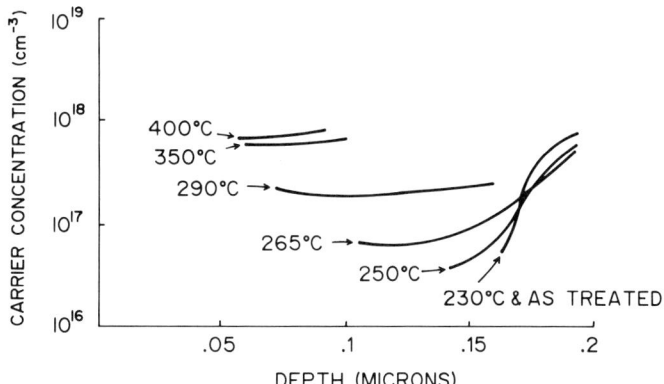

FIG. 12. Isochronal annealing after hydrogen plasma exposure on Si-doped GaAs. [From Chung *et al.* (1985).]

using $CF_4 + O_2$ plasma. An increase in transconductance value, at about 10%, was also confirmed without showing threshold voltage shift in the SAINTFETs. These improvements in device characteristics are attributed to the removal of surface contaminants that were introduced during the dry etching process. Asakawa and Sugata (1985) examined the cleaning of air-exposure-induced oxygen and carbon contaminants on the GaAs suface using a radical beam gun with ECR plasma source. From AES measurements, it was confirmed that O and C contaminants are removed sequentially by hydrogen-radical beam exposure, followed by chlorine radical beam exposure, for GaAs wafers kept at 400°C. These cleaning methods are effective in establishing dry process technology. Further development in the cleaning process will be necessary.

To interconnect the first-level metal to the second-level metal, via-hole opening is necessary in LSI fabrication. Because of device size shrinkage, via-hole size falls in the one-micron region. Since low-resistive contact between the first-level metal and the second-level metal is vital in the LSI, it is of great importance to establish an accurate and reliable dry etching process. Inferior step coverage for the second-level interconnect is often observed on the vertical wall of a via hole. For this reason, it is desirable to produce a tapered or sloped hole profile. SiO_2 or Si_3N_4 films are usually used for the intermediate film between the first- and the second-level interconnects, as in the case for Si LSIs. Therefore, a dry etching process, developed for Si devices, may be applicable. There are a number of techniques for making tapered hole profiles, such as the use of controlled erosion of flowed resist (Bondur and Frieser, 1981; Crabbe and Sirkin, 1983). The other methods are summarized by Gorowitz and Saia (1984). Another approach, a via-metal filling, which is achieved by a lift-off method, is reported and employed in LSI fabrication (Eden and Welch, 1982; Asai et al., 1983).

Ion-beam milling can be applied routinely for the second-level interconnections, where Au-based metallizations, such as Au/Pt/Ti, Au/Ti, and Au/Cr are usually employed. However, in order to achieve fine pattern definition, optimization in milling process conditions is necessary, along with taking into account several factors such as incident ion-beam angle, sputtered material redeposition and relatively poor selectivity. Moreover, investigations of radiation damage is needed, even though sensitive semiconductor surfaces or interfaces are covered with a passivating dielectric layer.

III. Self-Aligned MESFET Technology

To realize high-speed GaAs LSIs, a device technology that gives a high GaAs MESFET transconductance (g_m) and a high K-value is indispensable. Between the observed g_m and the intrinsic g_m(g_{m0}), the well-known relationship,

$g_m = g_{m0}/(1 + R_S g_{m0})$, exists, where R_S denotes source series resistance. R_S has been reduced by shortening the source-gate spacing by means of various self-aligned fabrication techniques. This section describes some representative self-aligned MESFET technologies developed for high-speed GaAsIC fabrications.

4. REFRACTORY METAL GATE TECHNOLOGY

Refractory metal gate technology enables the highly doped n^+-regions to be placed close to the gate electrode, resulting in a reduction in the source series resistance (Yokoyama *et al.*, 1981). Figure 13 indicates the major stages in the fabrication of the refractory metal gate self-aligned GaAs MESFETs. First, an *n*-type GaAs active layer is prepared by selective ion implantation into a semi-insulating (S.I.) GaAs substrate by using photoresist as an implantation mask. Second, a high-temperature-stable Schottky gate is formed on the active layer by sputter deposition or a metal CVD technique and is delineated by a reactive ion etching. Next, a high dosage Si^+-implantation is performed with the gate acting as a mask. Fourth, annealing is performed at 800°C for 10 minutes to activate dopants and to form the self-aligned n^+ regions. Fabrication is completed by ohmic metallization. In this self-aligned MESFET, the n^+-regions significantly reduce parasitic-source series resistance. Also, the self-aligned MESFET is superior in packing density, because it does not require gate alignment. The key point

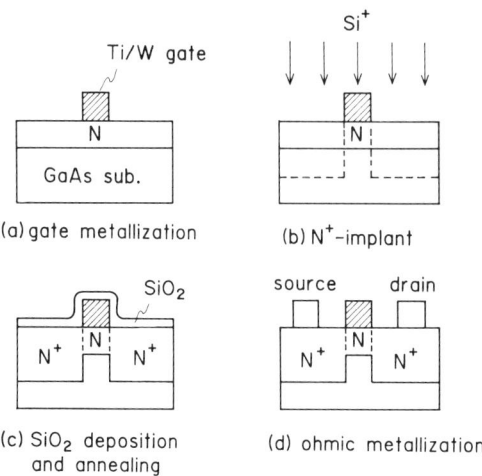

FIG. 13. Major steps in the refractory metal gate self-aligned GaAs MESFET fabrication process. (a) Gate metallization; (b) n^+ implant; (c) SiO_2 deposition and annealing; (d) ohmic metallization. [From Yokoyama *et al.* (1981).]

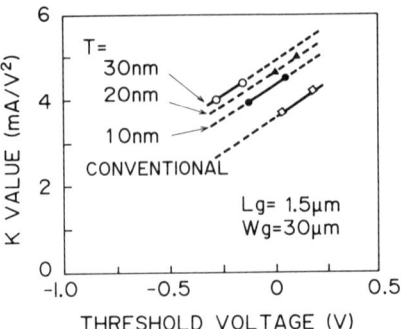

FIG. 14. Averaged K-value as a function of the threshold voltage with the parameter of AlN film thickness: (○) 30 nm, (▲) 20 nm, (●) 10 nm, (□) conventional. [From Kawata et al. (1985).]

in this technology is the use of a high-temperature-stable (refractory) Schottky gate. Several kinds of the refractory metals have been proposed and applied to the self-aligned MESFET fabrications, such as W, WSi, TiW, WN, TiWSi, Mo, and TaSi$_2$ (see Part II). Further improvement in g_m and K-value[1] can be achieved by introducing a shallow channel with a high carrier concentration (n_D).

Onodera et al. (1984) developed a through-AlN implantation technology for fabricating extremely thin active layers. Figure 14 (Kawata et al., 1985) shows the K-values versus V_T characteristics measured with the AlN film thickness as a parameter. Ion implantation energies for the active layer are 30 keV and 59 keV for bare-implantation and for through-implantation, respectively. WSi gate n^+ self-aligned technology was employed for device fabrication. It can be seen in Fig. 14 that K-values for through-implanted FETs are larger than those for conventional FETs and that K-values increase as the AlN-film thickness increase. The increased K-values are attributable to the decrease in the channel layer thickness. Another advantage of through-implantation technology is the improvement in the threshold voltage uniformity, which comes from the fact that the effective doping concentration is increased in the case of through-implantation; therefore, the threshold voltage is less sensitive to residual impurities.

An alternative for this refractory metal gate technology is to introduce MOCVD n^+-regions on the active layer (Imamura et al., 1984; Uetake et al., 1985), where the short-channel effect is reduced because the n^+-regions are not embedded in the active layer. Figure 15 shows the FET fabrication

---

[1] K-value is defined from the expression for saturated drain current I_{DS}, $I_{DS} = K(V_G - V_T)^2$. K is expressed by $K = \varepsilon \mu W_g / 2dL_g$, where ε is the dielectric constant of GaAs, μ the drift mobility of electrons in the channel layer, d the effective thickness of the channel layer, W_g the gate width and L_g the gate length.

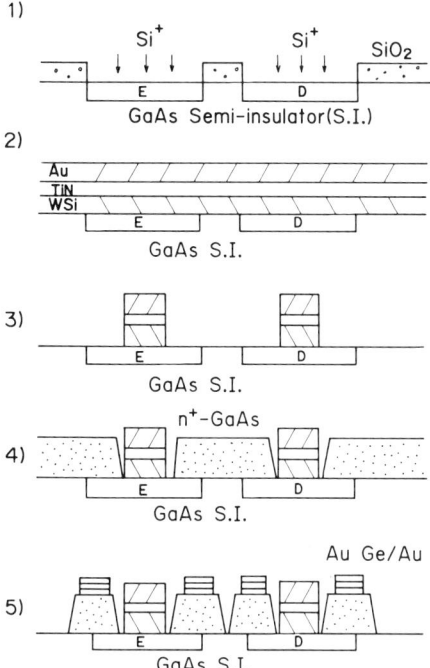

FIG. 15. Fabrication process of Au/TiN/WSi gate self-aligned E and D MESFETs. [From Imamura et al. (1984).]

procedure. First, an *n*-type active layer is prepared by Si$^+$ implantation into the substrate and subsequently annealed at 850°C for 15 min. with a SiO$_2$ encapsulant. Next, 200-nm-thick WSi$_x$, 40-nm-thick TiN and 200-nm-thick Au films are successively deposited on the GaAs surface. Au and TiN films are patterned by Ar$^+$ ion-milling using photoresist as a mask. WSi$_x$ is reactive-ion-etched using CF$_4$ gas. The source and drain regions are selectively grown by MOCVD under the conditions shown in Table III. The

TABLE III
MOCVD GROWTH CONDITIONS[a]

	System 1	System 2
Dopants	S	Se
Ga source	(C$_2$H$_5$)$_2$GaCl$_2$	TMG
As source	AsH$_3$	AsH$_3$
Growth temperature	580°C	620°C

[a] From Imamura et al. (1984).

nominal resistivity for a Au/TiN/WSi gate is 1.8×10^{-5} Ω-cm, which is approximately one order of magnitude smaller than that for a WSi gate. Sheet resistance for the selectivity formed n^+-GaAs layers is 33 Ω/□ for the S-doped 400-nm-thick layer, and 20 Ω/□ for the Se-doped layer. Annealing temperature dependences of barrier height Φ_B and the ideality factor n for the Au/TiN/WSi gate Schottky diodes show that the TiN acts as a good barrier metal between Au and WSi at temperatures up to 700°C, which is sufficiently high to prevent a metallurgical interaction during n^+-GaAs MOCVD.

Another approach to the reduction of the gate metal resistivity is to use W/WSi bilayer film (Kanamori et al., 1985). Figure 16 shows the average resistivities for $W/WSi_{0.4}$ films, before and after annealing at 800°C, as a function of $WSi_{0.4}$ thickness, while retaining 0.5 μm total film thickness. In the bilayer film structure, resistivity reduction is marked. For example, 18 μΩ-cm average resistivity, which is one order of magnitude lower than that for a $WSi_{0.4}$ single layer, is obtained for 800°C annealed $W/WSi_{0.4}$ film with 0.1-μm-thick $WSi_{0.4}$ film. Stable characteristics at the bilayer gate/GaAs interface are confirmed for W of 0.05 μm thickness (0.5 μm in total).

Recently, lower gate resistance as small as 3 μΩ-cm was achieved using Au/WSiN refractory metal gate technology (Onodera et al., 1988). Au/WSiN submicron gate patterns were delineated by Ar ion milling for the Au, followed by RIE for the WSiN. Because of the amorphous state of WSiN film even after high-temperature annealing, which results in suppression of Au penetration and As and Ga outdiffusion into WSiN, stable characteristics at the interface between WSiN and GaAs were confirmed (Asai et al., 1988). Along with shallow channel formation, higher electrical characteristics, such as a maximum transconductance of 630 mS/mm, were achieved in 0.2-μm gate length devices.

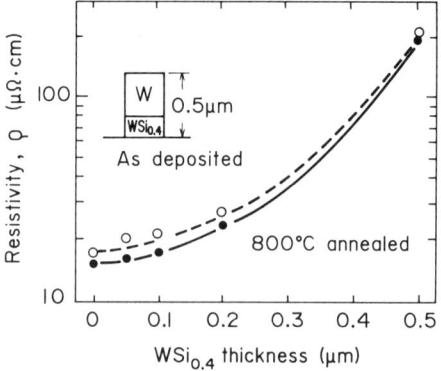

FIG. 16. Average $W/WSi_{0.4}$ bilayer film resistivities, (○) before and (●) after annealing at 800°C, as a function of WSi_4 film thickness. [From Kanamori et al. (1985).]

In the self-aligned MESFET technologies shown in Figs. 13 and 15, the n^+-source and drain regions are not intentionally separated, resulting in a relatively small gate breakdown voltage and a large gate input capacitance. Further improvements from this point of view are accomplished by SAINT, SWAT, or LDD, which are described in the following sections.

5. SAINT (Self-Aligned Implantation for n^+-Layer Technology)

SAINT (Yamasaki *et al.*, 1982a) makes it possible to arbitrarily control the spacing between the n^+-implanted layer and the gate contact by a dielectric lift-off process utilizing a multilayer resist with an undercut wall shape. The FET fabrication process utilizing SAINT is shown in Fig. 17. First, Si^+ ions are selectively implanted into S.I. substrate by use of a photoresist mask. After implantation, the GaAs surface is covered by a plasma-enhanced CVD SiN film of 0.15-μm thickness. Next, the multilayer resist scheme consists of bottom resist, sputtered SiO_2 and top photoresist. The bottom resist is FPM (polytetrafluoropropyl methacrylate), which possesses the following features: high dissolubility in organic solvent; thermal stability against SiO_2 sputter deposition and the top photoresist baking; and selective etchability for precise patterning. The top photoresist pattern can be replicated in the intermediate SiO_2 layer by RIE using $CF_4 + H_2$. The bottom FPM layer is undercut with respect to the upper SiO_2-layer pattern by RIE in O_2 discharge. The top photoresist is also etched during the bottom FPM etching, whereas the intermediate patterned SiO_2 is resistant to this etching. In the next stage, high-dose Si^+ ion implantation is made through the PCVD SiN-film for n^+-layer formation, where the intermediate SiO_2 film acts as an ion-stopping mask. Next, the second SiO_2 layer is deposited by RF magnetron sputtering as shown in Fig. 17d. The SiO_2 adhering to the sidewall of the multilayer resist can be easily removed in buffered HF solution. Finally, the second SiO_2 layer deposited on top of the resist is lifted off in acetone solvent, leaving the second SiO_2 layer on the ground SiN (Fig. 17e). Electrical activation of the n and n^+ implants is made at this stage, using the ground SiN film as an encapsulant during high-temperature annealing. After activation, source and drain contact windows in SiO_2/SiN are opened and ohmic metal is lifted off. To provide the gate electrode, the ground SiN located on the gate channel is plasma etched. The etching rate for SiO_2 is about one-fifth that for SiN; therefore, the Schottky barrier contact window can be opened only for the gate channel as shown in Fig. 17g. Gate metal is evaporated and delineated by the lifting-off technique. The gate length is therefore determined by separation distance between the second SiO_2 films.

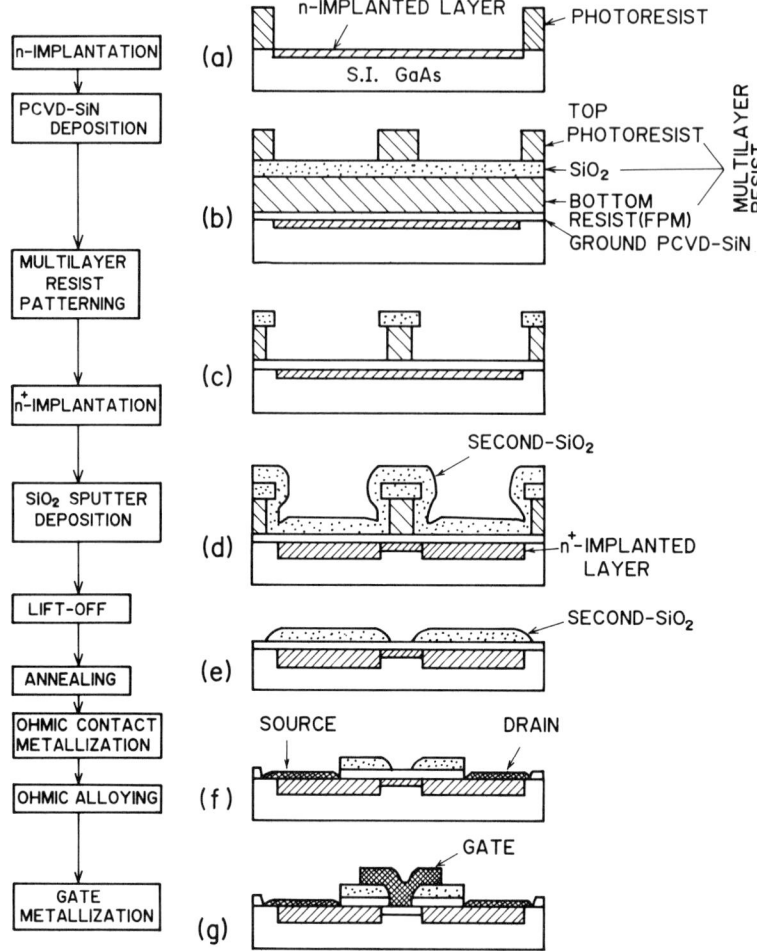

FIG. 17. SAINT process sequence. [From Yamasaki et al. (1982d).]

One important feature of SAINT is that it can determine the spacing between n^+-regions and gate electrode by the undercut amount. Optimization of n^+-gate spacing is of great importance from the viewpoint of FET switching speed. Too close spacing results in an increase in gate input capacitance because of the lateral spreading of n^+-implants caused by ion recoiling. On the other hand, too remote placement of an n^+ source from the gate causes an increase in external resistance. According to a two-dimensional numerical simulation (Yamasaki et al., 1982c), the capacitance increase due to the n^+-embedding is effectively restrained by the n^+-to-gate

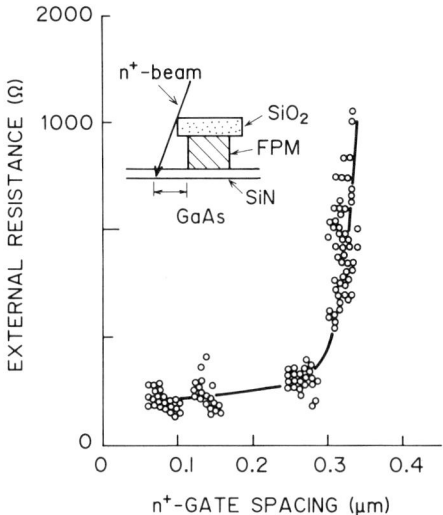

FIG. 18. Gate–source external series resistance vs. spacing between n^+-implant projection edge and gate electrode edge. Gate width is 9 μm. [From Yamasaki *et al.* (1982b).]

spacing, and the total charge movement in case of zero spacing is 1.6 times greater than that for 0.2-μm spacing. The capacitance increase is much more severe in practical devices because thermal diffusion of the n^+-implants takes place. Figure 18 shows gate–source external series resistance versus spacing between n^+-implant projection edge and gate electrode edge (Yamasaki *et al.*, 1982b). The drastic improvement in external resistance ceases at about 0.3 μm n^+-gate spacing, which corresponds to the lateral spreading of n^+ implants (Yamasaki *et al.*, 1982b). From the viewpoints of the gate charge movement and of the external resistance, the optimum distance between n^+ and gate electrode is about 0.3 μm.

Another advantage resulting from using the undercut technique is realizing a shorter gate length than the photoresist mask. Therefore, the submicron gate can be easily obtained by using the conventional photolithography technique.

The next important feature of SAINT is that the gate metal is not subjected to high-temperature annealing. Therefore, the gate metal is not necessarily restricted to a high-temperature-stable (refractory) metal. Figure 19 shows the MASFET (metallic amorphous silicon gate FET) fabricated by SAINT. A contact beween amorphous Si–Ge–B and GaAs exhibits a Schottky barrier height as large as 0.9–1.0 V, which is larger than the metal–GaAs Schottky barrier height of 0.7–0.9 V. A large barrier height enlarges DCFL circuit logic swing, which is restricted by the clamped voltage of gate barrier height, and

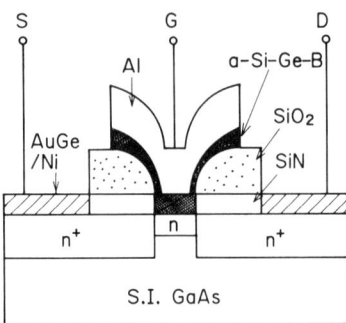

FIG. 19. Cross-sectional structure of MASFET (metallic amorphous silicon gate FET). [From Suzuki et al. (1984).]

results in the realization of fully operational GaAs LSIs with FET threshold voltage dispersion of 40–80 mV. Additionally, high-speed operation can be expected, because the FET maximum saturation current is large on account of the larger built-in voltage. The resistivity of a-Si–Ge–B is as great as 9 Ω-cm. Therefore, Al film is overlaid on it to reduce the gate electrode resistance as shown in Fig. 19 (Suzuki et al., 1983, 1984).

Terada et al. (1983) have proposed a unique self-aligned MESFET technology utilizing a pattern inversion technique as used in SAINT. Figure 20 shows an outline of the MESFET fabrication process.

(1) After an active channel layer is formed by selective Si^+ ion implantation, the 1.5-μm-long, 1-μm-thick SiO_2 "dummy gate" is formed by RIE using Al "cap metal" as an etching mask (Fig. 20a).

(2) Si^+ ion is selectively implanted to form n^+ source and drain regions using Al/SiO_2 as an implantation mask. Then the side-wall of SiO_2 is etched-off by 0.2 μm. As a result, the dummy gate length becomes approximately 1.1 μm, and the spacing between the dummy gate and the n^+-implanted layer becomes 0.2 μm (Fig. 20b).

(3) After removing the Al-cap metal, the 0.15-μm-thick SiO_2 is deposited as an encapsulant film for annealing (Fig. 20c).

(4) AuGe/Au ohmic contacts are formed and photoresist is then coated on using a resist spinner. The resist is etched by RIE until the top of the SiO_2 dummy gate is exposed (Fig. 20d). Once the SiO_2 is exposed, it is etched off by NH_4F + HF etchant. As a result, the inverted pattern is obtained. The resultant 1.1-μm-long and about 0.8-μm-deep hole remains in the resist film.

(5) Pt gate metal is evaporated and lifted off using the remaining resist. In this step, the SiO_2 dummy gate is replaced by a Pt gate. Figure 20e shows a cross-sectional view of the FET structure.

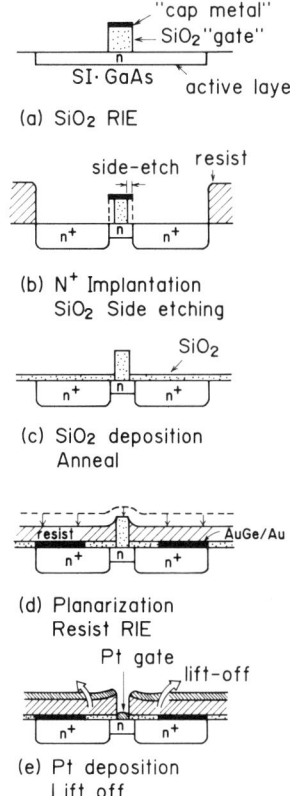

FIG. 20. Schematic diagram of self-aligned Pt-buried gate GaAs FET fabrication process steps: (a) SiO$_2$ RIE; (b) n^+ implantation and SiO$_2$ side etching; (c) SiO$_2$ deposition and annealing; (d) planarization and resist RIE; (e) Pt deposition and lift-off. [From Terada et al. (1983).]

One of the features of this FET structure is the usage of a Pt gate. Pt–GaAs solid-phase reaction consumes the interfacial layer in the annealing process, resulting in a good Schottky contact.

6. SWAT (Side-Wall Assisted Self-Alignment Technology)

Separation of source and drain regions from the gate electrode is performed with thin dielectric films formed on both sides of the gate electrode in SWAT (Higashisaka et al., 1983). SWAT is primitively applied to the fabrication of the closely spaced electrode FET. Figure 21 shows a schematic diagram for GaAs MESFET fabrication process by a side-wall-assisted closely spaced electrode technology. First, a 0.5-μm-thick Al gate electrode is formed on the

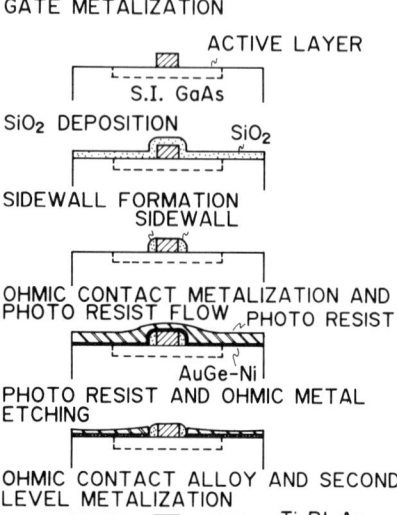

FIG. 21. Fabrication procedure of a closely spaced electrode FET with SWAT: (a) gate metallization; (b) SiO₂ deposition; (c) sidewall formation; (d) ohmic contact metallization and photoresist flow; (e) photoresist and ohmic metal etching; (f) ohmic contact alloy and second metallization. [From Higashisaka et al. (1983).]

active layer prepared by a selective ion implantation of Si^+ into S.I. GaAs substrate. The steepness of the cross-sectional structure of the gate electrode, which is important to achieve a reproducible fabrication of the sidewall, is realized by a relatively large amount of Al undercutting. Second, a 0.2-μm-thick SiO_2 film is deposited by chemical vapor deposition. The third step is side-wall formation. An anisotropic dry etching with CF_4 gas is employed to fabricate the SiO_2 side-wall on both sides of the gate electrode. The thickness of the side-wall, which determines the spacings between gate and source and between gate and drain, is almost the same as that for deposited SiO_2. This comes from both the anisotropic property of SiO_2 etching and the steep profile of Al-gate cross-sectional structure. The fourth step is ohmic metal evaporation. A AuGe/Ni film is selectively evaporated on the gate, source and drain regions using a conventional lift-off technique. To separate the gate electrode from the source and drain electrodes, the ohmic metal on top of the gate must be removed. To accomplish this objective, a unique technology was developed that utilizes the phenomenon wherein the spin-spread photoresist is thinner at the top of the gate electrode than in the bottom field. Anisotropic dry etching with CF_4 is again employed to etch

3. DEVICE FABRICATION PROCESS TECHNOLOGY

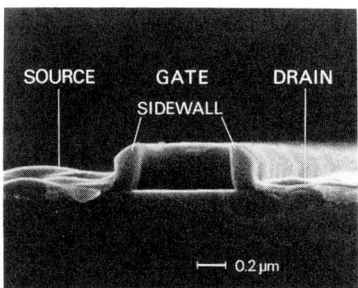

FIG. 22. Cross-sectional SEM view of a SWAT FET.

down the photoresist to reveal the ohmic metal on top of the gate. Ion milling with Ar^+ follows to remove the revealed, unnecessary ohmic metal on the gate. Closely spaced electrode GaAs MESFET fabrication is completed by alloying the ohmic metal at 420°C for 30 sec. Figure 22 shows a cross-sectional scanning electron microscope (SEM) view of a fabricated GaAs MESFET. The feature of this technology is that the gate and ohmic electrodes are closely spaced by a thin dielectric film (0.17 μm in Fig. 22), giving small parasitic resistances and, therefore, high g_m values. Typical FET performance dependence on gate length (L_g) is shown in Fig. 23. Because no

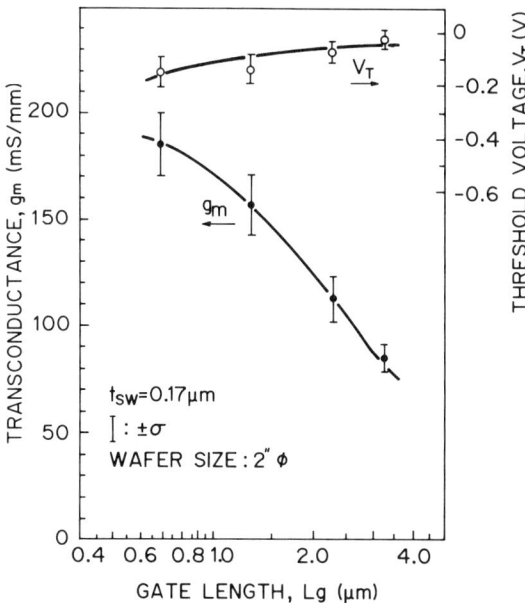

FIG. 23. Gate-length dependence of SWAT FET performance.

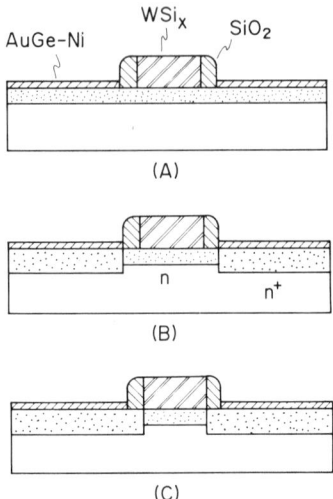

FIG. 24. Cross-sectional structure of SWAT FETs (A) is without an n^+-contact region, (B) and (C) are with n^+-implanted layers. [From Ueno et al. (1985a).]

n^+-contact regions are employed, the so-called short-channel effect is negligible in this device (Furutsuka et al., 1984).

SWAT can be used in combination with the refractory metal gate n^+-self-alignment technology and realizes a further reduction in source series resistance. Figure 24 shows SWAT FETs having n^+-implanted layers outside (B) and inside (C) the SiO_2 side-wall. For structure B device fabrication, another SiO_2 side-wall formation is needed before the second (n^+) ion implantation, where the side-wall serves as an implantation mask. For structure C, the side-wall is formed after n^+ ion implantation (Ueno et al., 1985a). A comparison concerning the gate-length (L_g) dependences of source series resistance R_s and g_m is shown in Fig. 25. The figure indicates that the V_T dependences on L_g for structures A and B are almost the same. This means that the lateral diffusion of n^+-implanted impurities has no influence on the channel region beneath the gate electrode in structure B, because of the existence of 0.2 μm-thick side-walls. On the other hand, in structure C, where n^+ regions are introduced adjacent to the gate electrode, a further increase in the g_m value is observed, which is due to the reduction in R_s under the sidewall and to the increase in the effective channel carrier density caused by lateral diffusion of the implanted n^+ impurities.

One of the important technical points of SWAT is how to recover any contamination and/or damage induced by the RIE process, which is needed for dielectric film (SiO_2 side-wall) delineation. Figure 26 shows XPS spectra

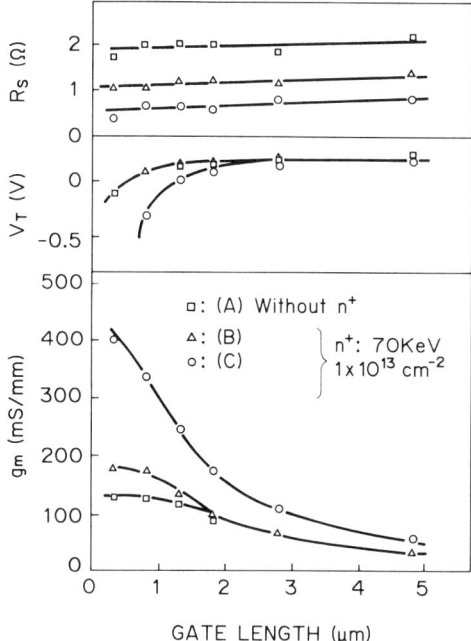

FIG. 25. Gate-length dependence of source series resistance (R_S), threshold voltage (V_T) and transconductance (g_m) for three SWAT FETs (n: 30 keV, 3.2×10^{12} cm^{-2}). □, structure A, without n^+; △, structure B, and ○, structure C, both with n^+: 70 keV, 1×10^{-13} cm^{-2}. [From Ueno et al. (1985a).]

for the GaAs surface just after RIE (A), after HCl cleaning (B) and after heat treatment (C). It can be seen in the figure that fluorine and/or fluorine-related polymer contamination–induced RIE can be fairly removed by HCl cleaning for relatively higher gas pressure (26 Pa). However, elimination of the surface contamination is not sufficient in the case of 8-Pa RIE. Therefore, with respect to the surface cleaning, the higher-pressure RIE is more advantageous; however, the gas pressure also must be optimized from the viewpoint of an etching direction, because anisotropic etching is sacrificed for higher gas pressure conditions. Even in the case of 8-Pa RIE, the remaining contamination can be removed by heat cleaning in N_2/H_2 atmosphere at 300°C for 3 minutes (Uetake et al., 1985).

7. LDD (Lightly Doped Drain) Technology

In place of space formation between the n^+ layer and the gate metal, LDD technology (Asai et al., 1986; Shimura et al., 1986), whose feature is the insertion of an intermediate carrier concentration region between the n^+

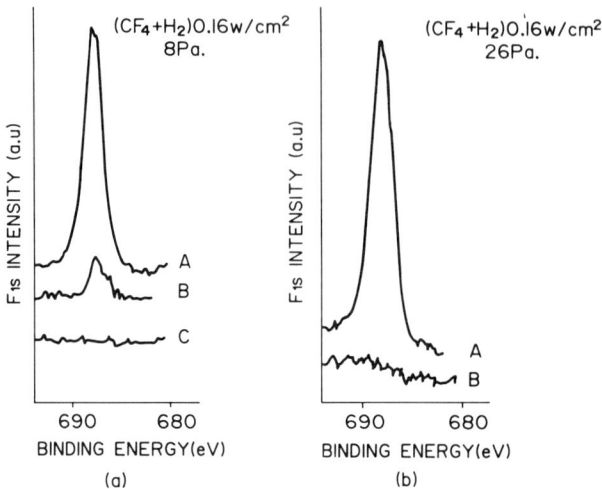

FIG. 26. XPS spectra of GaAs surface just after RIE (A), after HCl cleaning (B), and after heat treatment (C). RIE conditions were 8 Pa (30% in $CF_4 + H_2$) for (a), and 26 Pa for (b). [From Uetake et al. (1985).]

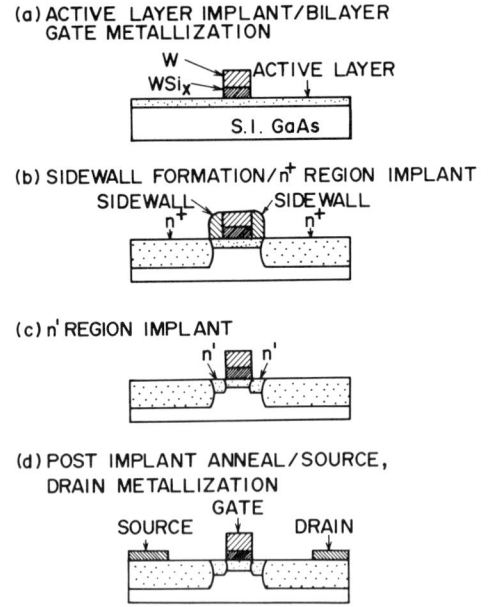

FIG. 27. LDD GaAs MESFET fabrication process: (a) active-layer implant and bilayer gate metallization; (b) sidewall formation and n^+-region implant; (c) n'-region implant; (d) post-implant anneal and source and drain metallization. [From Asai et al. (1986).]

layer and the gate metal, was investigated. Figure 27 shows the fabrication process sequence. First, an active n-layer is formed by selective Si ion implantation into semi-insulating GaAs substrates. Gate-metal film is deposited and reactively ion-etched to form a gate electrode. After that, the wafer surface is coated by a SiO_2 film and anisotropically dry-etched for making side-walls. Using the gate and side-walls together as a mask, Si ions are implanted for contact n^+ regions. Implantation for the intermediate carrier concentration regions (n'-regions) is performed after removing the sidewalls. To activate implanted impurities, high-temperature annealing (800°C, 20 min.) is carried out. Finally, ohmic electrodes are made by alloying AuGe–Ni film.

The effect of n'-region insertion was examined by varying ion doses for n'-regions, while maintaining 0.3 μm side-wall thickness and 50 keV ion energy. The results for 0.3 μm gate length devices are shown in Fig. 28. All the data shown are average values within a 2-inch wafer. It is clearly seen that K reaches its maximum at $\Phi(n') = 7$–10×10^{12} cm^{-2}. The maximum value is as high as 256 mS/V·mm, which is about 40% larger, than that for a conventional structure FET, which was fabricated by n^+ implantation with gate electrode as a mask. The transconductance value at 0.6 V gate bias was 270 mS/mm for the average threshold voltage of 0.08 V. Gate breakdown voltage (BV_G) and the R_S value decrease monotonically as $\Phi(n')$ increases.

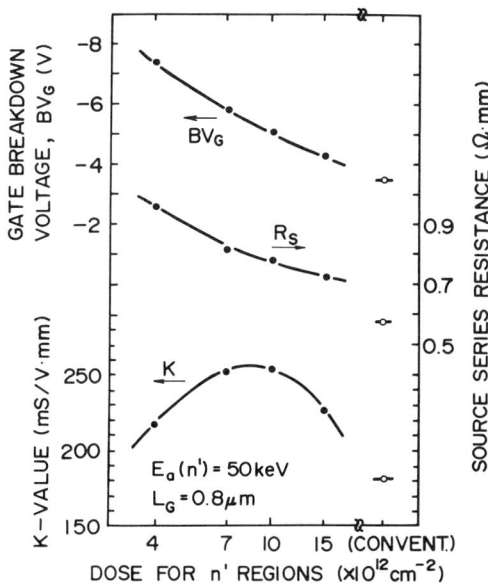

FIG. 28. LDD FET parameter dependence on dose for n' region. [From Asai et al. (1986).]

At $\Phi(n') = 7 \times 10^{12}$ cm^{-2}, the BV_G becomes up to -5.8 V, an improvement of about 60% over that for conventional structure FET. R_S value at that $\Phi(n')$ is 0.82 $\Omega \cdot$mm ($V_T = 0.08$ V). The K-value degradation in the lower $\Phi(n')$ region is mainly caused by R_S increase due to lower carrier concentration in the n'-region. In a higher $\Phi(n')$ region, a possible reason for decreasing K-value is the active layer thickness increase caused by lateral impurity diffusion from the n' region.

In Fig. 29, FET parameter dependences on gate length are compared for LDD and conventional structure MESFETs. The $\Phi(n')$ is 7×10^{12} cm^{-2}. A marked difference is that the K-value for the LDD FETs continues to increase as gate length is reduced to 0.8 μm, while in the conventional FET case, it tends to saturate where gate lengths are less than 2 μm. Considering the strong V_T value shift seen in Fig. 29, the K-value saturation behavior for the conventional FET is thought to be brought about by an active layer thickness increase caused by lateral impurity diffusion from the n^+ region. Since the lateral impurity diffusion effect is greatly suppressed with the LDD technology, the short-channel effect is sufficiently reduced. The threshold voltage shift is as small as -40 mV, when gate length varies from 4.9 to

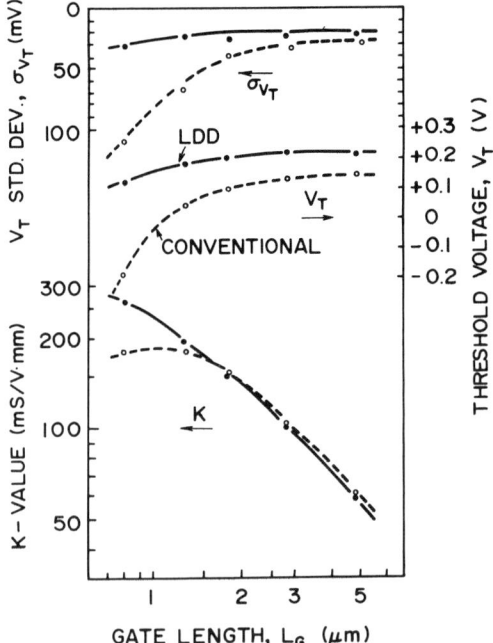

FIG. 29. Comparison of FET parameter dependence on gate length between LDD FET and conventional structure FET. [From Asai et al. (1986).]

0.8 μm. It is also noted that the V_T standard deviation value for the LDD FET exhibits very small dependence on gate length.

Similar results were also reported by Shimura *et al.* (1986), where the n' region was formed simultaneously by n^+ implantation through gate metal overcoating Si_3N_4 film.

IV. Device Technology Effective for Improving Characteristics in Submicron Gate MESFETs

Owing to improvements in device fabrication technologies and device structures, high performance characteristics have been confirmed in GaAs LSIs, such as a 16-kbit SRAM (Ishii *et al.*, 1984) and a 2-k gate array (Toyoda *et al.*, 1985). In these LSIs, enhancement-mode FETs having gate lengths ranging from 1.0 to 1.5 μm were employed as driver and switching FETs. In pursuit of higher performance and higher packing density, gate length must be reduced to the submicron range. However, along with the gate-length shortening, the so-called short-channel effects and the orientation effect emerge as the most crucial factors in preventing device-performance improvement. Therefore, it is very important to establish technologies effective in alleviating short-channel effects and the orientation effect. Sections 8 and 9 decribe results achieved for the suppression of short-channel effects and the orientation effect, respectively.

8. ALLEVIATION OF SHORT-CHANNEL EFFECTS

Short-channel effects include the following features: (1) Threshold voltage shift to the negative side; (2) threshold voltage dependence on source-drain voltage; (3) poor current cutoff in the subthreshold region; (4) reduction in K-value; and (5) increase in drain conductance.

The short-channel effects impose a limit on high-speed performance as well as leading to difficulty in achieving the designed threshold-voltage value.

To date, several mechanisms for the short-channel effects have been presented.

(1) Two-dimensional potential and current distributions, owing to the fact that the gate-length to channel-thickness ratio is no longer much greater than 1.

(2) Piezoelectric effect due to the stress induced by a gate material and/or overlayer dielectric film.

For n^+ self-aligned FETs, which are one of the most promising device structures for high performance GaAs LSIs as described in Part III, the following factors are involved.

(3) The change in channel layer carrier profile due to lateral diffusion of n^+-layer impurity.

(4) Space-charge-limited current between n^+ source/drain regions, which flows through the semi-insulating GaAs substrate under the channel layer.

It is difficult to determine the dominant factors in causing short-channel effects, since they depend on many factors such as device structures, processing parameters, gate and dielectric materials, etc.

Several investigations were carried out to determine ways to alleviate short-channel effects. In this section, process technologies that are effective in reducing the influence of mechanisms (1), (3), and (4) in short-channel effects are presented and discussed. Results and discussions on the piezoelectric effect, which is also responsible for the orientation effect in threshold voltage, are described in Section 9a.

a. Use of Shallow Channel Layer with Higher Doping Concentration

Using Monte Carlo particle simulation, Awano *et al.* (1983) reported a negative shift of the threshold voltage in the submicrometer-gate MESFETs, and improvements in device characteristics by use of a higher doping concentration channel. However, the examined doping concentrations, which were less than 1×10^{17} cm^{-3}, are lower than the doping concentration in actual GaAs LSIs.

Daembkes *et al.* (1984) showed by numerical simulations and experiments that the use of higher carrier concentration—up to 1×10^{18} cm^{-3}—in the channel layer is very effective to suppress the threshold voltage shift to the negative side. The numerical simulations used a one-dimensional model, which takes into proper account the two-dimensional electric field distribution along the channel and the current displacement into the substrate.

The threshold voltage V_T was calculated as the gate-to-source voltage, which reduces the drain current below 0.5 mA for a 300 μm-wide device. The shift in the threshold voltage, ΔV_T, was defined as

$$\Delta V_T = V_T - V_{Tnom} \qquad (1)$$

with

$$V_{Tnom} = V_{bi} - N_D a^2 q/2\varepsilon_s, \qquad (2)$$

where V_{Tnom}, V_{bi}, N_D, a, q, and ε_s are nominal threshold voltage for long-gate devices, built-in potential, channel doping concentration, channel thickness, electronic charge and semiconductor permittivity, respectively. Channel thickness a, as a function of N_D, was chosen so that the nominal

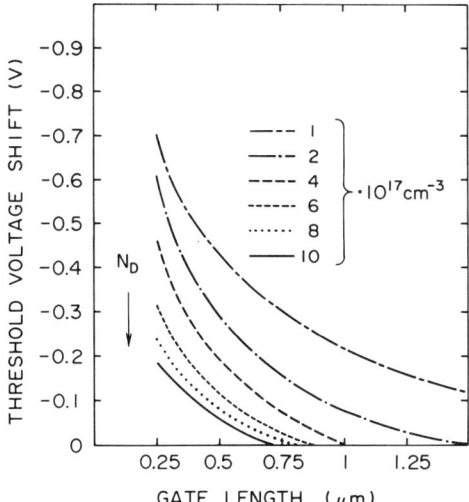

FIG. 30. Calculated threshold voltage shift versus gate length for different doping concentrations. [From Daembkes et al. (1984).]

pinchoff voltage, V_p, remains constant according to

$$a = [2\varepsilon_s V_p/(qN_D)]^{1/2} \quad (3)$$

The calculated results of threshold voltage shift are shown in Fig. 30. For the lower doping concentration, a pronounced shift appears for gates shorter than 1 μm. On the other hand, for the higher doping concentration, this negative shift becomes obvious only for $L_g < 0.5$ μm and is drastically reduced.

Ueno et al. (1985b) also reported an improvement in negative V_T shift by the use of an MBE-grown epitaxial layer with a higher doping concentration. Three kinds of epitaxial layers having different doping concentrations and thicknesses were prepared. They were 3.6×10^{18} cm^{-3} (20 nm), 8.6×10^{17} cm^{-3} (41 nm), and 2.5×10^{17} cm^{-3} (78 nm). Epitaxial layer thicknesses were determined in order to obtain the same V_T value in the long-gate FETs. Actually, V_T values ranging from 0.08 to 0.20 V were obtained. The V_T and K-value dependences on L_g are shown in Fig. 31, where ΔV_T was the threshold voltage shift from V_T for long-gate devices. As is clearly observed in Fig. 31, improved characteristics, that is, ΔV_T decrease and K-value increase, were achieved with increasing doping concentration in the channel layer.

From a practical viewpoint, optimization of doping concentration will be necessary in the use of highly doped channels, since a doping-concentration

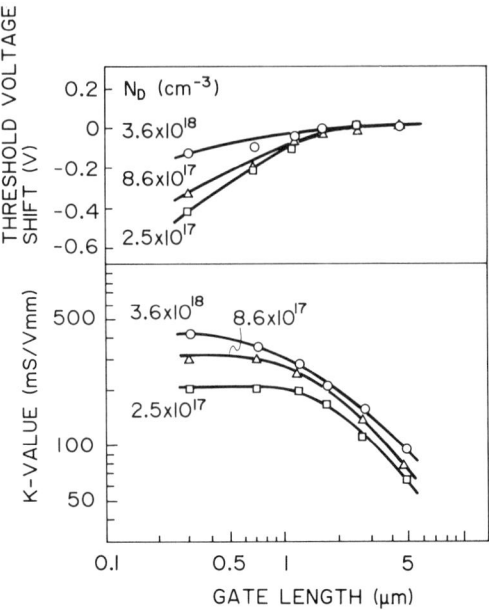

FIG. 31. Threshold voltage shift and K-value dependences on gate length for different doping concentrations: (○) 3.6×10^{18}, (△) 8.6×10^{17}, and (□) 2.5×10^{17} cm^{-3}. [From Ueno et al. (1985b).]

increase leads to an increase in gate capacitance and a degration in breakdown voltage. Moreover, in order to form a LSI-grade shallow channel layer having a higher doping concentration, advanced technologies should be developed. One promising method is the through-dielectric-film implantation technology. Onodera et al. (1984) reported the results on high-transconductance WSi$_x$-gate self-aligned GaAs MESFETs using a reactively sputtered AlN film as a through-dielectric-film for Si ions and as an annealing encapsulant. In addition, satisfactory results for the threshold voltage variation were experimentally confirmed, caused by the small value of the AlN film thickness variation—that is, about 3% over the 2-inch wafers. Another method is the low-energy ion-implantation technique. Sano et al. (1985) reported Si-implanted shallow channel layer formation using ion energies as low as 20 keV. Using 30 keV, Si ion implantation and SiO$_2$ encapsulation annealing under AsH$_3$ pressure, a high transconductance value with reduced short-channel effects was experimentally confirmed in W·Al-gate self-aligned FETs.

Recently, Onodera et al. (1988) reported shallower channel formation using 10 keV Si-ion implantation followed by rapid thermal annealing. Higher performance is achieved by 0.2-μm gate-length Au/WSiN refractory metal gate devices, as already described in Section 4.

b. Refinement of n^+-Layer Formation

As described in Part III, the n^+ self-aligned MESFET technology is expected to provide high performance FETs. In these FETs, the n^+-layer edge does not coincide with the n^+-implantation mask edge because of lateral stretch of the n^+ layer, which results from ion scattering during ion implantation and thermal diffusion during annealing. If the n^+ layer overlaps with the gate electrode, the effective gate length, L_{eff}, which determines the device characteristics, is not the length of a gate electrode but the distance between n^+-layers. Since L_{eff} is smaller than L_g, it is reasonable that an increased negative shift for V_T is observed in the V_T plot versus L_g. Therefore, in order to obtain meaningful insight regarding V_T behavior in the submicron-gate region, evaluations on the basis of V_T plot versus L_{eff} are necessary, even though the L_{eff} estimation is not easy. From a device manufacturing viewpoint, this n^+-layer–gate overlap is not advantageous, since such an overlap leads to an increase in gate capacitance and to degradation in gate breakdown voltage.

Control of n^+-Layer–Gate Gap. Kato *et al.* (1983) reported the effect of n^+-layer–gate gap control on alleviation of short-channel effects. Using the SAINT process, they fabricated two kinds of MESFETs having different n^+-layer–gate gaps, 0.15 and 0.3 μm. The control of n^+-layer–gate gap was achieved by varying the undercut for the bottom resist in a multilayer resist system. n^+ implantation was carried out through a 0.15 μm SiN film at 120 keV ion energy with 1.6×10^{14} cm^{-2} dose. Figure 29 shows the results of V_T dependence on L_g, where it is clear that the V_T shift in 0.3 μm undercut devices was substantially improved, as compared with that for devices with 0.15 μm undercut.

Since the lateral stretch of n^+ impurity is inevitable in the n^+-layer self-aligned FETs, as described previously, it is necessary to take into account this phenomenon in the discussion of short-channel effects.

In the SAINT FET, lateral stretch of n^+-impurity was experimentally examined by Yamasaki *et al.* (1982b), as described in Section 5. The gate-to-source external series resistance was measured against spacing between the n^+-implant projection edge and the gate electrode edge. The spacing was intentionally varied, using a shadow of n^+ implants caused by the multilayer resist wall in connection with slight tilting in the incident Si ion beam. Si ions for n^+-layer formation were implanted at 200 keV through a 0.15 μm PCVD

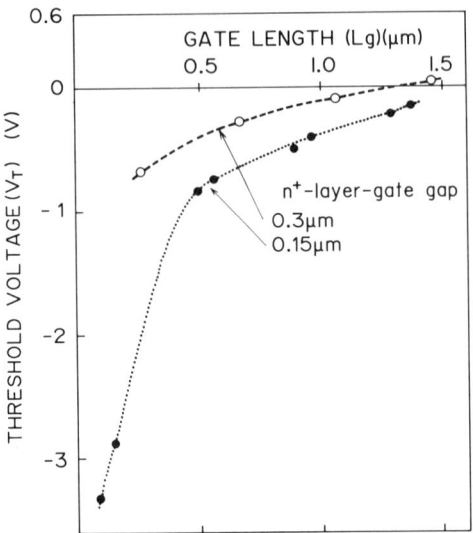

FIG. 32. Threshold voltage dependence on gate length for (●) 0.15 and (○) 0.3 μm n^+-layer–gate gaps. [From Kato et al. (1983).]

SiN film with a 4×10^{13} cm^{-2} dose. The results, shown in Fig. 18, indicate that about a 0.3 μm lateral stretch of n^+-impurity took place, caused by the ion scattering in GaAs during implantation and by redistribution during a 20-min thermal annealing at 800°C.

In the FETs whose results are shown in Fig. 32, the lateral stretch was not 0.3 μm, since the ion-implantation conditions for the n^+ layer were different. The gate-to-source series resistance, R_S, in FETs fabricated with a 0.3 μm n^+-layer–gate gap, was reported to be 0.11–0.13 Ω·mm. This value is very small, even though a slight increase was observed over the R_S value in FETs fabricated with a 0.15 μm n^+-layer–gate gap (0.085–0.105 Ω·mm). Therefore, it is reasonable to consider that the n^+-layer edge, even in FETs fabricated with a 0.3 μm n^+-layer–gate gap, extends to the gate electrode edge. Thus, an almost 0.3 μm L_{eff} difference in both FETs is probable. If this L_{eff} difference is taken into account for the results shown in Fig. 32, the difference in V_T behavior versus L_{eff} for both FETs becomes very small. This indicates that the differences in the short-channel effects for FETs fabricated with different n^+-layer–gate gap values originate from the difference in L_{eff}. This fact is reasonable, if the variations in the spacing beween the n^+-implant projection edge and the gate electrode edge lead to the same variations in the L_{eff}.

In actual devices, it is necessary to suppress gate capacitance increase and degradation in gate breakdown voltage. It is also necessary to achieve series resistance reduction. Therefore, it is a key point in process technology to control the separation between the n^+-implant projection edge and the gate electrode edge. From the same standpoint, refractory-metal-gate n^+ self-aligned technology, using T-gate structure, has already been reported (Lee et al., 1982; Sadler and Eastman, 1983a; Nakamura et al., 1983). Another approach utilizing dielectric side-walls was reported by Ueno et al. (1985a). The separation of n^+-implant projection edge and WSi_x gate-metal edge was made by a 0.2-μm-thick SiO_2 side-wall. It was experimentally confirmed that the V_T behavior versus L_g is almost the same as that for MESFETs fabricated without n^+ implantation.

Recently, in place of space formation between the n^+ layer and the gate metal, LDD technolgy, whose feature is the insertion of an intermediate carrier concentration region between the n^+ layer and the gate metal, was investigated. Experimental results concerning improvements in short-channel effects are described in Section 7.

Reduction in n^+-Layer Thickness. Using two-dimensional simulation, including lateral spreading of n^+-layer impurity during ion implantation, Yamasaki et al. (1982c) confirmed that thinning of the n^+ layer by reduction of the implantation energy is effective to alleviate short-channel effects.

Nakamura et al. (1983) also reported similar experimental results. W·Al-gate n^+ self-aligned MESFETs were fabricated with changing n^+ implantation energy from 60 to 180 keV, while still retaining the 2×10^{13} cm^{-2} implantation dose. Results are shown in Fig. 33. It is clear that reduction of

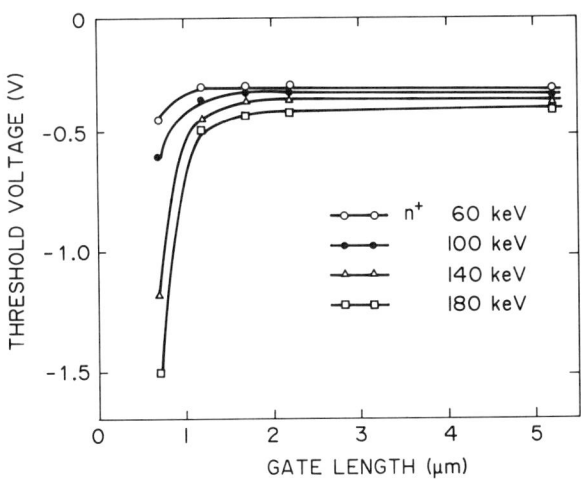

FIG. 33. Threshold voltage dependence on gate length for different n^+ ion energies: (○) 60 keV, (●) 100 keV, (△) 140 keV, and (□) 180 keV. [From Nakamura et al. (1983).]

the implantation energy has an advantage in reducing short-channel effects. In the device fabrication, a T-gate structure, as described previously, was employed. Since side etching of the gate metal was less than 0.1 μm, an overlap between gate metal and n^+ layer would occur, even in a FET fabricated with 60 keV n^+-implantation energy. Therefore, L_{eff} will decrease with increasing n^+-implantation energies, corresponding to the increase in n^+-layer depth. This fact is responsible for the results shown in Fig. 33. Another possible reason is a substrate current between adjacent n^+-source/drain regions, since substrate current may increase with increasing n^+-layer depth.

A decrease in n^+-layer depth leads to an increase in n^+-layer sheet resistance, resulting in an increase in parasitic resistance. Thus, optimization of n^+-implantation energy is necessary in order to suppress short-channel effects and to achieve high performance in the device.

Suppression of n^+-Layer Lateral Stretch during Post-implantation Annealing. The lamp annealing method was experimentally confirmed to be very effective for reduction in n^+-impurity diffusion during post-implantation annealing, according to Ohnishi *et al.* (1984). Successful results using the lamp annealing method were already well known in the formation of an ion-implanted conductive layer with negligible impurity diffusion (see Chapter 1). Using WSi$_x$-gate n^+ self-aligned technology, lamp-annealed FETs (LAFETs) and conventional furnace annealed FETs (FAFETs) were fabricated. Si implantation for the n^- and n^+ layers was performed at 59 keV with 1.0×10^{12} cm^{-2} and at 175 keV with 1.7×10^{13} cm^{-2}, respectively. Annealing conditions in n^- and n^+-layer formation for LAFETs were 1000°C, 12 sec and 960°C, 5 sec, respectively. For FAFETs, the annealing conditions were 850°C, 900 sec and 750°C, 900 sec, respectively.

Figure 34 shows V_T dependence on L_g, where marked improvement in negative V_T shift was observed for LAFETs. Improvements in N_g value[2] and increase in K-value in the submicron gate region were also confirmed for LAFETs, as is clear in Figs. 35 and 36. Suppression of short-channel effects in LAFETs is attributed to the reduction in the n^+-layer lateral stretch.

The lateral stretch of the n^+ layer, estimated from a comparison between experimental results concerning V_T dependence on L_g and one-dimensional simulation results, was 0.24 μm and 0.11 μm for FAFETs and LAFETs, respectively. From these values for n^+-layer lateral stretch, the L_{eff} difference between two FETs is estimated to be 0.26 μm. Even though this L_{eff}

---

[2]N_G is the subthreshold parameter defined from the expression for drain current in the subthreshold region, i.e., $I_{DS} \propto \exp(qV_G/N_G kT)$, where q is the elementary charge, V_G the gate bias, k the Boltzmann constant and T the temperature.

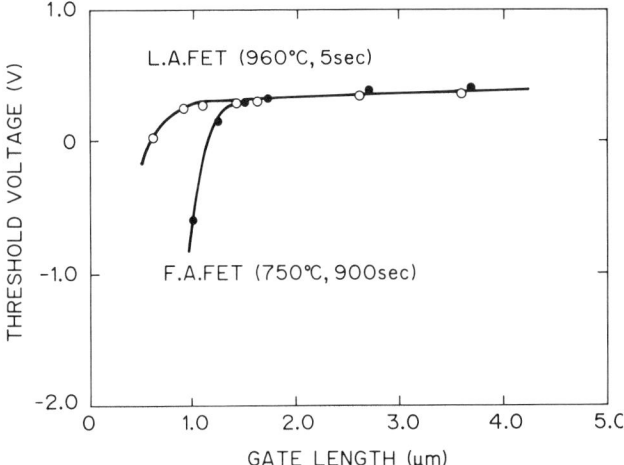

FIG. 34. Threshold voltage dependence on gate length in LA and FAFETs. [From Ohnishi et al. (1984).]

difference is taken into account in the results shown in Fig. 34, short-channel effects are still greater for FAFETs than for LAFETs, indicating enhancement in short-channel effects in FAFETs. Although the cause is not clear, a possible reason is the change in impurity concentration in the channel layer due to the impurity profile broadening at the n^+-layer tail region.

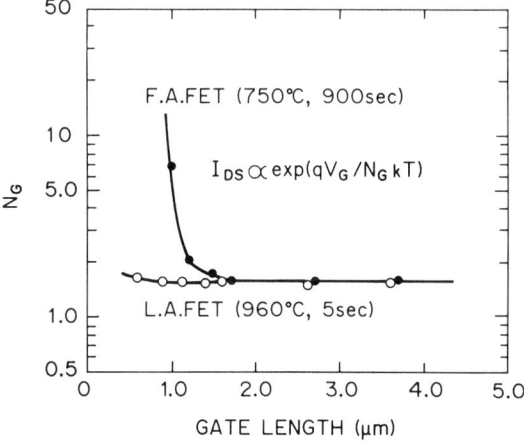

FIG. 35. N_g-value dependence on gate length in LA and FAFETs. Subthreshold current I_{sub} is expressed as $I_{sub} \exp(qV_g/N_g kT)$. [From Ohnishi et al. (1984).]

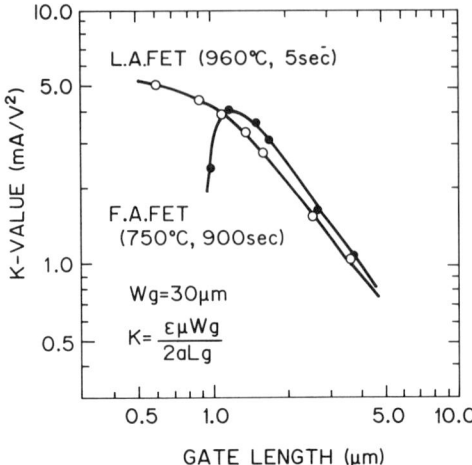

FIG. 36. K-value dependence on gate length in LA and FAFETs. [From Ohnishi et al. (1984).]

The possibility of change in impurity concentration in the channel layer was directly confirmed for W-gate n^+ self-aligned FETs, which were fabricated without channel layer implantation (Matsumoto et al., 1982). To form the n^+ layer, Se ions were implanted at 250 keV with 2×10^{13} cm^{-2} using W gate metal as an implantation mask. Post-implantation annealing was performed at 900°C for 10 minutes, without a cap, under an arsenic pressure. Even though no channel layer implantation was carried out, FET characteristics were observed, as shown in Fig. 37. These results support the presence of Se-ion lateral spreading and associated formation of an effectively very short active region. The length over which Se ions spread is at least one-half the length of the longest gate, i.e., 0.4 µm. This value is rather large. However, using the lamp-annealing method, it was experimentally confirmed that the lateral spread for the Se impurity was reduced to about 0.12 µm (Matsumoto et al., 1984a).

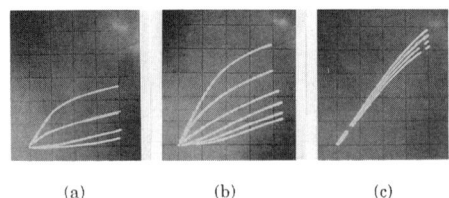

FIG. 37. FET characteristic dependence on gate length; (a) for $L_g = 0.8$ µm, (b) for $L_g = 0.6$ µm, (c) for $L_g = 0.4$ µm. [From Matsumoto et al. (1982).]

c. Buried p-Layer Formation underneath the Channel Layer

One of the main causes for short-channel effects is a substrate current that flows in the semi-insulating substrate between adjacent source/drain n^+ layers. To clarify substrate effects on FET characteristics, Monte Carlo simulation was carried out using an n^+ self-aligned FET as a model (Matsumoto et al., 1984b). The channel length, which was the same as the gate length, was 0.25 μm. Doping concentrations for the channel and the n^+ layer, both of which were 0.1 μm thick, were 2×10^{17} cm^{-3} and 8×10^{17} cm^{-3}, respectively. Figure 38 shows I-V characteristics for the FETs (a) with and (b) without a substrate. As compared to characteristics of FETs without a substrate, an increase in V_T absolute value, together with an increase in drain conductance, was observed for FETs with substrate, indicating the existence of substrate current.

In actual devices, to prevent substrate current, formation of a buried p layer under the channel layer is very effective. Matsumoto et al. (1984b) examined electrical characteristics for W-gate n^+ self-aligned FETs having a Mg-implanted p layer. Channel layer and buried p layer were made by ion implantation using 4×10^{12} cm^{-2} Se at 160 keV and 2×10^{12} cm^{-2} Mg at 250 keV, respectively. As a comparison, conventional FETs were also fabricated, in which 2×10^{12} cm^{-2} Se at 160 keV was used for channel-layer formation. In both FETs, 1.2×10^{13} cm^{-2} Se at 160 keV was used for n^+-layer formation. Implanted layers were annealed in an infrared lamp furnace at 950°C for 25 sec without a cap, under arsenic pressure. Figure 39 shows I-V characteristics for both FETs. Even though the FET with a buried p layer has a smaller L_g than that for the FET without a p layer, the drain conductance is almost

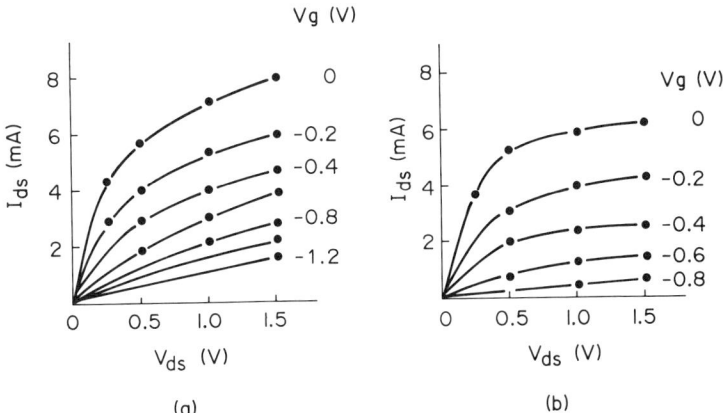

FIG. 38. I-V characteristics obtained by Monte Carlo simulation for self-aligned FETs. (a) with substrate and (b) without substrate. [From Matsumoto et al. (1984b).]

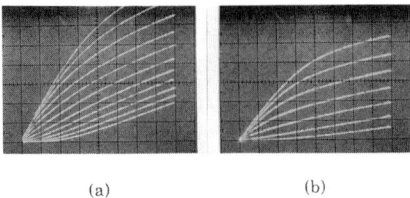

FIG. 39. *I–V* characteristics for FETs (a) with and (b) without a buried *p* layer: (a) gate length $L_g = 0.33\,\mu\text{m}$, gate bias $V_g = +0.4\text{--}-1.6$ V; (b) $L_g = 0.23\,\mu\text{m}$, $V_g = 0\text{--}-1.2$ V. 1 mA/division, 0.2 V/division. [From Matsumoto *et al.* (1984b).]

half as much. Marked improvements in drain current cutoff characteristics in the subthreshold region were also confirmed for FETs with a buried *p* layer. Figure 40 shows g_m dependence on L_g. The g_m value for a FET with a buried *p* layer shows a steep increase with decreasing L_g, as compared to the results for an FET without a *p* layer. However, when the L_g is reduced further, a decrease in the g_m value was observed. Two possible reasons were proposed. One is the lateral diffusion of the n^+ impurity into the channel layer, as described previously. The other is the increase in the ratio between channel thickness and gate length.

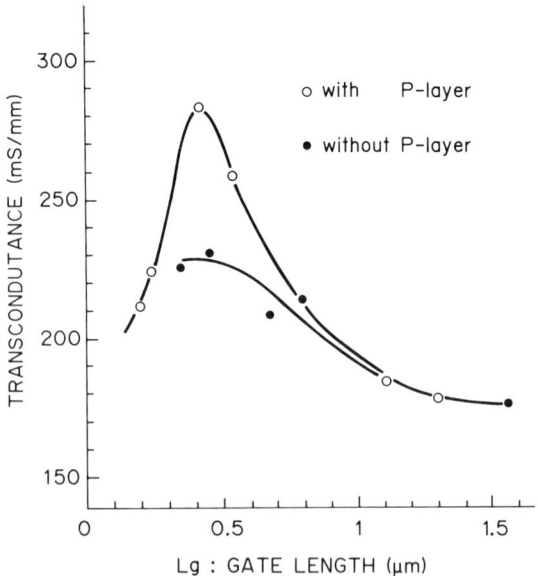

FIG. 40. Transconductance dependence on gate length for FETs (○) with and (●) without a buried *p* layer. [From Matsumoto *et al.* (1984b).]

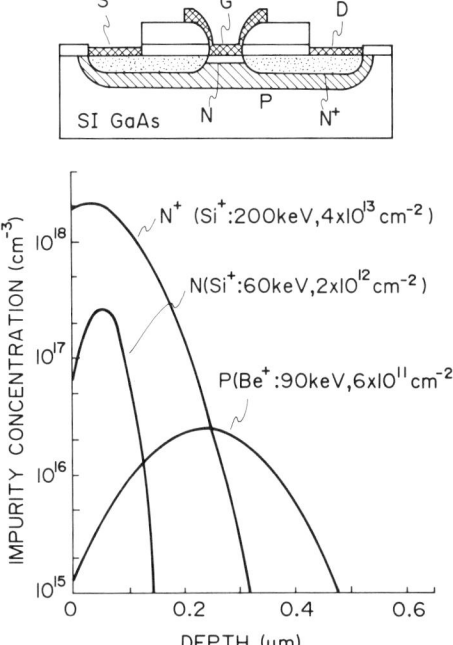

FIG. 41. Buried p-layer SAINT FET cross-sectional view and implanted ion profiles. [From Yamasaki et al. (1984).]

Yamasaki et al (1984) also reported improvement in short-channel effects, using a Be-implanted buried p layer. Figure 41 shows a schematic cross-sectional view of a buried p-layer SAINT FET. 2×10^{12} cm^{-2} Si at 60 keV, 4×10^{13} cm^{-2} Si at 200 keV and 6×10^{11} cm^{-2} Be at 90 keV were used for n^--, n^+-, and buried p-layer formation, respectively. In the n^+-layer formation, through implantation using 0.15 μm SiN film was employed. The Be implantation dose was chosen so that the p layer would be completely depleted by the built-in potential. Electrical activation of n^-, n^+, and p layers was simultaneously carried out by thermal annealing with a SiN cap at 800°C for 20 min. It was confirmed that the activation coefficient for implanted Be ions is approximately 100% and thermal diffusion is negligibly small in the small concentration range used. Threshold voltage shift due to Be$^+$ implantation in long-gate-length FETs is almost equal to the value estimated under the conditions that all implanted Be ions become shallow acceptors and that the p layer is completely depleted. Figure 42 shows the dependence of ΔV_T and the subthreshold parameter N_g on L_g, where satisfactory improvements were observed for FETs with a buried p layer. In addition, for FETs with

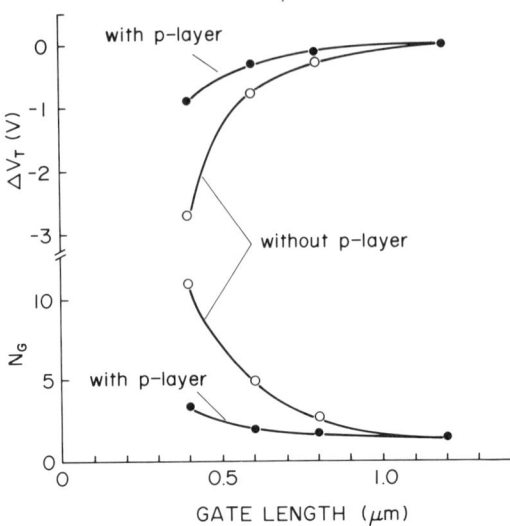

Fig. 42. Threshold voltage shift ΔV_T and subthreshold parameter N_g dependence on gate length for SAINT FETs, (●) with and (○) without buried p layer. [From Yamasaki et al. (1984).]

a buried p layer, an increase in drain current by a factor of about two was observed, compared to conventional FETs with the same V_T value. This is caused by an improvement in channel-layer profile in the tail region (Nozaki and Ohata, 1976) and a reduction in channel-layer effective thickness.

As is clear from these results, a buried p layer formed under the channel layer is very effective to alleviate short-channel effects. Since ion implantation technology can be applied for p-layer formation, controllability and reproducibility, which are vital factors in LSI fabrication, will be achieved accompanied by substrate quality improvement.

d. Formation of n^+ Selective Epitaxial Layer on Source and Drain Region

In order to achieve higher performance in device characteristics, a shallow channel layer having a higher carrier concentration is required. To reduce short-channel effects, n^+-layer thinning is also necessary, which leads in turn to an increase in series resistance. To overcome these restrictions, the technology for an n^+ selective epitaxial layer formation on source and drain regions is one solution.

Uetake et al. (1985) reported the characteristics of a sidewall assisted self-alignment GaAs MESFET having selectively grown n^+ contact regions. Using WSi_x gate-electrode metal as a mask, MOCVD-grown n^+ epitaxial layers were selectively formed on source and drain regions. A separation

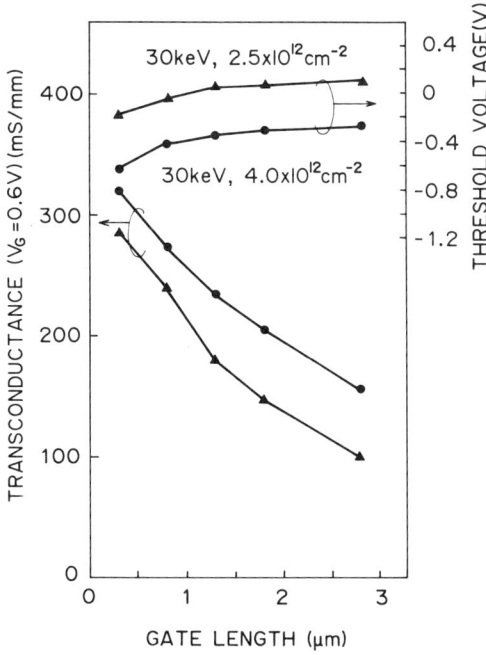

FIG. 43. Threshold voltage and transconductance dependence on gate length. [From Uetake *et al.* (1985).]

between gate electrode and epitaxially grown n^+ layers was made by 0.1-μm-thick CVD SiO_2 sidewalls, which were formed on the gate electrode sides. Figure 43 shows the V_T and g_m dependence on L_g. The negative shift of V_T in the submicron gate region is very small, almost comparable to that for a SWAT FET fabricated without ion-implanted n^+ layers (Higashisaka *et al.*, 1983). In addition, increases in the g_m value, corresponding to an L_g decrease, are clearly observed. This fact may indicate a smaller parasitic resistance.

In applying this technology to LSI fabrication, several technologies must be developed as well. One is establishing low-resistive and reliable properties in the n^+/n interface. Since several processing steps are carried out before the n^+ epitaxial growth, degradation in the GaAs surface, caused by contamination and damage that are introduced by the deposition and dry etching process, is unavoidable. Therefore, suitable techniques capable of removing contamination and damage, such as cleaning, low-temperature annealing and a slight amount of wet etching, must be established. The other is the establishment of controllability and reproducibility in the selective epitaxy. This is very important, since epitaxial layer thickness and selectivity are influenced by device pattern size and spacing in LSIs.

9. Suppression of the Orientation Effect

It has been reported by many authors that short-channel FETs, which have different gate-finger directions in $\langle 100 \rangle$ wafers, show different electrical characteristics such as V_T, σ_{V_T} and I_{dss}. The so-called orientation effect leads to a restriction in FET arrangement in LSIs. Normally, gate fingers are arranged in one direction (Hirayama *et al.*, 1984). This restriction is disadvantageous to realizing high-performance and high-packing-density GaAs LSIs. Therefore, it is necessary to clarify the origin of the orientation effect. It is also necessary to establish process technologies or device structures that are capable of reducing the orientation effect.

The origin of the orientation effect is not perfectly clear at this stage. However, the following factors have been presented as possible reasons.

(1) Piezoelectric effect due to the stress induced by the overlayer dielectric film and/or the gate-metal film.
(2) Preferential lateral diffusion of n^+-layer impurity.

Many reports on orientation effects have been published. However, the definition of crystalline direction to the gate finger was different in each study. In order to avoid confusion, a fixed definition for the gate-finger direction is necessary. Figure 44 shows the notations of the six kinds of FETs A, B, C, D, E, and F, whose gate fingers have different orientations with respect to the $\langle 100 \rangle$ GaAs substrates. FETs reported in each paper correspond to some of these FETs. Hereafter, these notations are fixed in this

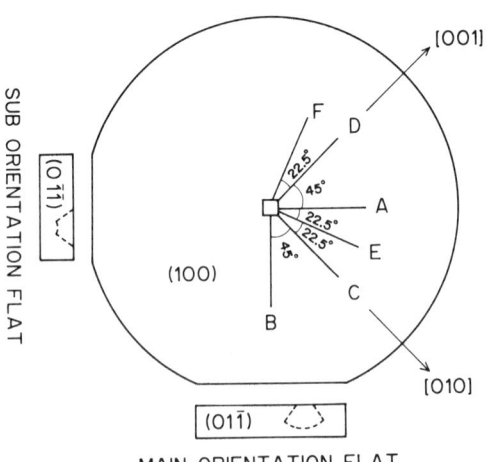

FIG. 44. Gate-finger orientations for six kinds of FETs, A, B, C, D, E, and F, in $\langle 100 \rangle$ GaAs. Shapes of the etched grooves in main and sub orientation flat are also shown by dashed lines.

section. The shapes of the etched grooves in two different [011] directions, which correspond to the main and sub orientation flats, are shown by the dashed lines. The difference in the etched pattern enables identification of the two different [011] directions.

a. Piezoelectric Effect

Recently, the piezoelectric effect has been presented as the physical mechanism for orientation effects. Chang et al. (1984) examined V_T shift as a function of overlayer Si_3N_4 film thickness for FETs having different gate-finger directions. FETs having six different gate-finger directions, as shown in Fig. 44, were fabricated using 3-inch semi-insulating LEC GaAs substrates with ⟨100⟩ surfaces. The *n*-type channel layer under the gate was doped by a Se implant, while the deeper n^+ regions under the source and drain had an additional Si implant to improve the ohmic contact. Gate length and width were 1 μm and 50 μm, respectively. The spacing between the n^+ region and the gate was approximately 0.75 μm. Device fabrication processes were similar to those for planar GaAs integrated circuits (Welch and Eden, 1977).

Figure 45 shows V_T shifts observed for FETs having different thicknesses of the overlayer dielectric films. The overlayer Si_3N_4 film was thinned by CF_4 plasma etching. As is clear in Fig. 45, the absolute value of the threshold voltage V_T for FETs A and E increased with decreasing overlayer Si_3N_4 film thickness, while V_T for FETs B and F decreased. The greatest changes occurred for FETs A and B. V_T difference in these two FETs was decreased

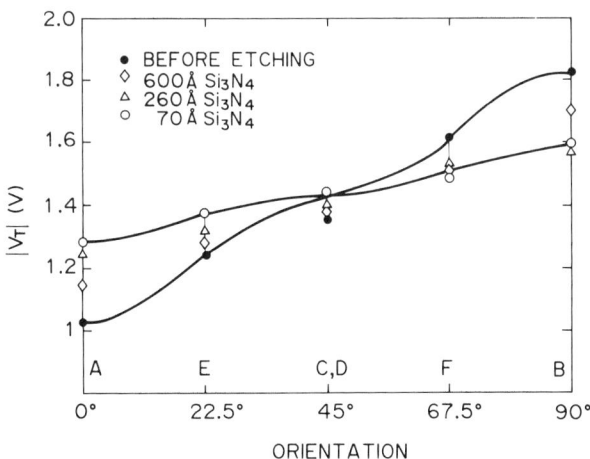

FIG. 45. Threshold voltage for FETs A, B, C, D, E, and F, as suface Si_3N_4 film was thinned. (●) Before etching; (◇) 600 Å, (△) 260 Å, and (○) 70 Å Si_3N_4. [From Chang et al. (1984).]

with decreasing overlayer Si_3N_4 film thickness. On the other hand, FETs C and D had nearly no change in V_T. These orientation effects, which depend on the thickness of the overlayer dielectric film, were explained by the existence of a piezoelectric charge density ρ_{pz} induced by stress in the overlayer Si_3N_4 film. Because of the anisotropy of GaAs crystals, ρ_{pz} induced in the channel for FETs A is of opposite sign to that in FET B, and it vanishes for FETs C and D. In the channels of FETs A and B, the magnitude of ρ_{pz} is proportional to the thickness of the dielectric overlayers and approximately inversely proportional to the square of the length of the film openings in the gate area (Asbeck *et al.*, 1984).

Figure 46 shows $|V_T|$ for FETs A, B, and C, as a function of their positions on a 3-inch GaAs wafer. $|V_T|$ showed strong radial dependence for FETs A and B. However, this radial dependence of device characteristics is opposite in these two FETs. The average V_T and σV_T are indicated on the figure, where large σV_T was observed for FETs A and B because of the radial dependence of V_T. For large-diameter GaAs wafers, radial dependences of processing parameters are sometimes difficult to avoid. The deposited dielectric films are generally thinner around the wafer edge and thicker in the center of the substrate. In addition, the amount of undercut for dielectric openings made by plasma etching also often appears to be radially dependent. Combining

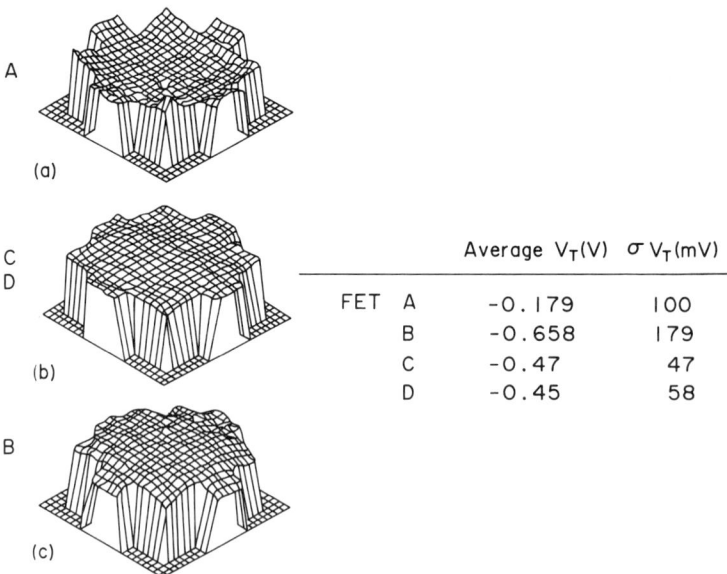

FIG. 46. V_T of FETs as a function of their position on a 3-inch GaAs wafer: (a) for FET A, (b) for FET C, and (c) for FET B. [From Chang *et al.* (1984).]

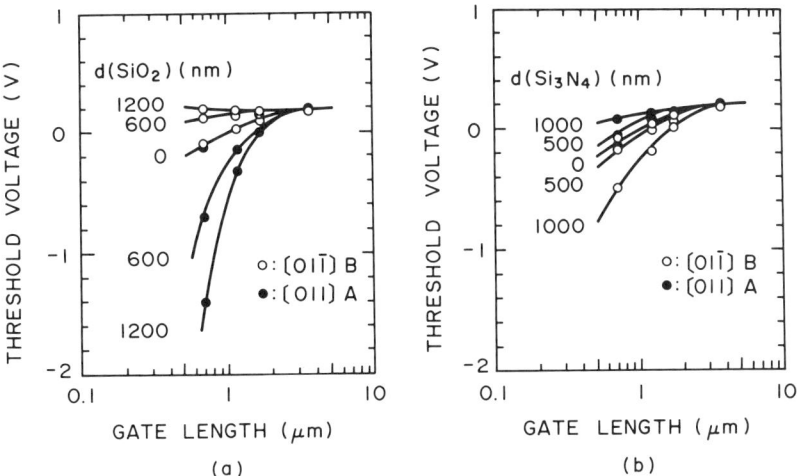

FIG. 47. Gate length dependence of threshold voltage for FET A and B: (a) FETs with SiO_2 film thickness, $d(SiO_2)$, as a parameter; (b) FETs with Si_3N_4 film thickness, $d(Si_3N_4)$, as a parameter. [From Ohnishi *et al.* (1985).]

the results shown in Fig. 45, it can be explained that these variations in the dielectric film thickness and in the film openings size for gate deposition are responsible for the radial dependence of the FET characteristics.

Ohnishi *et al.* (1985) also reported experimental results that support the piezoelectric effect model of the orientation effect. WSi_x-gate n^+ self-aligned FETs A and B, having four different gate lengths with 20 μm gate width, were fabricated on a ⟨100⟩ semi-insulating GaAs substrate. The *n*-channel layer was formed using Si implantation at 59 keV with 9×10^{11} cm^{-2}, followed by furnace annealing at 800°C for 15 min. Self-aligned n^+ layers were formed by Si-ion implantation with 0.4-μm thick WSi_x gate as a mask at 175 keV with 1.7×10^{13} cm^{-2} and activated by furnace annealing at 750°C for 15 min. The encapsulation film for annealing after both n and n^+ implantation was 100-nm-thick CVD SiO_2. Au/Au·Ge (280 nm/20 nm) ohmic source/drain electrodes were lifted off with dielectric overlayers as spacers and alloyed at 450°C for 1 min. For dielectric overlayers, two kinds of films, 1200-nm-thick CVD SiO_2 and 1000-nm-thick plasma-CVD Si_3N_4, were used. Achieving intermittent chemical etching, the V_T values for 40 or more FETs were measured and averaged in 2-inch-diameter wafers.

Figure 47 shows the gate-length dependence of V_T for FETs A and B with SiO_2 film thickness (a), and with Si_3N_4 film thickness (b), as a parameter. From the results shown in Fig. 47, the following features can be pointed out:

(1) V_T dependence on L_g coincides for FETs A and B, when the dielectric overlayer thickness is equal to zero.

(2) Orientation effects begin to be observed when dielectric overlayer films exist.

(3) V_T shifts for FETs A and B were opposite for SiO_2 and Si_3N_4 film.

(4) Differences in V_T value for FETs A and B increased with increasing dielectric overlayer film thickness.

These results, observed in FETs having dielectric overlayers, were explained by the piezoelectric effect. The SiO_2 and Si_3N_4 films showed different signs of internal stress in the film. The SiO_2 film has compressive stress with a magnitude of about mid 10^9 dyn/cm^2 and the Si_3N_4 film has tensile stress with a magnitude of about 10^9 dyn/cm^2. Therefore, it is reasonable that different orientation effects are observed for both FETs. Increase in the V_T differences for FETs A and B with increasing thickness of dielectric overlayer films can be explained by the fact that the magnitude of ρ_{pz} is proportional to the dielectric overlayer film thickness. In the studies reported by Chang et al. (1984), there is no description of the Si_3N_4 film stress sign and magnitude. If the sign of the Si_3N_4 film stress is assumed to be the same as in the films used by Chang et al. and by Ohnishi et al., the threshold voltage differences in FETs A and B are qualitatively identical in both studies. Similar results were also reported by Yamasaki et al. (1982d) for SAINT FETs, where Si_3N_4 film exists ajacent to the gate electrode, and by Ueno et al. (1985a) for SWAT FETs, where SiO_2 film was used as an overlayer film.

When the thickness of the dielectric overlayer film was equal to zero, no differences in V_T dependence on L_g were observed for FETs A and B, as is clear in Fig. 47. This fact means that the stress in the gate material WSi_x is negligibly small, since a piezoelectric effect also should be observed if the gate material possesses stress. As described in Section 8, several factors induce in negative shifts V_T in the submicron-gate region. From the fact that no difference in V_T negative shifts was observed for FETs A and B, it is reasonable to assume that the effects of these factors are isotropic.

From the results shown in Fig. 47, it is also clear that enhanced short-channel effects were observed for FETs A with a SiO_2 overlayer and FETs B with a Si_3N_4 overlayer. Thus, in actual devices, the piezoelectric effect becomes one of the main mechanisms in short-channel effects, since a dielectric overlayer film is necessary to isolate the first and second interconnections. On the other hand, for FETs A with a Si_3N_4 overlayer and FETs B with a SiO_2 overlayer, V_T shift in the submicrometer-gate region becomes very small, if an appropriate overlayer film thickness is employed. These results originate from the compensation effect due to the piezoelectric effect.

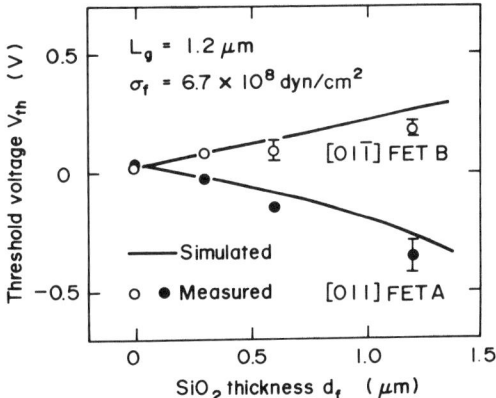

FIG. 48. Simulated (solid line) and measured threshold voltages as a function of the SiO_2 overlayer thickness for (●) FET A and (○) FET B. [From Onodera et al. (1985a).]

Using a two-dimensional device simulator including the effect of piezoelectric charges, Onodera et al. (1985a, b) examined, experimentally and theoretically, the FET-characteristics variation due to piezoelectric effects. Figure 48 shows V_T variation against SiO_2 overlayer thickness for 1.2 μm gate-length WSi_x-gate devices. Assuming the internal stress of the SiO_2 overlayer to be 6.7×10^8 dyn/cm², good agreement in theoretical and experimental results is found. They also reported the coincidence in experimental results and theoretical calculation for K-value improvements due to piezolectric effects.

Recently, Ueno et al. (1988) reported the disappearance of orientation effects when using ⟨111⟩ B substrates in place of conventional ⟨100⟩ substrates. Figure 49a and Fig. 49b show $I-V$ characteristics for 0.5 μm gate-length WSi_x-gate self-aligned MESFETs on ⟨100⟩ and ⟨111⟩ B substrates, respectively. As compared to ⟨100⟩ substrates, almost the same FET characteristics were observed for ⟨111⟩ B substrates, exhibiting good transconductance values as high as 300 mS/mm ($V_T = -0.65$ V). These results will open reconsideration of substrate orientation from the viewpoint of realization of higher-packing-density ICs.

b. Preferential Lateral Diffusion of n^+-Layer Dopant

As described in Section 7b, because of lateral diffusion of Se impurities, which were implanted at 250 keV with 6×10^{12} cm^{-2} to form n^+ layers, a channel layer was actually formed for 0.6 μm gate-length FETs fabricated without channel-layer implantation (Matsumoto et al., 1982). In these devices, a larger saturation drain current and a larger pinch-off voltage were

(a)

(b)

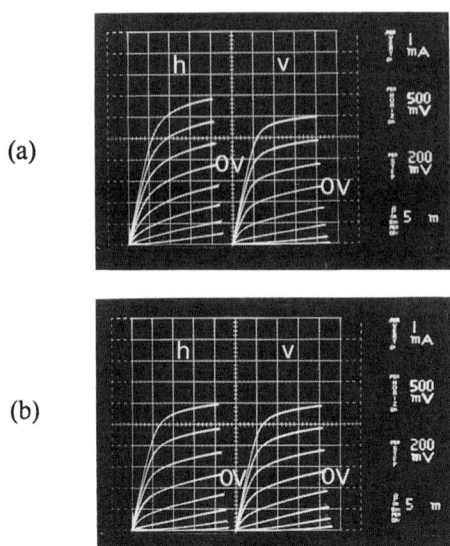

FIG. 49. I–V characteristics for FET A and B on different orientation substrates: (a) ⟨100⟩ substrate and (b) ⟨111⟩ substrate. [From Ueno et al. (1988).]

observed for FET B as compared to FET A. The authors have reported that the preferential lateral diffusion of Se impurities is a possible reason for the anisotropic characteristics. However, the piezoelectric effect cannot be ruled out. Even though the implantation for the channel was not carried out, the channel layer was definitely formed by lateral diffusion of Se impurities. Therefore, it is possible that anisotropy in FET characteristics originate from the piezoelectric effect, as discussed in Section 9a.

At first, anisotropic diffusion of n^+-layer dopant was proposed as a possible reason for the orientation effect (Lee et al., 1980; Yokoyama et al., 1983b; Sadler and Eastman, 1983b; McLaughlin and Birrittella, 1984). However, reexamination of the experimental results will be necessary, along with taking into account the piezoelectric effect.

At present, the gate-finger orientation is arranged in one direction in LSI fabrication (Hirayama et al., 1984). This is, of course, one way to avoid the orientation effect. However, this restriction is undesirable to realize high performance and high packing density in LSIs. Therefore, further work is needed to overcome the orientation effect. Since the piezoelectric effect is a main reason for the difference between FETs A and B in Fig. 44, reduction of metal and dielectric film stress should be effective. The sign of stress in the film as well as its magnitude depends on many factors, such as the type of materials, film thicknesses, deposition parameters and annealing conditions. Therefore, to reduce the stress, it is necessary to optimize these factors in a device structure used in LSIs. To do so, clarification of the origins of stress should be made. The other method is to use FETs C and D. Since FETs C and D show nearly the same characteristics due to disappearance of the piezoelectric effect, this method seems to be a good choice. However, optimizing the gate-finger direction, including choice of a one-direction arrangement, in actual LSIs should be determined by considerations of device characteristics and manufacturing ease, including parameters such as performance, chip size and production yield.

V. Summary

The work reported here reflects the current state of the art in device fabrication process technologies for achieving high-performance GaAs LSIs. In order to maximize the advantages of the material properties of GaAs in device fabrication, developments in device structures and process technologies, which are directed to high-performance FET characteristics, are essential. Among the developed novel device structures, the n^+ self-aligned gate structure is very promising, since high performance of FETs is realized by reducing parasitic elements such as source-to-gate resistance. Two kinds of n^+ self-aligned gate structures have been investigated, as described in Part III. One uses a refractory-metal gate and the other is SAINT (self-aligned implantation for n^+-layer technology). By using both structures, GaAs LSIs such as 4k and 16 kbit SRAM have been fabricated (Yokoyama *et al.*, 1985; Ishii *et al.*, 1984).

In order to achieve further improvements in device characteristics and to achieve controllable and reproducible processes, it is very important to establish process technologies that are effective in suppressing short-channel effects and in controlling orientation effects, both of which emerge in the submicron-gate region.

To suppress short-channel effects, several technologies have been developed. First, it is very important to establish an ion-implantation technology suitable for forming a shallow channel layer with high doping concentration. The through-AlN implantation or low-energy-implantation technology is one of the most promising candidates. In the formation of n^+ source/drain regions, control of the spacing between n^+ layer and gate is essential, since an overlap between gate electrode and n^+ layer leads to a decrease in effective gate length, degradation in gate breakdown voltage and an increase in gate capacitance. As well as decreasing the active-layer thickness, it is also necessary to reduce n^+-layer depth, in order to suppress short-channel effects. This, in turn, leads to an increase in parasitic resistance. Therefore, from the trade-off between suppression of short-channel effects and parasitic resistance reduction, some limitation might occur in n^+-layer formation by ion implantation. To overcome this limitation, LDD technology and selective epitaxial growth of the n^+ layer in source and drain regions, in combination with buried p^--layer formation underneath the channel layer, will become a useful technology.

Another problem in submicron-gate MESFETs is the orientation effect. Because of the piezoelectric property of GaAs, device characteristics are modified by the piezoelectric charge in the channel that is induced by internal stress in a gate-metal film and/or a dielectric overlayer film. The sign and the magnitude of the piezoelectric charge differ according to the gate-finger orientation, resulting in different device characteristics. In fact, in the combination of the gate-finger orientation and the kind of dielectric overlayer film used, enhanced short-channel effects are observed. On the other hand, a suitable choice of the gate-finger orientation and the kind of dielectric overlayer film eliminates short-channel effects. Usually, to avoid the orientation effect, gate fingers are arranged in one crystal orientation. Since this restriction is undesirable to realize high performance and high packing density in LSIs, further work is necessary to overcome the orientation effect on the optimization in film deposition processes and device structures.

In order to fabricate high-performance GaAs LSIs with good yield, it is very important to establish a controllable and reproducible total device fabrication process. Stable characteristics of the interface between a gate metal and GaAs are essential. In n^+ self-aligned structure with refractory metals such as WSi_x, $TaSi_2$, $W \cdot Al$, and WN_x film as a gate, stable characteristics of the barrier height and the ideality factor are obtained in initial devices. However, in order to obtain long-term stability, further study may be needed to optimize process technologies. In the SAINT FETs, a gate is formed after opening the gate area by removing a SiN film with dry etching. In this structure, many kinds of gate metals can be employed, since n^+ layers are already formed before gate-metal deposition. This is a useful point for the SAINT structure. However, it is further necessary to develop a process technology to minimize damage and contamination during dry etching.

Reduction of parasitic resistance is essential to maximize the extrinsic transconductance. In addition to the n^+-layer formation, it is very important to form a low resistive ohmic contact uniformly across a wafer. Among many kinds of ohmic-contact metals for n-type, $Au \cdot Ge$ eutectic metals, which realize low specific contact resistance (down to 1×10^{-6} Ω-cm^2) after the alloying process, are useful. In view of demands for device-size shrinkage, further work is needed, including the discovery of new ohmic-contact metals to form an ohmic contact as close as possible to a gate and to make very fine patterns suitable for submicron device structures.

To achieve device-size shrinkage in GaAs LSIs, progress in dry etching technologies is necessary for fine pattern definition in areas such as first- and second-level interconnection and contact-hole opening. Some process technologies developed in Si LSIs can be applied to GaAs LSIs as well. However, owing to the inherent characteristics of GaAs, technologies particularly suitable for GaAs devices should be developed.

Finally, in manufacturing high-performance GaAs LSIs with satisfactory yield, it should be noted that a total fabrication process should be developed with advances in each elemental process that are directed to high-performance device characteristics and to high-yield device fabrication.

References

Asai, K., Kurumada, K., Hirayama, M., and Ohmori, M. (1983). *Int. Solid-State Circuits Conf., Dig. Tech. Papers*, p. 46.

Asai, S., Goto, N., Kanamori, M., Tanaka, Y., and Furutsuka, T. (1986). *Conf. Solid State Devices and Materials, Tokyo, Japan*.

Asai, K., Sugahara, H., Matsuoka, Y., and Tokumitsu, M. (1988). *J. Vac. Sci. Technol.* **B6**, 1526.

Asakawa, K., and Sugata, S. (1985). *Proc. Symp. Dry Process, Tokyo, Japan*, p. 138.

Asbeck, P. M., Lee, C. P., and Chang, M. F. (1984). *IEEE Trans. Electron Devices* **ED-31**, 1377.

Awano, Y., Tomizawa, K., Hashizume, N., Kawashima, M., and Kanayama, T. (1983). *Tech. Dig.-Int. Electron Devices Meet*, p. 617.

Bondur, J. A., and Frieser, R. G. (1981). *Proc. Symp. Plasma Etching Deposition, Pennington, New Jersey*, p. 180.

Braslau, N. (1981). *J. Vac. Sci. Tech* **19**, 803.

Chang, M. F., Lee, C. P., Asbeck, P. M., Vahrenkamp, R. P., and Kirkpatrick, C. G. (1984). *Appl. Phys. Lett.* **45**, 279.

Chung, Y., Langer, D. W., and Look, D. (1985). *IEEE Trans. Electron Devices* **ED-32**, 40.

Crabbe, E., and Sirkin, E. R. (1983). *Ext. Abstr., Int. Symp. Plasma Process*, p. 261.

Daembkes, H., Brockerhoff, W., Heime, K., and Cappy, A. (1984). *IEEE Trans. Electron Devices* **ED-31**, 1032.

Eden, R. C., and Welch, B. M. (1982). *In* "VLSI Electronics Microstructure Science" (N. G. Einspruch, ed.), Vol. 3, Chap. 4. Academic Press, New York.

Furutsuka, T., Tsuji, T., Katano, F., Higashisaka, A., and Kurumada, K. (1981). *Electron. Lett.* **17**, 944.

Furutsuka, T., Takahashi, K., Ishikawa, M., Yano, S., and Higashisaka, A. (1984). *Tech. Dig.—Int. Electron Devices Meet.*, p. 344.

Gorowitz, B., and Saia, R. J. (1984). *In* "VLSI Electronics Microstructure Science" (N. G. Einspruch and D. M. Brown, eds.), Vol. 8, Chap. 10. Academic Press, New York.

Higashisaka, A., Ishikawa, M., Katano, F., Asai, S., Furutsuka, T., and Takayama, Y. (1983). *Conf. Solid State Devices and Materials, Tokyo, Japan*.

Hirayama, M., Ino, M., Matsuoka, Y., and Suzuki, M. (1984). *Int. Solid-State Circuits Conf., Dig. Tech. Papers*, p. 46.

Hojo, A., Toyoda, N., Mochizuki, M., Mizoguchi, T., and Nii, R. (1981). *GaAs IC Symp., San Diego, California*.

Ida, M., Uchida, M., Shimada, K., Asai, K., and Ishida, S. (1981). *Solid-State Electron.* **24**, 1099.
Imamura, K., Yokoyama, N., Ohnishi, T., Suzuki, S., Nakai, K., Nishi, H., and Shibatomi, A. (1984). *Jpn. J. Appl. Phys.* **20**, L342.
Ishii, Y., Ino, M., Ida, M., Hirayama, M., and Ohmori, M. (1984). GaAs *IC Symp., Boston, Massachusetts.*
Kanamori, M., Nagai, K., and Nozaki, T. (1985). GaAs *IC Symp., Monterey, California.*
Kato, N., Matsuoka, Y., Ohwada, K., and Moriya, S. (1983). *IEEE Electron. Device Lett.* **EDL-4**, 417.
Kawata, H., Onodera, H., Shinoki, T., Yokoyama, N., and Nishi, H. (1985). *Conf. Solid-State Devices and Materials, Tokyo, Japan.*
Kohn, E. (1979). *Tech. Dig.—Int. Electron Devices Meet.*, p. 469.
Kuzuhara, M., Nozaki, T., and Kohzu, H. (1985). *J. Appl. Phys.* **58**, 1204.
Lee, C. P., Zucca, R., and Welch, B. M. (1980). *Appl.. Phys. Lett.*, **37**, 311.
Lee, R. E., Levy, H. M., and Matthews, D. S. (1982). GaAs *IC Symp., New Orleans, Louisiana.*
Macksey, H. M. (1976). *Int. Symp. GaAs Relat. Compd., St. Louis, Missouri.*
Matsumoto, K., Hashizume, N., Atoda, N., Tomizawa, K., Kurosu, T., and Iida, M. (1982). *Int. Symp. GaAs Relat. Compd., Albuquerque, New Mexico.*
Matsumoto, K., Hashizume, N., and Atoda, N. (1984a). *Electron. Lett.* **20**, 940.
Matsumoto, K., Hashizume, N., Atoda, N., and Awano, Y. (1984b). *Int. Symp. GaAs Relat. Compd., Biarritz, France.*
Matsuura, H., Nakamura, H., Ishida, T., and Kaminishi, K. (1984). *Int. Conf. Solid-State Devices and Materials, Kobe, Japan.*
McLaughlin, K. L., and Birrittella, M. S. (1984). *Appl. Phys. Lett.* **44**, 252.
Miyazaki, M., Yasuda, T., Masuki, J., and Hashimoto, N. (1984). *Proc. Symp. Dry Process, Tokyo, Japan*, p. 25.
Mozzi, R. L., Fabian, W., and Piekarski, I. J. (1979). *Appl. Phys. Lett.* **35**, 337.
Mukherjee, S. D., Morgan, D. V., and Howes, M. J. (1979). *J. Vac. Sci. Tech.* **16**, 138.
Muraguchi, M., Ohwada, K., and Hirayama, M. (1984). *Proc. Symp. Dry Process, Tokyo, Japan*, p. 37.
Nakamura, H., Sano, Y., Nonaka, T., Ishida, T., and Kaminishi, K. (1983). GaAs *IC Symp., Phoenix, Arizona.*
Nozaki, T., and Ohata, K. (1976). *Int. Conf. Solid-State Devices, Tokyo, Japan.*
Ogawa, M. (1980). *J. Appl. Phys.* **51**, 406.
Ogawa, M., Shinoda, D., Kawamura, N., Nozaki, T., and Asanabe, S. (1970). *Int. Symp. GaAs Relat. Compd., Aachen, West Germany.*
Ohnishi, T., Yokoyama, N., Onodera, H., Suzuki, S., and Shibatomi, A. (1983). *Appl. Phys. Lett.* **43**, 600.
Ohnishi, T., Yamaguchi, Y., Onodera, T., Yokoyama, N., and Nishi, H. (1984). *Int. Conf. Solid-State Devices and Materials, Kobe, Japan.*
Ohnishi, T., Onodera, T., Yokoyama, N., and Nishi, H. (1985). *IEEE Electron. Device. Lett.* **EDL-6**, 172.
Onodera, H., Kawata, H., Yokoyama, N., Nishi, H., and Shibatomi, A. (1984). *IEEE Trans. Electron Devices* **ED-31**, 1808.
Onodera, T., Ohnishi, T., and Nishi, H. (1985a). *Int. Symp. GaAs Relat. Compd., Karuizawa, Japan.*
Onodera, T., Ohnishi, T., Yokoyama, N., and Nishi, H. (1985b). *IEEE Trans. Electron. Devices* **ED-32**, 2314.
Onodera, K., Tokumitsu, M., Sugitani, S., Yamane, Y., and Asai, K. (1988). *IEEE Electron. Device Lett.* **EDL-9**, 417.

Pang, S. W. (1984). *Solid State Technol.* **27**, 249.
Piotrowska, A., Guivarich, A., and Pelous, G. (1983). *Solid-State Electron.* **26**, 179.
Sadler, R. A., and Eastman, L. F. (1983a). *IEEE Electron Device Lett.* **EDL-4**, 215.
Sadler, R. A., and Eastman, L. F. (1983b). *Appl. Phys. Lett.* **43**, 865.
Sakaki, H., Sekiguchi, Y., Sum, D. C., Taniguchi, M., Ohno, H., and Tanaka, A., (1981). *Jpn. J. Appl. Phys.* **20**, L107.
Sano, Y., Egawa, T., Nakamura, H., Ishida, T., and Kaminishi, K. (1985). *Pap. Tech. Group, TGSSD-84-114, IECE Japan*, p. 81 (in Japanese).
Shimura, T., Noda, M., Hosogi, K., Tanino, N., Nishitani, K., and Otsubo, M. (1986). *Conf. Solid State Devices and Materials, Tokyo, Japan.*
Stall, R., Wood, C. E. C., Board, K., and Eastmann, L. F. (1979). *Appl. Phys. Lett.* **35**, 951.
Suzuki, M. (1983). *Jpn. J. Appl. Phys.* **22**, L709.
Suzuki, M., Murase, K., and Hirayama, M. (1984). *Int. Conf. Solid-State Devices and Materials, Kobe, Japan.*
Terada, T., Kitaura, Y., Mizoguchi, T., Mochizuki, M., Toyoda, N., and Hojo, A. (1983). *GaAs IC Symp., Phoenix, Arizona.*
Toyoda, N., Mochizuki, M., Mizoguchi, T., Nii, R., and Hojo, A. (1981). *Int. Symp. GaAs Relat. Compd., Oiso, Japan.*
Toyoda, N., Uchitomi, N., Kitaura, Y., Mochizuki, M., Kanazawa, K., Terada, T., Ikawa, Y., and Hojo, A. (1985). *Int. Solid-State Circuits Conf., Dig. Tech. Papers*, p. 206.
Tseng, W. F., Zhang, B., Scott, D., Lau, S. S., Christou, A., and Wilkins, B. R. (1983). *IEEE Electron Device Lett.* **EDL-4**, 207.
Uchitomi, N., Kitaura, Y., Mizoguchi, T., Ikawa, Y., Toyoda, N., and Hojo, T. (1984). *Int. Conf. Solid-State Devices and Materials, Tokyo, Japan.*
Ueno, K., Furutsuka, T., Kanamori, M., and Higashisaka, A. (1985a). *Conf. Solid-State Devices and Materials, Tokyo, Japan.*
Ueno, K., Furutsuka, T., Toyoshima, H., Kanamori, M., and Higashisaka, A. (1985b). *Tech. Dig.—Int. Electron Devices Meet.*, p. 82.
Ueno, K., Hida, H., Ogawa, Y., Tsukada, Y., and Nozaki, T. (1988). *Tech. Dig.—Int. Electron Devices Meet.*, p. 846.
Uetake, K., Katano, F., Kamiya, M., Misaki, T., and Higashisaka, A. (1985). *Int. Symp. GaAs Relat. Compd., Karuizawa, Japan.*
Van Tuyl, R. L., and Liechti, C. A. (1974). *IEEE J. Solid-State Circuits* **SC-9**, 269.
Welch, B. M., and Eden, R. C. (1977). *Tech. Dig.—Int. Electron Devices Meet.*, p. 205.
Yamagishi, H. (1984). *Jpn. J. Appl. Phys.* **23**, L895.
Yamasaki, K., Asai, K., Mizutani, T., and Kurumada, K. (1982a). *Electron. Lett.* **18**, 119.
Yamasaki, K., Yamane, Y., and Kurumada, K. (1982b). *Electron. Lett.* **18**, 592.
Yamasaki, K., Asai, K., and Kurumada, K. (1982c). *Int. Conf. Solid-State Devices, Tokyo, Japan.*
Yamasaki, K., Asai, K., and Kurumada, K. (1982d). *IEEE Trans. Electron. Devices* **ED-29**, 1772.
Yamasaki, K., Kato, N., and Hirayama, M. (1984). *Electron. Lett.* **20**, 1029.
Yokoyama, N., Mimura, T., Fukuta, M., and Ishikawa, H. (1981). *Int. Solid-State Circuits Conf., Dig. Tech. Papers*, p. 218.
Yokoyama, N., Ohnishi, T., Onodera, H., Shinoki, T., and Ishikawa, H. (1983a). *Int. Solid-State Circuits Conf., Dig. Tech. Papers*, p. 44.
Yokoyama, N., Onodera, H., Ohnishi, T., and Shibatomi, A. (1983b). *Appl. Phys. Lett.* **42**, 270.
Yokoyama, N., Ohnishi, T., Onodera, H., Shinoki, T., Shibatomi, A., and Ishikawa, H. (1983c). *IEEE J. Solid-State Circuits* **SC-18**, 520.
Yokoyama, N., Onodera, H., Shinoki, T., Ohnishi, H., and Nishi, H. (1985). *IEEE Trans. Electron Devices* **ED-32**, 1797.

CHAPTER 4

GaAs LSI Circuit Design

M. Ino

NTT LSI LABORATORIES
NIPPON TELEGRAPH AND TELEPHONE COMPANY
ATSUGI, JAPAN

T. Takada

NTT ELECTRONICS TECHNOLOGY CORPORATION
ATSUGI, JAPAN

I.	INTRODUCTION	160
II.	DEVICE ANALYSIS	160
	1. Macroscopic FET Analysis	160
	2. Particle Simulation	166
	3. Gate Capacitances and Cutoff Frequency	168
III.	FET MODEL	170
	4. Circuit Simulation Program	170
	5. Simple Physical Model	171
	6. Practical Model	174
IV.	OTHER ELEMENTS	187
	7. Diodes	187
	8. Resistors	188
	9. Capacitors	188
	10. Interconnection Lines	188
V.	BASIC CIRCUIT CONFIGURATIONS	190
	11. DCFL (Direct Coupled FET Logic)	191
	12. BFL (Buffered FET Logic)	195
	13. SDFL (Schottky Diode FET Logic)	196
	14. SCFL (Source-Coupled FET Logic)	198
VI.	LSI DESIGN	202
	15. Memory	202
	16. Full-Custom Logic	209
	17. Gate Array	215
	18. Frequency Divider	216
VII.	TESTING TECHNOLOGY	223
	19. High-Frequency Probe Card	223
	20. Package	226
VIII.	SUMMARY	230
	REFERENCES	230

I. Introduction

For transistors, GaAs material has a speed potential two to three times higher than that of Si. However, this potential can only be realized in large-scale integration (LSI) performance by advanced designs that also accommodate an increase in both function complexity and integration scale. In this chapter, several design techniques including fabrication and testing results are described for several kinds of GaAs LSIs.

Specifically, Part II presents methods and results for the analysis of GaAs field effect transistor (FET) devices, and the superiority of GaAs material characteristics is made clear from the viewpoint of FET intrinsic response time. Next, FET models including both simple physical and practical models are discussed for use in GaAs LSI design and circuit simulation (Part III). Then, Part IV outlines other elements, such as diodes, resistors, capacitors, and interconnection lines.

In Part V, various basic circuit configurations and their features are presented for GaAs LSIs. By adopting these basic circuits, the actual LSI design approaches for memory and logic circuits are described in Part VI. In addition, these circuits are tested and their characteristics are given. Finally, in Part VII, high-speed testing technology and package design methods are introduced.

II. Device Analysis

The evaluation of elementary device characteristics is essential for IC design. Device analysis is particularly effective in estimating effects of material, structures, and size on device characteristics as they change with the progress of the fabrication process. In this part, a two-dimensional MESFET (metal semiconductor field effect transistor) transient analysis is described first to estimate the intrinsic response delay time of the FET itself, as well as to effectively compare effects of different materials (Si, GaAs, and InP) (Ino and Ohmori, 1980). The particle model FET simulation results, including drift velocity overshoot, are then described.

1. MACROSCOPIC FET ANALYSIS

a. Analysis Method

The device structure for the analysis is a planar type MESFET as shown in Fig. 1. The normally-off-type FETs are briefly considered here. A semi-insulating substrate is omitted in this analysis for simplicity. Generally, two dimensional transient analysis requires more CPU time, as the channel impurity concentration becomes higher. Therefore, the impurity density is

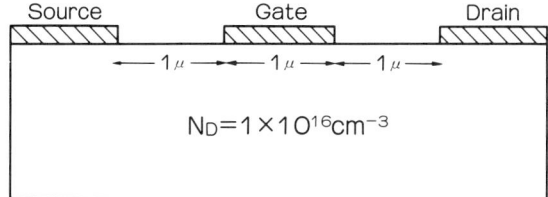

FIG. 1. MESFET model for two-dimensional analysis.

set to 1×10^{16} cm^{-3}, which is relatively low for normal FETs used in GaAs ICs. The basic equations for the analysis are Poisson's equation and the continuity equation, expressed as

$$-\nabla^2 \varphi = \frac{q}{\varepsilon}(N_D - n), \qquad (1)$$

$$\mathbf{E} = -\nabla \varphi, \qquad (2)$$

$$\frac{\partial n}{\partial t} = \nabla(n\mathbf{v} + D\nabla n), \qquad (3)$$

and

$$\nabla \mathbf{J}_{\text{tot}} = \nabla\left(qn\mathbf{v} + qD\nabla n + \varepsilon\frac{\partial \mathbf{E}}{\partial t}\right) = 0. \qquad (4)$$

In these equations, φ is potential, q the electronic charge, N_D the donor density, n the electron density, $\mathbf{E}$ the electric field, $\mathbf{v}$ the drift velocity and D the electron diffusivity. Nonequilibrium effects of drift-velocity overshoot and ballistic carrier transport are not included in this analysis; additionally, diffusivity is assumed to be isotropic.

The conventional differential method similar to a Si MESFET analysis by Reiser (1973) is used to solve the above equations. As shown in Fig. 2, a device area is divided into a $100(x) \times 15(y)$ mesh. The objective grid point is named O, and its neighbors are named P, Q, R, and S as shown in Fig. 3. Using the

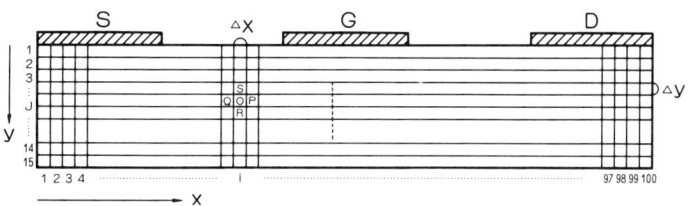

FIG. 2. Mesh for two-dimensional FET analysis. Numbers of meshes are 100 for the x-direction and 15 for the y-direction.

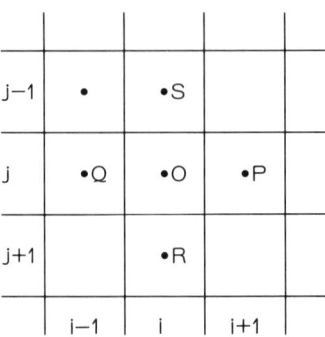

FIG. 3. Mesh point and its neighbors for differential equations.

$$O = [i\,\Delta x, j\,\Delta y]$$
$$P = [(i + 1)\,\Delta x, j\,\Delta y]$$
$$Q = [(i - 1)\,\Delta x, j\,\Delta y]$$
$$R = [i\,\Delta x, (j + 1)\,\Delta y]$$
$$S = [i\,\Delta x, (j - 1)\,\Delta y]$$

variables at these points, Eqs. (1)–(4) are transformed into differential forms as follows:

$$\frac{-\varphi_P + 2\varphi_O - \varphi_Q}{\Delta x^2} + \frac{-\varphi_R + 2\varphi_O - \varphi_S}{\Delta y^2} = \frac{q}{\varepsilon}(N_{DO} - n_O); \quad (5)$$

$$E_{xO} = \frac{-\varphi_P + \varphi_Q}{2\,\Delta x}, \quad E_{yO} = \frac{-\varphi_R + \varphi_S}{2\,\Delta y}. \quad (6)$$

The right-hand side of the continuity equation (3) was initially transformed and set at z, with the following equation subsequently given:

$$z(n, v_x, v_y) = \frac{n_P v_{xP} - n_Q v_{xQ}}{2\,\Delta x} + D_O \frac{n_P - 2n_O + n_Q}{\Delta x^2}$$
$$+ \frac{n_R v_{yR} - n_S v_{yS}}{2\,\Delta y} + D_O \frac{n_R - 2n_O + n_S}{\Delta y^2}. \quad (7)$$

From the field at the O point of $E_O = \sqrt{E_{xO}^2 + E_{yO}^2}$, the drift velocity, v_O, is given by the relation

$$v_O = \frac{\mu E_O + v_s(E_O/E_C)^4}{1 + (E_O/E_C)^4}. \quad (8)$$

The x and y components of v_O are v_{xO} ($= v_O E_{xO}/E_O$) and v_{yO} ($= v_O E_{yO}/E_O$).

The diffusivity, D_O, is expressed as

$$D_O = \frac{kT}{q}\frac{v_O}{E_O} + 1.5\tau v_O^2. \tag{9}$$

Here, mobility μ is 5500 cm$^2\cdot$V$^{-1}\cdot$sec^{-1}, saturation velocity v_s is 8×10^6 cm sec^{-1}, T is 300 K, τ is 0.1 ps, and the parameter E_C is set at 3.764 kV cm^{-1} so that Eq. (8) has a peak velocity at $E_O = 3.5$ kV cm^{-1}. The carrier concentration, $n^k = n(k\,\Delta t)$, where Δt is a unit time interval (mesh), is given by

$$\frac{n^{k+1} - n^k}{\Delta t} = z(n^k, v_x^k, v_y^k). \tag{10}$$

A more accurate n^{k+1} is given by

$$\frac{n^{k+1} - n^k}{\Delta t} = \frac{z(n^{k+1}, v_x^k, v_y^k) + z(n^k, v_x^k, v_y^k)}{2}. \tag{11}$$

Let us now take a look at the calculation procedures for the step response to applied gate voltage. As an example, when gate-to-source voltage, V_{GS}, changes from 0.1 V to 0.6 V (off-on) with constant drain–source voltage, V_{DS}: (1) distributions of the potential φ^0 and the electron concentration n^0 are calculated at the equilibrium state ($V_{GS} = 0.1$ V) at the first step. This is a time-independent calculation, from which the static characteristics are obtained. Following step (1), whenever the time advances as $t = t + \Delta t$ ($k = k + 1$), calculation steps (2) and (3) described below are iterated until convergence is reached. The convergence conditions for n^k and φ^k are: $n_{i,j}^k(m+1)/n_{i,j}^k(m) - 1 < \varepsilon_n$ and $\varphi_{i,j}^k(m+1)/\varphi_{i,j}^k(m) - 1 < \varepsilon_\varphi$, where m is an iteration number, and ε_n and ε_φ are convergence error, e.g. 0.001. In step (2), $\varphi^{k+1}(i)$ is obtained from n^k by solving Eq. (5). The drift velocity and diffusivity are then calculated from Eqs. (6), (8), and (9). In step (3), $n^{k+1}(m)$ is obtained from Eq. (11), and, using this value, $\varphi^{k+1}(m+1)$ is calculated by employing Eq. (5). When both n and φ converge, the drain current $\mathbf{J}_{tot}$ is derived from Eq. (4). The successive-over-relaxation (SOR) method is adopted for solving the matrix.

b. Analysis Results

MESFETs made from three kinds of materials—GaAs, Si, and InP—are analyzed for comparison. The Schottky barrier heights employed are 0.8 V for GaAs and Si and 0.5 V for InP, which represent typical experimental values. FETs having gate lengths of 1 μm and 0.5 μm are calculated, whose active layer thicknesses are set to satisfy the threshold voltage of $(V_T) = 0$ V at each gate length under the constant $N_D = 1 \times 10^{16}$ cm^{-3}.

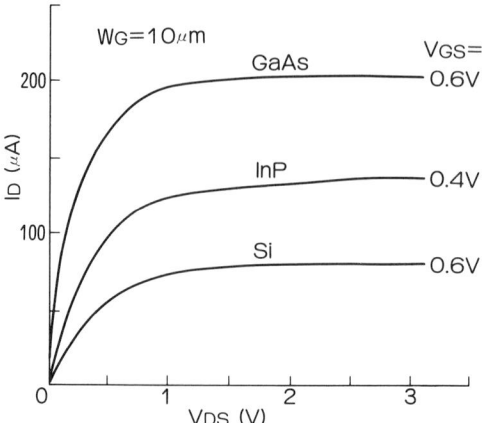

FIG. 4. Comparison of I-V characteristics for three kinds of normally-off MESFETs made from GaAs, InP, and Si.

The static characteristics are calculated first. The I_D-V_{DS} characteristics of 1-μm gate FETs with a gate width (W_G) of 10 μm at on-state are shown in Fig. 4. As can be seen, on current (I_{on}) of the Si FET is the smallest because it has the lowest drift velocity among the three materials. The I_{on} of InP is smaller than that of GaAs with the same V_T, because the InP active layer corresponding to lower barrier height is thinner than that of GaAs. Since the threshold voltages for the three FETs are 0 V, the average transconductance (g_m) was obtained by I_{on}/V_{GS}(on). The g_m for the GaAs FET is comparable with that for the InP FET and is about 2.5 times higher than that of the Si FET. Using the relationship of gate capacitance $C_{GS} \propto (V_{bi} - V_G)^{-1/2}$, the cutoff frequency, $f_T \propto g_m/C_{GS}$, of the GaAs FET is estimated to be higher than that of the InP FET with the first approximation adopted here.

The transient characteristics are calculated to obtain the FET intrinsic response time (t_{int}). The step response of the drain current I_D to change of V_{GS} with a constant V_{DS}, t_{int} is calculated as an average of the t_{on} and t_{off} — where t_{on} is defined as the time needed for a 90% change in I_D when V_{GS} is changed in a stepwisely manner from a low level V_L to a high level V_H(on-stage), and t_{off} is defined as the time required for a 90% change in I_D for off-stage. Here, considering application to DCFL (direct coupled FET logic), the V_H is near the forward-bias clamp voltages of 0.6 V for GaAs and Si and 0.4 V for InP, while V_{DS} is equal to the typical experimental barrier height of 0.8 V for GaAs and Si and 0.5 V for InP.

A result of the calculation of transient characteristics for a 1-μm-gate GaAs FET is shown in Fig. 5. From the figure, both t_{on} and t_{off} are found to be about 10 psec. The calculated transient characteristics for 1-μm-gate

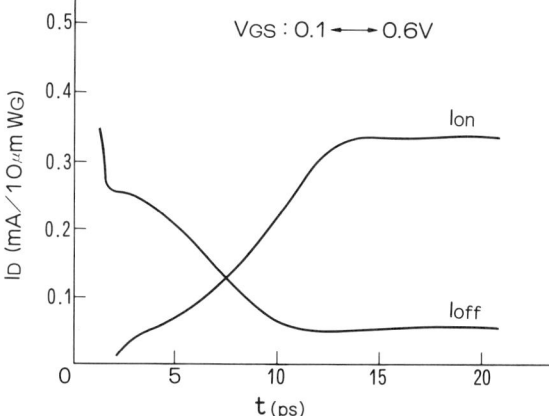

FIG. 5. Calculated transient response of drain current when switched from off-state to on-state and vice versa.

FETs are summarized in Table I, where t_{int} is in the order of GaAs, InP, Si. From the calculated drift velocity profiles in the channels shown in Fig. 6, the average velocity is found to be approximately 5×10^6 cm·sec^{-1} in Si and 1×10^7 cm·sec^{-1} in GaAs. Assuming a channel length equal to the gate length of 1 μm, the channel transit time is estimated to be 20 psec for Si and 10 psec for GaAs, which are almost equal to the values for t_{int}. This result corresponds to the well-known relationship $C_G/g_m \propto L/v$. Although InP has a peak velocity about 1.6 times higher than that for a GaAs FET, the InP FET is slower than the GaAs, because the channel field does not reach the peak field ($E_p \sim 10$ kV·cm^{-1}) of InP at $V_{DS} = 0.5$ V.

Similar calculations were made for the 0.5-μm-gate FETs. The results are listed in Table II. The InP FET, which has a 35% larger t_{int} than the GaAs FET when the gate length is 1 μm, shows speed comparable to a GaAs FET with a 0.5 μm gate, because the channel field is high for a short gate length,

TABLE I

CALCULATED CHARACTERISTICS OF TRANSIENT RESPONSE FOR 1 μm GATE FETS

Material	V_{DS} (V)	V_{GS} (V)	I_{on} (μA)	I_{off} (μA)	t_{on} (psec)	t_{off} (psec)	P_{dis}[a] (μW)	t_{epi} (μm)
GaAs	0.8	0–0.6	194.3	3.5	12.4	10.4	79	0.4
Si	0.8	0–0.6	71.2	1.1	22.1	21.3	29	0.4
InP	0.5	0–0.4	107.4	2.5	16.8	14.0	28	0.32

[a] $P_{dis} = V_{DS}(I_{on} + I_{off})/2$.

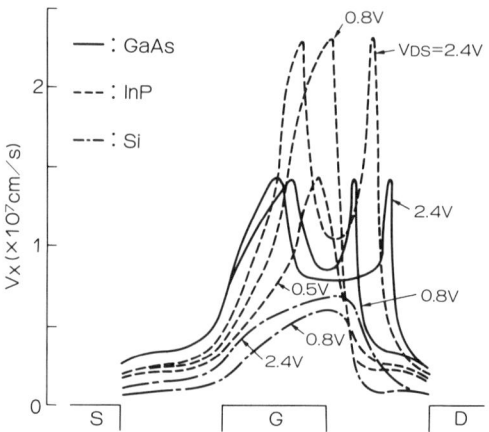

FIG. 6. Profiles of electron drift velocity in a channel at on state for (———) GaAs, (– – –) InP, and (— – —) Si.

consequently reaching the E_p. For high-speed ICs at present technology, GaAs is the most advantageous because it exhibits a peak velocity at low field of about $3.5 \text{ kV} \cdot \text{cm}^{-1}$. It has higher switching speed at lower voltage or lower power dissipation. As seen in Fig. 6, low field mobility is also important to determine t_{int}. In an actual FET, the series resistance outside a gate, which is not taken into account in the aforementioned simulation, has an important effect on switching speed, especially if a surface depletion increases series resistance and causes degradation of switching speed. Keeping all these points in mind, GaAs can be concluded to be superior to InP and Si, mainly due to its highest mobility, even in 0.5-μm-gate FETs.

2. Particle Simulation

Along with advances in process technologies, submicron or sub-half-micron FETs will be applied to GaAs ICs in the near future to improve IC performances. For such short-channel FETs, the aforementioned macroscopic model

TABLE II

Calculated Characteristics of Transient Response for 0.5 μm Gate FETs

Material	V_{DS} (V)	V_{GS} (V)	I_{on} (μA)	I_{off} (μA)	t_{on} (psec)	t_{off} (psec)	P_{dis} (μA)	t_{epi} (μm)
GaAs	0.8	0–0.6	168.9	4.1	10.0	7.5	69	0.3
Si	0.8	0–0.6	67.2	2.5	17.0	14.1	28	0.3
InP	0.5	0–0.4	106.9	3.7	9.0	8.1	28	0.24

by continuity approximation using drift and diffusion is insufficient to accurately describe carrier motion. Particle simulation is far more useful for this purpose.

In particle simulation, one particle represents several electrons, and each particle is traced both in k-space and real-space. In k-space, random numbers are used to determine the energy-dependent scattering of particle. This calculation method is referred to as a Monte Carlo method. Two dimensional Monte Carlo simulation by using particle models was first accomplished by Hockney *et al.* (1974). The calculation techniques are not described in this section. Only the results obtained by Tomizawa *et al.* (1983) will be discussed.

The calculated results for drift velocity overshoot by particle simulation are shown in Fig. 7. The overshoot effect becomes remarkable for less than 0.5 μm transit-length in GaAs and for less than 0.1 μm transit-length in Si. The reason for overshoot is as follows. The momentum relaxation time is in general shorter than the energy relaxation time. That is, when an electric field is applied stepwise, the carrier velocity immediately responds to a field change, while the energy response is delayed. Therefore, carriers drift with a low scattering rate corresponding to low energy, which, in turn, results in

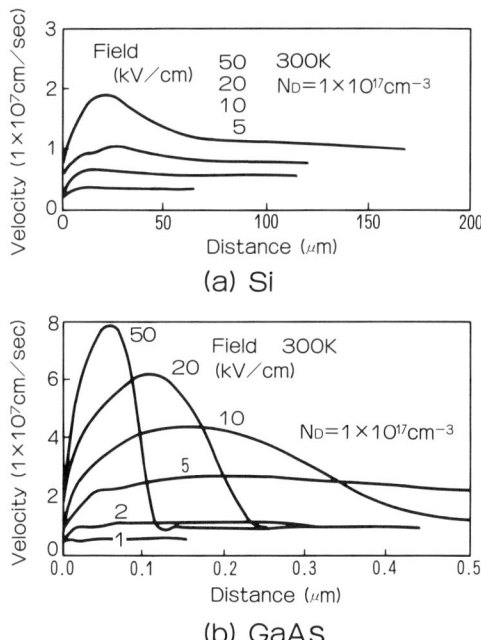

FIG. 7. Drift velocity versus transit distance calculated by particle simulation (see text) (a) for Si and (b) for GaAs.

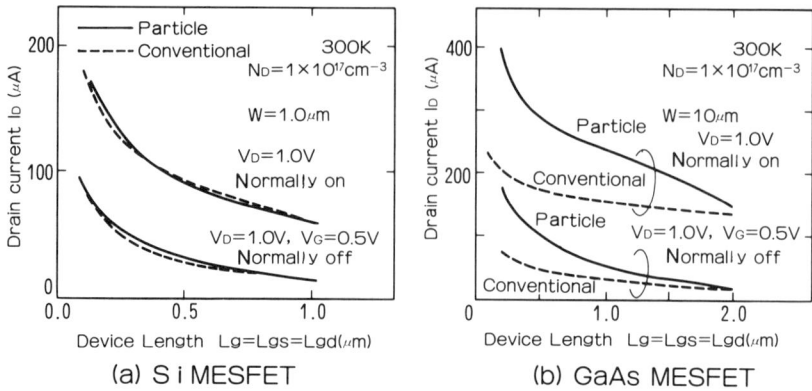

FIG. 8. Drain current versus device length calculated by particle simulation (solid lines) and conventional simulation (broken lines) for Si MESFET (a) and GaAs MESFET (b). For both MESFETs, the temperature was 300 K and N_D was 1×10^{17} cm^{-3}.

a velocity overshoot. Energy gradually increases with time, however, and converges at the equilibrium state. Even in GaAs that features a large overshoot effect, the peak velocity increases with an increase in the applied field, while carriers immediately transfer to the equilibrium state because of the decrease in the energy relaxation time. This result confirms that the effective device length is shortened by the overshoot effect in a higher electric field.

The calculated results using the conventional macroscopic model and the particle model are compared for GaAs and Si MESFETs. As indicated in Fig. 8, there is almost no difference between the models for the Si MESFETs having a gate length of 0.1 μm or longer. For GaAs FETs having a gate length less than 1 μm, on the other hand, a drain current derived by particle simulation is larger than that by conventional simulation, with this tendency being enhanced for a gate length less than 0.5 μm. Consequently, it is expected that the velocity overshoot should have a striking effect on characteristics of submicron-gate GaAs FETs.

3. Gate Capacitances and Cutoff Frequency

Parameters important for IC design are gate-to-source capacitance C_{GS}, gate-to-drain capacitance C_{GD} and cutoff frequency f_T, which determine the operational speed of the IC. The C_{GS} and C_{GD} calculated by using the two-dimensional analysis of static FET characteristics are shown as a function of gate voltage in Fig. 9. Considered here, is a 1 μm gate planar-type FET with an impurity concentration of 1×10^{17} cm^{-3} with a 0.09 μm active layer thickness, and $V_T = 0.03$ V. From Fig. 9, it is clear that C_{GS} has a peak

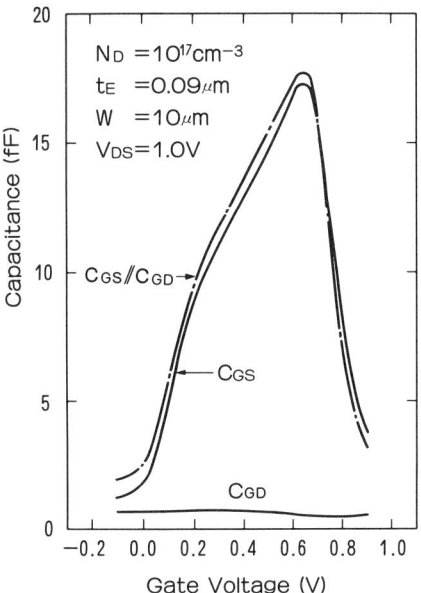

FIG. 9. Gate capacitances, C_{GS} and C_{GD}, versus gate voltage calculated by conventional two-dimensional analysis. $C_{GS}//C_{GD}$ is the sum of C_{GS} and C_{GD}.

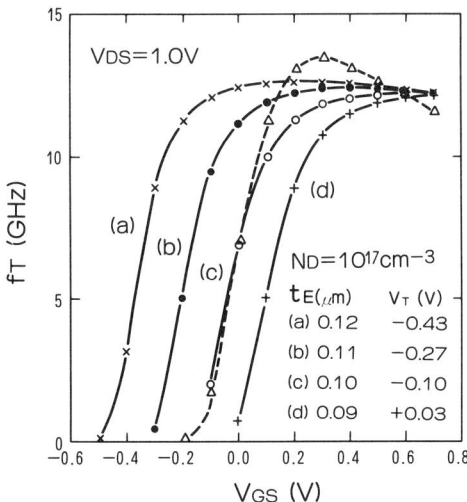

FIG. 10. Cut-off frequency f_T versus gate–source voltage V_{GS} with active layer thickness t_E as a parameter. V_T is threshold voltage.

around $V_{GS} = 0.6$ V and is considerably smaller at the near pinchoff of $V_{GS} = 0.2$ V. Since V_{DS} is sufficiently high at 1 V for a drain pinchoff, C_{GD} is small and almost constant, similar to C_{GS} after gate pinchoff ($V_{GS} < V_T$). Details of the gate capacitance model will be described in a later section. The cutoff frequency f_T is calculated with active-layer thickness or V_T as a parameter as shown in Fig. 10. Here, f_T is defined at each V_{GS} as

$$f_T = \frac{g_m}{2\pi C_{GS} / C_{GD}}. \tag{12}$$

The normally-on FET has a higher peak f_T than the normally-off FET, as seen in this figure. For digital IC applications, however, switching speed is closely related to the average f_T over the dynamic range from the off to the on state. Therefore, switching speed increases as the logic swing becomes larger in both the normally-on and the normally-off FETs.

III. FET Model

4. Circuit Simulation Program

Circuit simulation programs are generally used in circuit design for performing DC, AC, transient, and statistical analysis. Among several simulation programs available at present, ASTAP and SPICE II are widely used by digital circuit designers.

ASTAP is effective for studying detailed circuit performance, since circuit designers can describe their own device model by using user functions. Furthermore, this program has a function for statistical analysis that provides distributions of a circuit performance when device parameters statistically deviate from the center values. This analysis is particularly important when standard deviations of fabricated FET parameters are large.

On the other hand, SPICE II can directly handle device models, which applies to a bipolar transistor, a MOSFET and a JFET. Thus, it is only necessary to input the network descriptions. In addition, the calculation time is less for SPICE II than for ASTAP. It can also handle larger maximum node-numbers in a circuit. Therefore, SPICE II is more convenient than ASTAP when large-scale circuits are analyzed. Two major problems, however, exist in SPICE II for the simulation of GaAs MESFET circuits. First, the JFET model implemented in SPICE II does not accurately represent the GaAs MESFET characteristics, especially in terms of the description of capacitance characteristics. Second, a statistical analysis is not available through the program.

For these reasons, both ASTAP and an improved version of SPICE II, in which a precise GaAs MESFET model can be expressed, are necessary for the design of present GaAs MESFET circuits. In the following sections, a simple physical model and a practical model that were especially developed for a GaAs-MESFET circuit simulation will be described. These models can simulate precise GaAs MESFET characteristics; in addition they use simplified expressions in order to shorten calculation time.

5. Simple Physical Model

A physical FET model is necessary to ensure effective IC design and to predict IC performance prior to the fabrication. In this section, a simple but accurate FET model for a circuit simulation developed by Ino *et al.* (1982a) will be described.

In the model, the parameters in the intrinsic equivalent circuit for a FET, enclosed by the broken lines in Fig. 11, are self-consistently calculated. The drift velocity is given as a function of electric field by Lehovec *et al.* (1970):

$$v = \frac{\mu E}{1 + \mu E/v_s}, \tag{13}$$

where E is the electric field and v_s, the saturation velocity of 1.5×10^7 cm·sec^{-1}. The mobility μ is assumed to be dependent on impurity concentration as the following empirical form, with μ_0 being 4000 cm^2 V^{-1} sec^{-1}

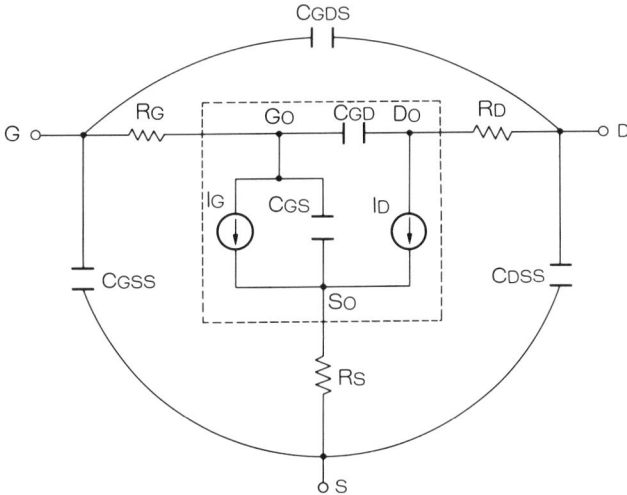

FIG. 11. Large signal equivalent circuit for GaAs MESFET. Intrinsic FET elements are inside the broken line.

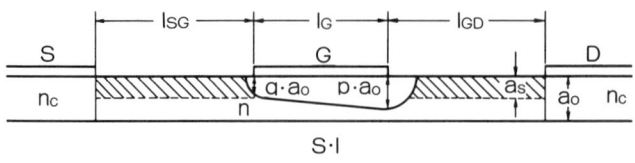

Fig. 12. Cross-sectional view of FET model.

at $N_D = N_0 = 1 \times 10^{17}$ cm^{-3}:

$$\mu = \mu_0 (N_D/N_0)^{-0.202}. \quad (14)$$

A cross-sectional view of a FET model is shown in Fig. 12. A stepwise impurity distribution is assumed even in active layer formed by ion implantation. The thickness a_0 is approximated by twice the projected range, $2R_p$. The carrier density is represented by an average donor density, $\bar{n}$. The threshold voltage V_T is related to $\bar{n}$ by

$$V_P = V_{bi} - V_T = \frac{e\bar{n}a_0^2}{2\varepsilon}. \quad (15)$$

The drain current is given as

$$I_D = I_{D0} \bigg/ \left[1 + \frac{\mu V_P}{v_s L_G}(p^2 - q^2)\right], \quad (16)$$

where

$$I_{D0} = \frac{e\bar{n}\mu a_0 W_G V_P}{L_G}[p^2 - q^2 - \tfrac{2}{3}(p^3 - q^3)], \quad (17)$$

$$q^2 = \frac{V_{bi} - (V_{G0} - V_{S0})}{V_P}, \quad (18)$$

and

$$p^2 = \frac{V_{bi} - (V_{G0} - V_{D0})}{V_P}. \quad (19)$$

Here, V_{G0}, V_{S0}, and V_{D0} are the potentials at nodes G_0, S_0, and D_0 in Fig. 11, and qa_0 and pa_0 are depletion width at the gate edge of the source and drain sides, respectively, as shown in Fig. 12.

In order to obtain C_{GS} and C_{GD}, the gate depletion charge is divided for convenience into the gate-source depletion charge, Q_{GS}, and the gate-drain depletion charge, Q_{GD}. The boundary between Q_{GS} and Q_{GD} lies at the gate edge of the drain side. The gate-source capacitance is approximated by

$$C_{GS} = \varepsilon L_G W_G / [0.5(p + q)a_0], \quad (20)$$

where, if $p > 1$, then set $p = 1$. The gate-drain capacitance is expressed as

$$C_{GD} = \frac{\pi}{2}\varepsilon W_G, \quad \text{when } p \leq 1,$$

$$C_{GD} = \frac{\pi}{2}\varepsilon W_G[1 - (\tan^{-1}\sqrt{p^2 - 1})/0.5\pi], \quad \text{when } p > 1, \quad (21)$$

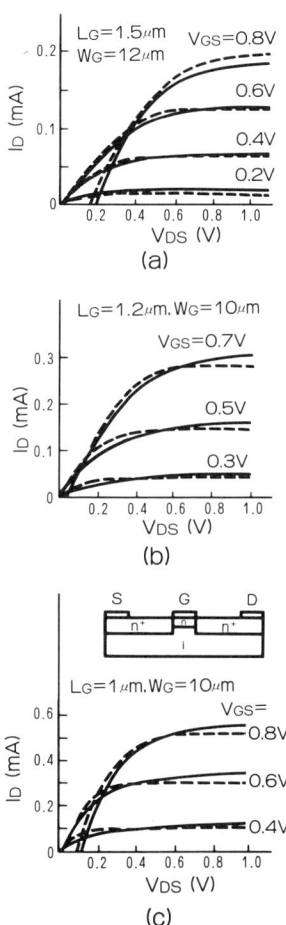

FIG. 13. I_{DS}-V_{DS} characteristics. Solid lines represent experimental results, and dashed lines represent calculated results. (a) Conventional FET with $L_G = 1.5\,\mu$m, $L_{SD} = 5\,\mu$m, $W_G = 12\,\mu$m, and $V_T = 0.10$ V. (b) Short source-drain FET with $L_G = 1.2\,\mu$m, $L_{SD} = 2.0\,\mu$m, $W_G = 10\,\mu$m, and $V_T = 0.10$ V. (c) SAINT FET with $L_G = 1\,\mu$m, $L_{SD} = 5\,\mu$m, $W_G = 10\,\mu$m, and $V_T = 0.17$ V.

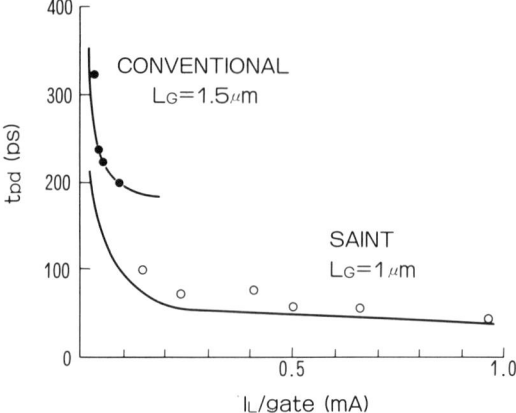

FIG. 14. Propagation delay time per gate (t_{pd}) as a function of load current. Curves represent the simulation results. Filled and open circles are experimental values for 15-stage E-D ring oscillator using conventional FETs ($L_G = 1.5\ \mu m$) and SAINT FETs ($L_G = 1\ \mu m$), respectively.

under the condition that the gate–drain depletion layer is approximated as a cylindrical shape having a radius of pa_0. The extrinsic elements of series resistance and parasitic capacitance are not expressed here.

To ensure the capability of the simulation model, the I–V characteristics were calculated for three types of MESFETs. In Fig. 13a, the I–V curve is shown for a conventional flat-type FET with $L_G = 1.5\ \mu m$ and the source–drain spacing $L_{SD} = 5\ \mu m$. In Figs. 13b and 13c, the I–V curves are shown for FETs fabricated by self-aligned technolgies. A FET with a short source–drain distance ($L_G = 1.2\ \mu m$ and $L_{SD} = 2\ \mu m$) is shown in (b), while a SAINT (self-aligned implantations for n^+-layer technology) FET with $L_G = 1\ \mu m$ and $L_{SD} = 5\ \mu m$ is shown in (c). The calculated results shown in these figures agree well with the experimental I–V curves for corresponding FETs.

Dynamic performance is also calculated by the same model. The propagation delay time of E-D-type ring oscillators is shown in Fig. 14 in comparison with the experimental results. The agreement is satisfactory. Therefore, it is concluded that the present FET model mainly based on the capacitance consideration is applicable to a large-scale-circuit simulation.

6. PRACTICAL MODEL

a. Intrinsic FET Model

I–V Characteristics. The intrinsic FET equivalent circuit is shown in Fig. 15. The I–V characteristics of the Shichmann–Hodges model, which is

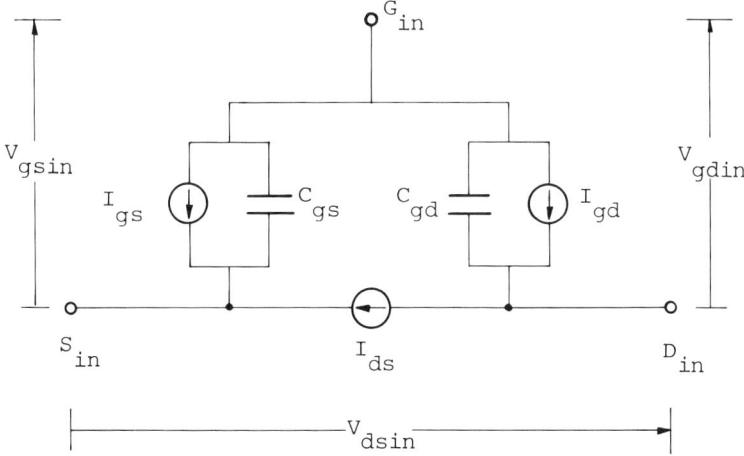

FIG. 15. Equivalent circuit of intrinsic GaAs MESFET. G_{in}, S_{in}, and D_{in} are internal electrodes for gate, source, and drain, respectively. I_{ds}, I_{gs}, and I_{gd} are drain-to-source, gate-to-source, and gate-to-drain currents. V_{dsin}, V_{gsin}, and V_{gdin} are intrinsic drain–source, gate–source, and gate–drain voltages. C_{gs} and C_{gd} are gate–source capacitance and gate–drain capacitance.

used for JFET in SPICE II, are expressed as

$$I_{ds} = 0 \qquad (V_{sat} < 0), \qquad (22)$$

$$I_{ds} = \beta_P V_{dsin}(2V_{sat} - V_{dsin}) \qquad (V_{dsin} < V_{sat}), \qquad (23)$$

$$I_{ds} = \beta_P V_{sat}^2 \qquad (V_{dsin} > V_{sat}), \qquad (24)$$

$$\beta_P = \beta(1 + \lambda V_{dsin}), \qquad (25)$$

$$V_{sat} = V_{gsin} - V_T. \qquad (26)$$

Here, I_{ds}, V_{dsin}, V_{sat}, V_{gsin}, β, λ, and V_T are drain current, drain–source voltage of the intrinsic FET, saturation voltage, gate–source voltage of the intrinsic FET, transconductance parameter, channel length modulation parameter and threshold voltage, respectively. This Shichmann–Hodges model is also applicable to the GaAs MESFETs. However, a GaAs MESFET with a submicron gate or a near-submicron gate has a large drain conductance and a triode-like I–V characteristic near the pinch off voltage because of short-channel effects, as indicated in Fig. 16. The FET shown in this figure has a gate length of 1.0 μm and is fabricated by Si ion implantation into a chromium-doped semi-insulating substrate.

The channel length modulation parameter λ can represent a finite drain conductance but not a triode-like characteristic. Experimental studies and a two-dimensional analysis show that a mechanism producing a finite drain

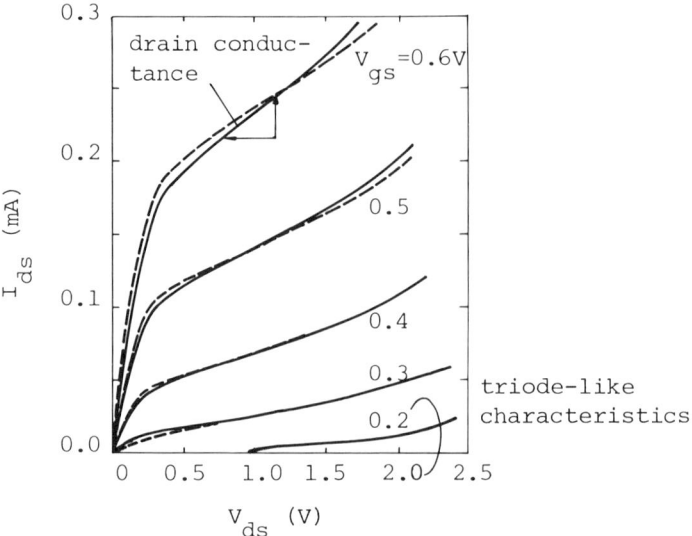

FIG. 16. Typical GaAs MESFET I-V characteristics (gate length is 1.0 μm, gate width is 9.0 μm). Solid and broken lines represent measured and simulated I-V characteristics, respectively.

conductance and a triode-like characteristic is not channel-length modulation but threshold voltage modulation resulting from changes in V_{dsin}. This phenomenon is caused by current flow in a semi-insulating substrate. V_T changes linearly with V_{dsin} in almost all GaAs MESFETs. Based on this fact, Takada et al. (1984a) introduced a channel thickness modulation parameter γ, and expressed V_T as

$$V_T = V_{T0} - \gamma V_{dsin}. \tag{27}$$

The parameter γ can express both a large drain conductance and a triode-like characteristic.

Gate-to-source current I_{gs} and gate-to-drain current I_{gd} are expressed by using the same equations as used for a Schottky diode:

$$I_{gs} = I_{sat}(e^{qV_{gsin}/nkT} - 1) \tag{28}$$

and

$$I_{gd} = I_{sat}(e^{qV_{gdin}/nkT} - 1), \tag{29}$$

where I_{sat}, n, and kT/q are the backward saturation current, the ideality factor and the thermal voltage, respectively.

The device parameters in Eqs. (22–27) are derived from measured I-V characteristics as follows.

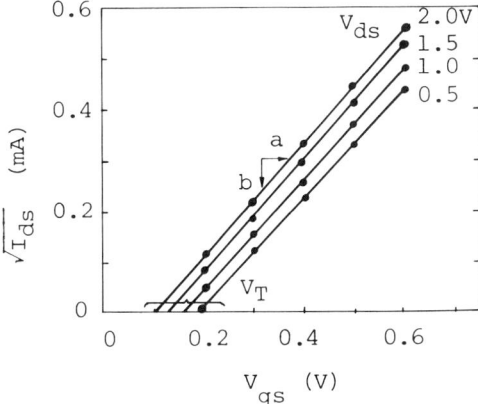

FIG. 17. Square root of drain current vs. gate-source voltage at various drain-source voltages.

First, the square root of I_{ds} is plotted as a function of V_{gs} with V_{ds} as a parameter from the measured I-V characteristics, as shown in Fig. 17 (this figure corresponds to Fig. 16). Here, V_{gs} and V_{ds} are the external gate-source voltage and the external drain-source voltage, respectively. The value of β_p for each V_{ds} is obtained from the gradient of each line (b/a) in Fig. 17 as

$$\beta_p = (b/a)^2. \tag{30}$$

The intersection with the horizontal axis gives V_T.

Second, β_p is plotted against V_{ds} as shown in Fig. 18. β can be obtained by extrapolating the line to $V_{ds} = 0$ (see the figure). λ is also derived using the gradient (d/c) of the line and β as

$$\lambda = \left(\frac{d}{c}\right)\bigg/\beta. \tag{31}$$

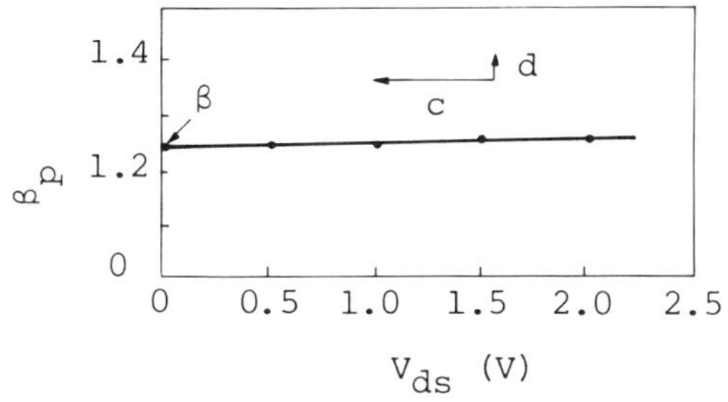

FIG. 18. β_p vs. drain-source voltage.

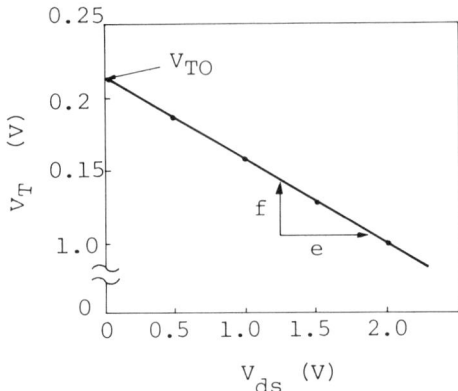

FIG. 19. Threshold voltage V_T vs. drain–source voltage.

Third, V_T is plotted against V_{ds} as shown in Fig. 19. Here, the intersection with the vertical axis gives V_{T0}, and the gradient (f/e) is equal to γ.

$$\gamma = f/e. \tag{32}$$

The simulated I-V characteristics using the device parameters obtained from the above procedure agree well with the experimental I-V characteristics when source and drain resistance (R_s and R_d) are negligibly small. (These resistances are discussed in Section 6b.) When R_s and R_d cannot be neglected, the iterative calculation of I-V characteristics by computer easily gives the precise device parameters. In this calculation, the parameters obtained from the aforementioned procedure and the measured values of R_s and R_d are used as the initial values for iterative calculation. For conventionally fabricated GaAs MESFETs, iterative I-V calculations modifying only one parameter of β are necessary. The simulated I-V characteristics using the modified β are shown by broken lines in Fig. 16. They agree well with the measured I-V characteristics throughout the entire region of V_{gs} and V_{ds}.

To determine I_{sat} and n in Eqs. (28) and (29), the following procedure is used. Gate current is measured under the short-circuited source and drain condition. Figure 20 shows the gate current as a function of applied voltage. The intersection of the extrapolated line with the vertical axis gives $2I_{sat}$. n is derived by

$$n = \frac{q}{kT} \frac{V_2 - V_1}{\ln I_2 - \ln I_1}, \tag{33}$$

where I_2 and I_1 should be selected from the straight line region. R_d and R_s

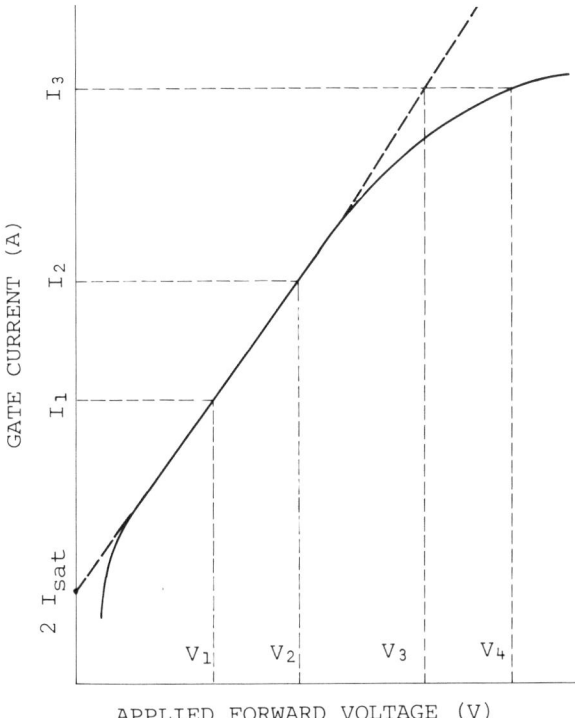

FIG. 20. Gate current vs. applied forward voltage.

are also graphically obtained using

$$R_s (\text{or } R_d) = 2(V_4 - V_3)/I_3. \tag{34}$$

The R_s and R_d are equal to each other for a symmetrical FET.

Gate Capacitances. Two kinds of capacitances are present in GaAs MESFETs. They are bias voltage-dependent junction capacitances and the parasitic capacitances present between electrodes and between an electrode and the ground plane. In this subsection, only the junction capacitances will be discussed.

We now consider GaAs MESFETs having a low donor density–thickness product, in which there is no effect of electron accumulation on the capacitance characteristics (Takada *et al.*, 1982). The model described here is obtained under this assumption. The capacitance characteristics for FETs having a large donor density–thickness product are described in Yokoyama's paper (1983). Although a uniform carrier profile is assumed in the active layer, the derived model can be applied to FETs having a Gaussian profile in the active layer with only minor change of expressions as described later.

Since a capacitance measurement is not easy as compared with a DC current measurement, a physical model is more convenient for circuit simulation than a direct measurement, which is better for characterizing I-V curves. Therefore, we shall set up a physical model for gate–source capacitance C_{gs} and gate–drain capacitance C_{gd}, which are to be expressed using gate length ℓ_g, gate width W_g, active layer carrier concentration N_d, thickness of active layer a, and built-in voltage V_{bi}.

The bias voltage-dependent C_{gs} can be expressed by classifying it into two or three bias-voltage regions as shown in Fig. 23. The two-region model is expressed by a before-pinchoff region (I) and an after-pinchoff region (II), which are represented as solid lines in Fig. 23. In the three-region model, the transition region (III) drawn as a broken line is added to the two-region model.

C_{gs} is derived as follows. First, the internal space charge distribution for the before-pinchoff bias condition (I) is simplified as shown in Fig. 21. Space charges Q_1, Q_2, and Q_3 for space charge regions A, B, and C are given by

$$Q_1 = \tfrac{1}{2} q N_d W_g \ell_g a \left\{ \left(\frac{V_{bi} - V_{gsin}}{V_p}\right)^{1/2} + \left(\frac{V_{bi} - V_{gdin}}{V_p}\right)^{1/2} \right\}, \quad (35)$$

$$Q_2 = \frac{\pi}{4} a^2 q N_d W_g \frac{V_{bi} - V_{gsin}}{V_p}, \quad (36)$$

$$Q_3 = \frac{\pi}{4} a^2 q N_d W_g \frac{V_{bi} - V_{gdin}}{V_p}, \quad (37)$$

where V_p is the pinchoff voltage represented by

$$V_p = q N_d a^2 / (2\varepsilon). \quad (38)$$

By differentiating total space charge Q ($Q = Q_1 + Q_2 + Q_3$) with respect to V_{gsin}, C_{gs} becomes

$$C_{gs} = \frac{W_g \ell_g}{2\sqrt{2}} \left(\frac{q N_d \varepsilon}{V_{bi} - V_{gsin}}\right)^{1/2} + \frac{\pi}{2} \varepsilon W_g. \quad (39)$$

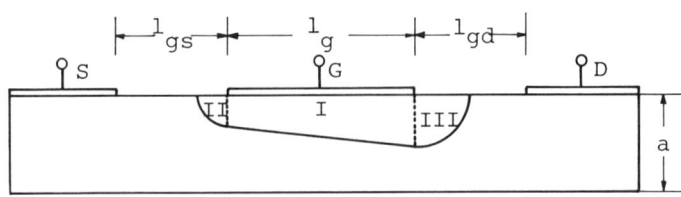

FIG. 21. Internal space charge distribution in before-pinchoff region. A, B, and C represent space charge regions; Q_1, Q_2, and Q_3 are the corresponding space charges.

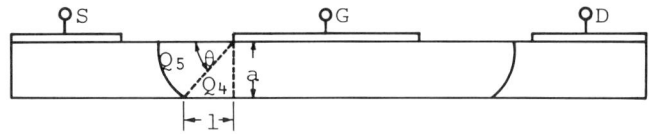

FIG. 22. Internal space charge distribution in after-pinchoff region.

The first and second terms in Eq. (39) correspond to the capacitances for regions A and B. The capacitance for region C is zero because the amount of this space charge is not varied with V_{gsin}.

When FETs are fabricated by ion implantation, a Gaussian profile of active layer is formed. In this case, N_d is replaced in the above equations by the effective carrier concentration, N_{de}, represented as

$$N_{\text{de}} = \frac{\varepsilon}{2q}(V_{\text{bi}} - V_T)/R_p^2, \tag{40}$$

where R_p is the projected range (Gibbons *et al.*, 1975) of ion implantation.

C_{gs} for the after-pinchoff region is obtained as follows. When V_{gsin} is decreased beyond the threshold voltage V_T, the internal space charge distributes as shown in Fig. 22. Here, V_T is expressed as

$$V_T = V_{\text{bi}} - V_p. \tag{41}$$

Space charges Q_4 and Q_5, indicated in the figure, are expressed as

$$Q_4 = qN_d W_g \frac{a}{2}\ell, \tag{42}$$

where the bottom length ℓ of the triangle space charge region in Fig. 22 is

$$\ell = \left\{\frac{2\varepsilon}{qN_d}(V_{\text{bi}} - V_{\text{gsin}}) - a^2\right\}^{1/2}, \tag{43}$$

and

$$Q_5 = \theta\varepsilon W_g(V_{\text{bi}} - V_{\text{gsin}}), \tag{44}$$

where

$$\tan\theta = \frac{a}{\ell} = \sqrt{\frac{V_{\text{bi}} - V_T}{V_T - V_{\text{gsin}}}}. \tag{45}$$

C_{gs} is derived by differentiating the total charge, $Q_4 + Q_5$, with respect to V_{gsin} as

$$C_{\text{gs}} = \varepsilon W_g \tan^{-1}\sqrt{\frac{V_{\text{bi}} - V_T}{V_T - V_{\text{gsin}}}}. \tag{46}$$

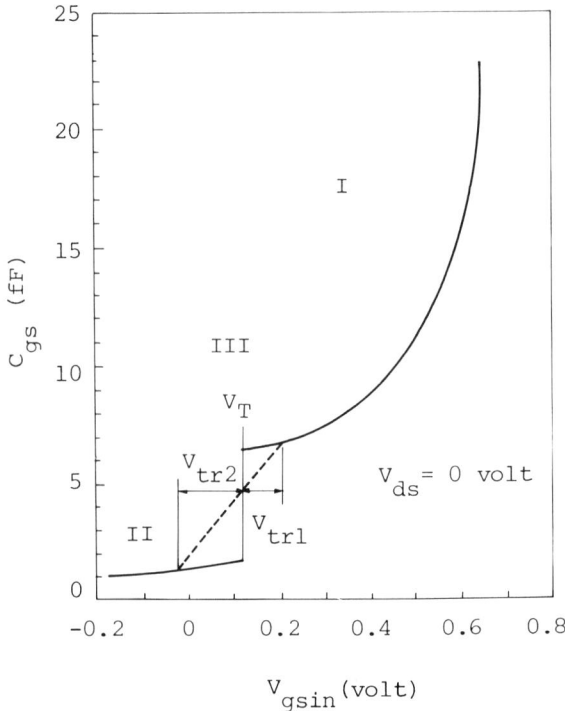

FIG. 23. Gate–source capacitance C_{gs} as a function of intrinsic gate–source voltage V_{gsin}. The curves can represent either a two-region or a three-region model. Regions I, II, and III correspond to before-pinchoff, after-pinchoff, and transition regions.

The two-region model is thus derived. Using Eqs. (39) and (46), the C_{gs} for the two-region model is calculated as shown by the solid line in Fig. 23 for a FET having the following parameters: $\ell_g = \ell_{gs} = \ell_{gd} = 1.0\,\mu\text{m}$, $a = 0.09\,\mu\text{m}$, $W_g = 10\,\mu\text{m}$, $N_d = 1.0 \times 10^{17}\,\text{cm}^{-3}$, $\varepsilon_s = 12.6$, and $V_{bi} = 0.7\,\text{V}$. However, C_{gs}, which is obtained from a two-dimensional analysis (solid line in Fig. 24) does not agree well with the C_{gs} shown as a solid line in Fig. 23 in the vicinity of the V_T value of V_{gs}. It is a fact that C_{gs} measured by Willing et al. (1978) and the present authors is closer to the two-dimensional analysis result than C_{gs} for the two-region model. Therefore, if a highly precise bias-dependent C_{gs} is required in circuit simulations, use of the following three-region model is preferable.

A disagreement near V_T is caused by an abrupt approximation at a depletion layer edge. Strictly speaking, since a carrier concentration does not change abruptly at a boundary of the depletion layer because of diffusion effect (Yamaguchi et al., 1976b; Grey, 1963), C_{gs} also does not change

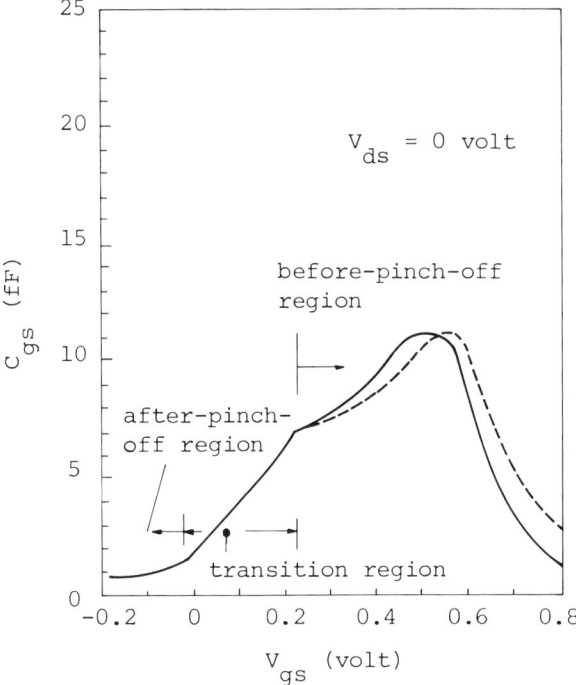

FIG. 24. Gate–source capacitance C_{gs} obtained by two-dimensional analysis. V_{gs} is external gate–source voltage.

abruptly with V_{gsin} but rather gradually. Since it changes almost linearly in the transition region in Fig. 24, C_{gs} in this region can be modeled as shown by a broken line in Fig, 23. C_{gs} then becomes

$$C_{gs} = \varepsilon W_g \tan^{-1} \sqrt{\frac{V_{bi} - V_T}{V_{tr2}}}$$
$$+ \left(\frac{\pi}{2} \varepsilon W_g + \frac{\ell_g W_g}{2\sqrt{2}} \sqrt{\frac{q N_d \varepsilon}{V_{bi} - (V_T + V_{tr1})}} - \varepsilon W_g \tan^{-1} \sqrt{\frac{V_{bi} - V_T}{V_{tr2}}} \right)$$
$$\cdot \frac{V_{gsin} - (V_T - V_{tr2})}{V_{tr1} + V_{tr2}}. \tag{47}$$

where V_{tr1} and V_{tr2} are approximately 0.1 V.

C_{gs} decreases when $V_{gs} > 0.5$ V in Fig. 24, while in Fig. 23 it increases monotonically. This disagreement results from the following reason. In the two-dimensional analysis (Fig. 24), C_{gs} is calculated as a function of the external gate–source voltage V_{gs}, including the voltage drop at R_s. Since the gate current begins to flow at about 0.5 V of V_{gs}, this voltage drop at R_s dominates V_{gs}. Because of this phenomenon, C_{gs} decreases in this region. C_{gs} calculated by using the three-region model is shown by a broken line in Fig. 24. In this case, a R_s of 200 Ω obtained from a and ℓ_{gs} is used. The curve agrees well with the two-dimensional analysis result (solid line). Consequently, Eq. (39) is concluded to be reasonable for representing C_{gs} in the before-pinchoff region.

Next, C_{gd} is described. Since the drain and the source are symmetrical with respect to the gate, C_{gd} is expressed by replacing V_{gsin} with V_{gdin} in Eqs. (39), (46), and (47). C_{gd} obtained by the three-region model is shown as a function of V_{dsin} in Fig. 25 by broken lines. The solid lines show C_{gd} calculated as a function of the external drain–source voltage V_{ds} by using the two-dimensional analysis. The difference between the two increases with V_{gs} because of the voltage drop at the drain resistance. Since the voltage drop increases for higher V_{gs}, the difference increases as V_{gs} increases. In principle, the two curves should be identical in a low V_{gs} range. Actually the C_{gd} obtained by the three-region model shows a good agreement with the two-dimensional analysis result when $V_{gs} < 0.2$ V, and therefore it is concluded that this model is reasonable for representing C_{gd}.

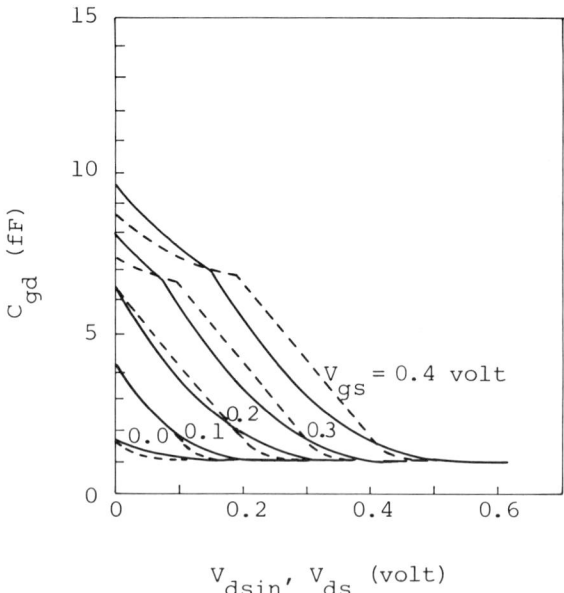

FIG. 25. Gate–drain capacitance C_{gd} as a function of external drain–source voltage V_{ds} (two-dimensional analysis, solid line) and of internal drain–source voltage V_{dsin} (three-region model, broken line).

4. GaAs LSI CIRCUIT DESIGN

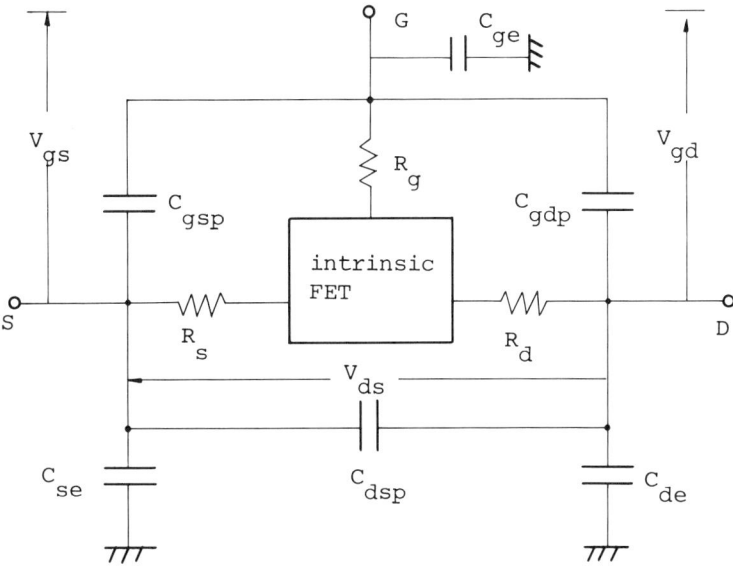

FIG. 26. Equivalent circuit of extrinsic GaAs MESFET.

b. Extrinsic FET Model

FET circuit performance depends on parasitic elements as well as intrinsic FET characteristics. The equivalent circuit for parasitic elements of an extrinsic MESFET is shown in Fig. 26. Since the source series resistance R_s and the drain series resistance R_d degrades g_m of an extrinsic FET, they must be made very small. R_s, in particular, decreases g_m. R_s and R_d consist of ohmic contact resistance and active layer resistance between gate and source (for R_s) and between gate and drain (for R_d). These resistances are considerably larger than the ones calculated from the mobility, the carrier concentration and the depth of the active layer. This is because of the effect of a surface depleted layer, which is induced by a surface potential of about 0.4 eV. Therefore, to decrease R_s and R_d, n^+ layers are formed in the offset regions of present FETs by self-alignment techniques. In such FETs with small series resistance, ohmic contact and n^+-layer resistances dominate in R_s and R_d. To decrease these resistances, new processes such as rapid thermal anealing and a nonalloying contact are being actively investigated. A typical R_s for a self aligned FET with 10 μm gate width is 50–100 Ω.

Gate resistance R_g, which is caused by the gate-metal resistance, is negligibly small as compared with the load resistance in digital circuits. Therefore, R_g is usually neglected in FET modeling.

Parasitic capacitances, which are independent of bias voltages, exist between electrodes (C_{gsp}, C_{gdp}, C_{dsp}), and between an electrode and the ground plane (C_{ge}, C_{se}, C_{de}), where C_{gsp}, C_{gdp}, C_{dsp}, C_{ge}, C_{se}, and C_{de} are gate-to-source, gate-to-drain, drain-to-source, gate-to-ground, source-to-ground, and drain-to-ground parasitic capacitances. A circuit simulation model that includes effects of all these capacitances is too complex to be executed in a practical application. Therefore, only dominant capacitances that affect circuit operational speed are selected and included in a model. The dominant capacitances depend on FET structures. Thus, circuit designers must reasonably simplify a model and evaluate dominant capacitances from the measured speed of simple circuits such as ring oscillators in conjunction with theoretical calculations.

The following is a typical case in which C_{ge}, C_{gsp}, and C_{gdp} are neglected. When a FET has a gate length of 1.0 μm, gate-to-source and gate-to-drain distances of 0.5 μm, and drain and source metal contact lengths of 7.0 μm (ohmic area is 7.0 × gate-width μm²), C_{dsp} is 0.14 pF per 1 mm gate width,

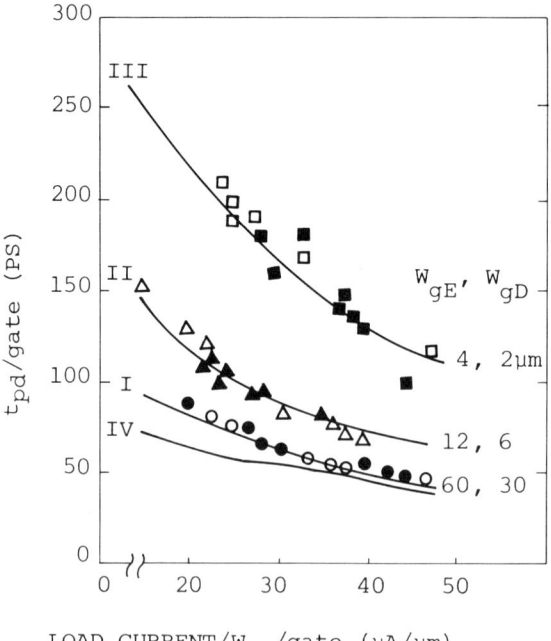

FIG. 27. Propagation delay time t_{pd} as a function of load current per unit gate width of DFET. W_{gE} and W_{gd} are gate width of EFET and DFET. C_{dsp} and C_{se} ($=C_{de}$) are 0.14 pF/mm and 10 fF for calculations of curves I, II, and III and are zero for curve IV. Gate length is 1.0 μm; supply voltage V_{DD} is 1.0 V.

and C_{se} and C_{de} are 10 fF. C_{se} and C_{de}, which are equivalently derived from fringing capacitance at each electrode (Takada *et al.*, 1984a) are independent of gate width.

In Fig. 27, experimental and simulation results are compared for propagation delay time t_{pd} for E/D DCFL ring oscillators. W_{gE} and W_{gD} are the gate-widths of the EFET and the DFET, respectively. The horizontal axis is normalized by load current for the gate-width of the DFET in one gate circuit. The solid lines represent simulation results. It can be seen that these two results agree very well.

The propagation delay time t_{pd} is inversely proportional to the value of the load current divided by a total capacitance including intrinsic and extrinsic capacitances and the capacitance of an interconnection line in a ring oscillator. The t_{pd} of the fastest ring oscillator indicated by curve IV, for which all parasitic capacitances are assumed to be zero, does not depend on gate width. This is because load current and intrinsic capacitance are proportional to the gate width. (The capacitance of an interconnection line, which is constant for each ring oscillator, is negligibly small.) On the other hand, actual devices have a slower speed caused by parasitic capacitances (I, II, and III). In addition, a large gate-width dependence of t_{pd} occurs. This results from the parasitic capacitances (C_{de}, C_{se}) that are independent of the gate width.

IV. Other Elements

In this part, we discuss circuit elements other than FETs in GaAs MESFET ICs: diodes, resistors, capacitors, and interconnection lines.

7. DIODES

Diodes are used mainly as switching elements and level shifters in various logic circuits. Schottky diodes are commonly used for these purposes because of their high-speed operation and ease of fabrication.

Diodes, which are easily realized by short-circuiting source and drain electrodes in FET structures, are usually used to simplify IC fabrication processes. However, the optimal diode structure depends on its function in the IC. Thus, Lee *et al.* (1982) optimally designed diode structures and employed implantation conditions for diode fabrication different from those for active-layer formations in SDFL (Schottky diode FET logic) circuits. They used the higher implantation energy to decrease capacitances of the diodes. A small series-resistance value is required for level-shift diodes in source-follower circuits, because series resistance degrades the capacitance drive capability in source-follower circuits. Accordingly, a diode having a highly doped active layer is preferable for this purpose. Reverse-biased diodes are also used as capacitors in CCFLs (capacitor-coupled FET logic) (Namordi and White, 1982) and CDFLs (capacitor diode coupled FET logic) (Eden, 1984).

8. Resistors

Resistors are mainly used as load resistances in ER-DCFLs (direct-coupled FET logic using enhancement FETs and resistors) and SCFLs (source-coupled FET logic). Usually, ion implantation into a semi-insulating substrate is employed in fabrication. In addition, sheet resistances ranging from 200 to 1000 Ω are usually used. This is because resistors with larger sheet resistances have larger standard deviations on a chip, and those with smaller sheet resistances require a larger resistor area. The standard deviation of sheet resistance is less than 1.8% through a wafer for an 880-Ω sheet resistance (Takada et al., 1985b).

Nonlinear resistors, in which the current saturates with increasing applied voltage, can also be utilized for a load in a circuit such as a SDFL (Lee et al., 1981). This nonlinear resistor provides higher transfer gain and higher circuit speed than the linear load resistor does. The nonlinear resistor is realized by forming a short active channel (less than 1 μm) between two electrodes.

A thin film metal with large resistivity such as cermet was also used by Lee et al. (1983) in a static RAM. Since this film's resistivity is as great as 68 MΩ ± 10%, it is effective for making small-size resistor patterns in low-current circuits.

9. Capacitors

Capacitors used for speed-up, coupling and decoupling capacitors in ICs are usually fabricated with MIM (metal–insulator–metal) structures. Si_3N_4 and SiO_2 thin films are employed for the insulator material. For speed-up and coupling capacitor use, reverse-biased diodes are often employed. Decoupling capacitors are used to minimize the supply voltage impedance.

10. Interconnection Lines

Interconnection lines play an important role in determining circuit performance. Line capacitance, cross capacitances and line resistance produce unfavorable decreases in circuit speed. Since the resistivity of a GaAs semi-insulating substrate is sufficiently high ($>10^6$ $\Omega \cdot$cm), propagations of slow-wave mode and skin-effect mode do not occur in the frequency ranges of digital circuits (Seki and Hasegawa, 1982). Therefore, the line capacitance

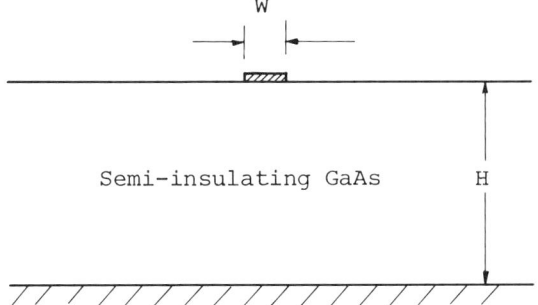

FIG. 28. Microstrip line structure formed on semi-insulating GaAs.

in the structure shown in Fig. 28 can be derived from the characteristic impedance of the microstrip line (Schneider, 1969; Yuan et al., 1982) as

$$C = 2\varepsilon_{\text{eff}}\varepsilon_0 / \ln\left(\frac{8H}{W} + \frac{W}{4H}\right), \quad W \leq H, \quad (48)$$

where

$$\varepsilon_{\text{eff}} = \frac{\varepsilon_r + 1}{2} + \frac{\varepsilon_r - 1}{2}\left(1 + \frac{10H}{W}\right)^{-1/2} \quad (49)$$

Here, ε_0, ε_r, and ε_{eff} are the permittivity of free space, the relative dielectric constant and the effective relative dielectic constant, respectively.

By using equation (48), the line capacitance becomes 50–60 fF for line widths W ranging from 1 to 3 μm with a semi-insulator thickness H of 400 μm. The dependence on W is very small because of the fringing effect (Higashisaka et al., 1980). In actual IC chips, multiple interconnection lines having small distances between each line are formed. In this case, line-to-line capacitance C_M (100–250 fF/mm) is larger than the line-to-ground capacitance (35–40 fF/mm). These values were calculated by using the two-dimensional simulator developed by Matsui and Moriya (1983). However, since the simulation for propagation delay that considers all of these C_M values is too complex, only a line-to-ground capacitance of 70–100 fF/mm is usually taken into account in a simulation.

Cross capacitances between first and second level lines are about 2 times larger than the capacitances calculated by assuming parallel-plate capacitance due to the fringing effect.

Crosstalk (Seki and Hasegawa, 1982) is also important because this reduces the circuit operation margin. In order to decrease crosstalk, Takada et al. (1984b) implemented a coplanar line for the output signal line in a high-speed frequency divider. Figure 29 shows the Q and $\bar{Q}$ output waveforms of

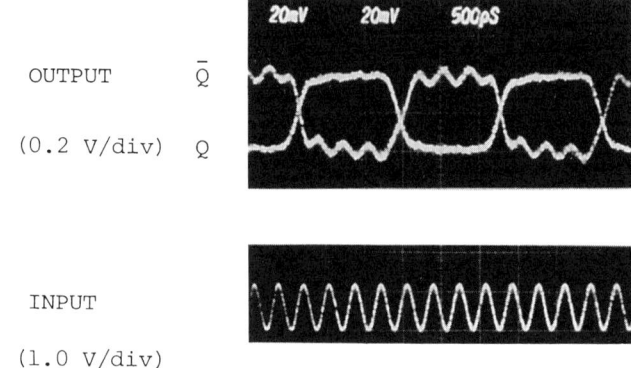

FIG. 29. Q and $\bar{Q}$ output and input waveforms of the 1/8 divider operated at 3.1 GHz.

the 1/8 divider at an input frequency of 3.1 GHz. The only Q is by coplanar line structure. Both output waveforms are clean at less than 2.7 GHz. However, at higher frequencies, the waveforms at the Q-signal line become cleaner because of the shielding effect.

Since interconnection line resistances as well as line capacitances decrease circuit speed, they must be included in circuit simulations. This can be accomplished by dividing the lines into small sections and assuming an equivalent circuit that consists of capacitance and resistances. Voltage drops resulting from current flows and line resistances should also be considered in layout design to ensure proper circuit noise margins. The voltage drop at the voltage-supply line must be less than 20 to 50 mV. These values are dependent on the magnitude of the logic voltage swings in implemented logic circuits.

Furthermore, the current density of interconnection lines must be considered to ensure the reliability, because when current density is too high, it induces electromigration and degrades reliability. Therefore, line widths should be determined in such a way that current density will be less than 5×10^4 A/cm^2 and 5×10^5 A/cm^2 for aluminum and gold interconnection metals, respectively.

V. Basic Circuit Configurations

Many kinds of logic circuit configurations have been so far proposed and studied for GaAs MESFET integrated circuits. In this part, the representative basic circuit configurations and these features are described.

11. DCFL (Direct Coupled FET Logic)

a. Features

Two-input NOR configurations of DCFL (direct coupled FET logic) are shown in Figs. 30a and 30b. The first one is called an E/D DCFL. This circuit consists of two enhancement mode FETs (EFETs), T1 and T2, as switching elements and a depletion mode FET (DFET), T3, as a load. The second is an E/R DCFL using a resistor as a load.

DCFLs are similar to E/D and E/R logics in Si-MOS technology. That is, direct connection of an output to an input terminal of a next stage circuit is available in the same way as for Si MOS logics. However, in high static voltage in the DCFL is not determined by the supply voltage V_{DD}, as it is MOS logics, but is a fixed, built-in voltage of about 0.8 V. This is because the load current flows into the next stage input terminals.

The supply voltage is normally 1.0 to 1.5 V. When very low power dissipation is required, it is less than 1.0 V. The logic voltage swing is normally 0.6 to 0.7 V. The threshold voltages V_T of EFETs and DFETs are usually designed to be 0.1 to 0.2 V and -0.4 to -0.8 V, respectively.

DCFL is considered to be the most suitable circuit for realizing LSI/VLSIs among the many available types of GaAs logic circuits. This is because it has three specific advantages: a small power dissipation of $10\,\mu\text{W}$ to 1 mW per gate, a small power-delay product of 1 fJ to 50 fJ, and a simple circuit configuration, featuring a small number of elements to construct a one gate circuit and a small gate cell area.

No DCFL ICs are commercially available at present, however, because of the narrow allowance for deviations of FET parameters, especially V_T, and

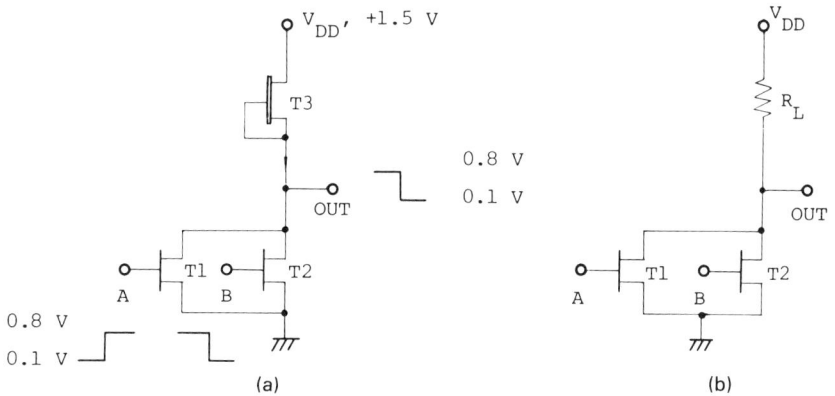

Fig. 30. (a) Two-input NOR configuration using E/D DCFL; (b) two-input NOR configuration using E/R DCFL.

the small noise margin. Therefore, strict controllability and high uniformity are required for fabrication. The FET fabrication technology and the quality of GaAs wafers are being improved to realize LSI/VLSI using DCFLs. Along with this development, feasibility studies have been performed on the 16 kb static RAM (Ishii et al., 1984) with 10^5 devices, which is the largest integration level achieved for GaAs LSIs. Moreover, 16×16 bit parallel multipliers (Nakayama et al., 1983) have been demonstrated.

b. Threshold Voltage Margin

One of the most significant problems for DCFL is its small tolerance for threshold voltage (V_T) fluctuation. In this subsection, the tradeoff between threshold voltage margin and speed is discussed by simulating E-R DCFL

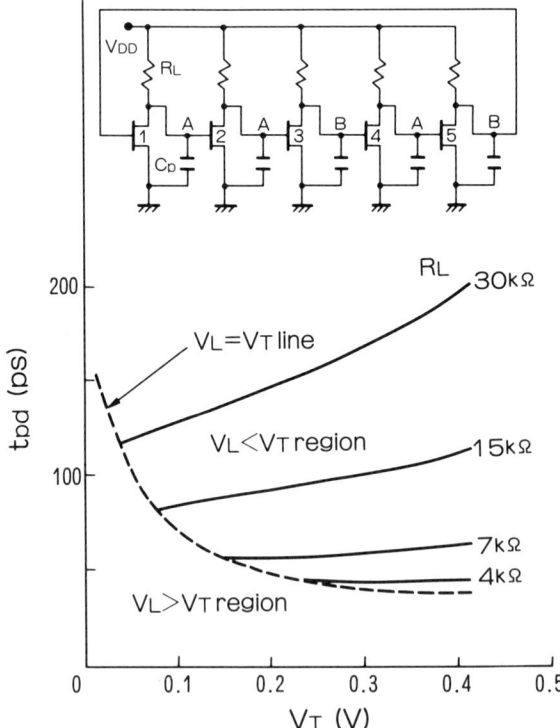

FIG. 31. Calculated propagation delay time per gate (t_{pd}) vs. threshold voltage (V_T) without dispersion. In the region above the broken line, criteria 1 (oscillation) and 2 ($V_L < V_T$) hold simultaneously. The inset is the circuit configuration of a five-stage E-R ring oscillator for the simulation. The output nodes of FETs 1, 2, and 4 are denoted by A, and those of FETs 3 and 5, by B.

ring oscillators (Ino *et al.*, 1981). To measure the normal inverter operation margin, the following two criteria are introduced for determining the deviation limit among threshold voltages: (1) the inverter chain is capable of ring oscillation; and (2) at any output node, the low level V_L settles below the V_T of the next gate ($V_L < V_T$). Criterion (2) assumes the condition that V_L does not exceed V_T at any stage in a ring oscillator, which is indispensable to the realization of DCFL LSIs. 0.5 μm gate FETs with 10 μm gate widths are assumed to calculate the tradeoff, using the aforementioned physical model. In Fig. 31, the calculated propagation delay time (t_{pd}) is plotted against V_T for the case in which the threshold voltage has no dispersion. If all FETs have the same V_T in a circuit, the criterion boundary for $V_L < V_T$ in this figure gives the fastest combination of V_T and R_L.

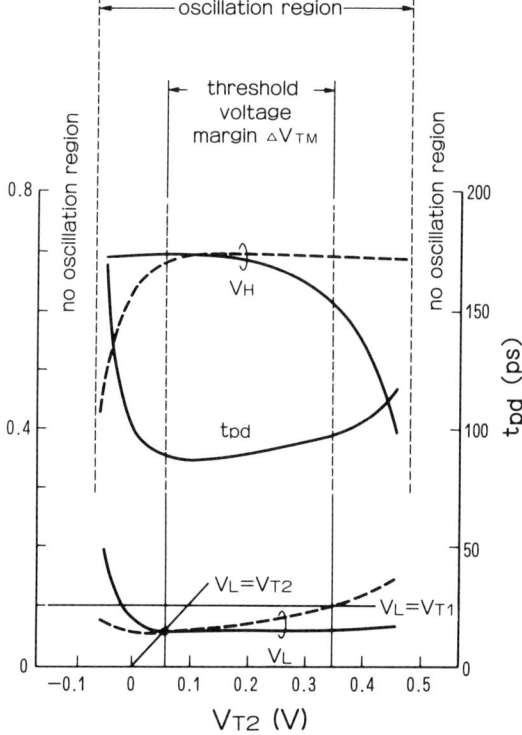

FIG. 32. V_{T2} dependence of oscillation characteristics of propagation delay time t_{pd} and output voltages V_H and V_L. In this figure, V_T of three FETs 1, 2, and 4 is fixed at $V_{T1} = 0.1$ V, while that for FETs 3 and 5 is variable V_{T2}. Load R_L is 15 kΩ. Solid lines for V_H and V_L represent voltages at nodes A, and broken lines represent those for nodes B. It is notable that ΔV_{TM} satisfying criterion 2 always lies within the oscillation region.

To demonstrate how the deviation of the threshold voltage affects the inverter output voltages V_H and V_L and the propagation delay time t_{pd}, it is assumed in the circuit shown in the inset of Fig. 31 that V_T values for FETs 1, 2, and 4 are constant at V_{T1}, while V_T values for FETs 3 and 5 have the same value, V_{T2}, which is varied. Furthermore, the load resistor R_L is assumed to be constant.

The calculated V_H, V_L, and t_{pd} are shown as a function of V_{T2} in Fig. 32 for the case of $V_{T1} = 0.10$ V and $R_L = 15$ kΩ. According to criterion 2, V_{T2} must be within a region ΔV_{TM} such that $V_L < V_{T1}$ at nodes B and $V_L < V_{T2}$ at nodes A. As shown in Fig. 32, the threshold voltage margin ΔV_{TM} always lies within the oscillation region.

Similar calculations were made for $V_{T1} = 0.1$–0.4 V with varying R_L. The summarized results are expressed in Fig. 33. Here, the threshold voltage margin ΔV_{TM} is plotted as a function of the average threshold voltage, $\langle V_T \rangle = (V_{T1} + V_{T2})/2$ with t_{pd} as a parameter. This figure also shows a quantitative tradeoff between speed and ΔV_{TM}. For large t_{pd} with large R_L, ΔV_{TM} is also large. The optimum $\langle V_T \rangle$ to be designed lies around 0.3 V for all t_{pd}.

In the first step in estimating the V_T tolerance, it is considered that V_T must satisfy $\langle V_T \rangle - \Delta V_{TM}/2 < V_T < \langle V_T \rangle + \Delta V_{TM}/2$, where ΔV_{TM} is inherently independent of the integration scale. In a specific circuit, the number of switching FETs N, the yield of the circuit Y, and ΔV_{TM} are related as

$$Y = [I(k\sigma)]^N. \tag{50}$$

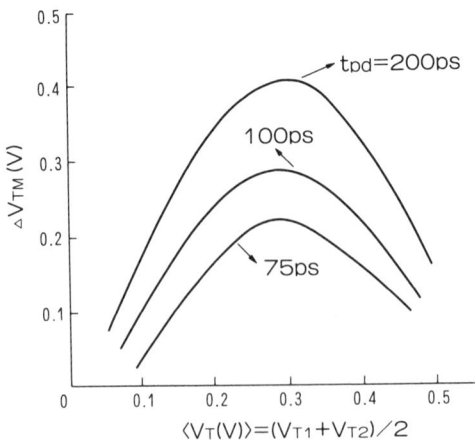

FIG. 33. Threshold voltage margin ΔV_{TM} as a function of $\langle V_T \rangle = (V_{T1} + V_{T2})/2$. The optimum value of $\langle V_T \rangle$ is about 0.3 V.

4. GaAs LSI CIRCUIT DESIGN

TABLE III
ALLOWABLE STANDARD DEVIATION OF THRESHOLD VOLTAGE FOR VARIOUS INTEGRATION SCALES WITH t_{PD} AS A PARAMETER

N (number of FETs)	σ (mV)		
	$t_{pd} = 75$ psec $\Delta V_{TM} = 220$ mV	$t_{pd} = 100$ psec $\Delta V_{TM} = 280$ mV	$t_{pd} = 200$ psec $\Delta V_{TM} = 407$ mV
5	101	128	187
10	76	97	141
100	46	58	84
1K	33	42	62
10K	29	37	54

$^a \langle V_T \rangle = 300$ mV, $Y = 20\%$.

Here, $I(k\sigma)$ is the probability V_T to satisfy for $\langle V_T \rangle - k\sigma V_T < V_T < \langle V_T \rangle + k\sigma V_T$ with $k\sigma = \Delta V_{TM}/2$, and σ is the standard deviation of V_T. If the value of Y is given, then the factor k increases as N increases. In Table III, tolerable standard deviations in relation to N are listed in the case of $Y = 20\%$, at $\langle V_T \rangle = 0.3$ V. For example, σV_T should be made less than 42 mV to achieve a 20% yield for a 1K gate LSI having a 100 psec gate delay time.

12. BFL (BUFFERED FET LOGIC)

A two-input NOR gate in a BFL circuit (buffered FET logic) (Van Tuyl et al., 1977) is shown in Fig. 34. It consists of depletion-type FETs (DFETs) and diodes. Since BFL uses DFETs as switching elements, a level-shift circuit is required to connect between two stages. A source-follower circuit employing diodes is used for this purpose. Two voltage sources, V_{DD} and V_{SS}, are necessary in BFL. The magnitude of supply voltage and the number of level shift diodes depend on the V_T of the FETs used. Usually, V_{DD} and V_{SS} are $+3$ to $+4.5$ V and -2 to -3.5 V, respectively, for V_T of -1.0 to -2.5 V.

One of the advantages of BFL is its high-speed performance because DFETs have a higher cutoff frequency f_T as compared with that of enhancement-type FETs (EFETs). The drive capability for interconnection line capacitances is also larger than that of the DCFL because of the small output impedance of the source-follower circuit.

In addition, the noise margin is larger because of the large logic voltage swing of 1.0 to 2.0 V. The large noise margin in conjunction with only DFET implementation makes BFL IC fabrication easier than in the case of DCFL. Therefore, BFL ICs are now commercially available.

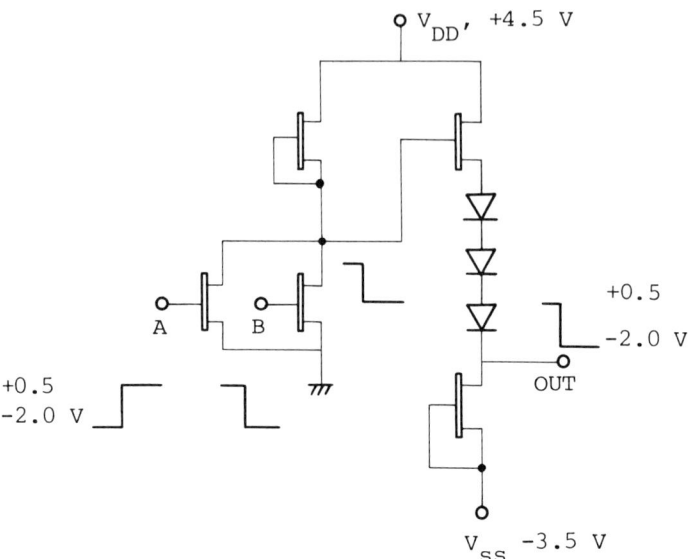

FIG. 34. Two-input NOR configuration using BFL.

On the other hand, the power dissipation (10 to 40 mW/gate) is too large to integrate a large number of gates on one chip. The maximum integration is limited to 50 to 200 gates if one assumes a power dissipation of 2 W is allowed on one chip.

In order to decrease the power dissipation, LBFL (low power BFL) using DFETs with a small negative V_T (-0.5 V) was proposed (Yamamoto et al., 1983). By using the LBFL, a 32-bit adder with 420 gates and power dissipation of 1 mW/gate was realized. Supply voltages are reduced to $+2.0$ and -1.0 V in this LSI. However, the complexity of circuit configuration remains, and the aforementioned advantages of normal BFL become smaller in LBFL.

CDFL (capacitor diode-coupled FET logic) (Eden, 1984) is also an improved version of BFL. The level shift circuit shown in Fig. 35 is employed in this circuit to achieve a higher speed with a lower power dissipation. reverse-biased Schottky diodes are used as speed-up capacitors.

13. SDFL (Schottky Diode FET Logic)

The two-input NOR configuration of SDFL (Schottky diode FET logic) (Eden, 1981) is shown in Fig. 36. SDFL consists of DFETs and diodes. Diodes, D1 and D2, and a current-source FET, T1, provide an OR function, and a switching FET, T3, and a load FET, T2, operate as an inverter,

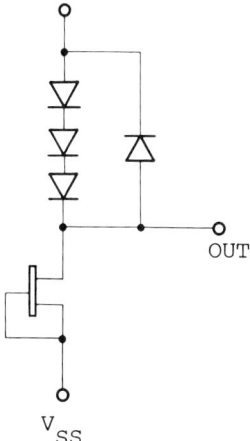

FIG. 35. Level shift circuit in CDFL.

providing a gain. A diode, D3, acts as a level shifter. The supply voltages V_{DD} and V_{SS} are 1.5 to 2.5 V and -1.0 to 2.0 V, respectively, and the logic voltage swing is 1.0 to 1.5 V. The DFET threshold voltage ranges from -1.0 to -1.5 V.

In SDFL, the current of the functional part, I_{SS}, is designed to be smaller than the load current I_{DD}. Thus, a small power dissipation (0.12–1.5 mW/gate) is obtained as compared with that of BFL. The gate cell area is also smaller than that in BFL because of the small diode area

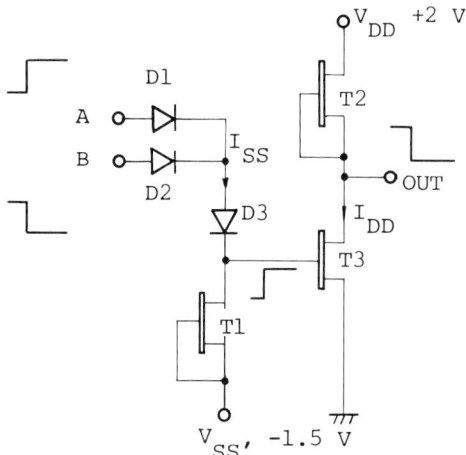

FIG. 36. Two-input NOR configuration using SDFL.

corresponding to a small I_{SS}. However, the speed decreases in accordance with the small power dissipation.

Therefore, SDFL can achieve a moderate level of integration (up to 1K gate) with relatively high speed performance. An 8 × 8 parallel multiplier including 1008 NOR gates was developed using SDFL (Lee *et al.*, 1982). This was the first LSI fabricated which integrates over 1K gates using DFETs in GaAs integrated circuits.

14. SCFL (Source-Coupled FET Logic)

SCFL (source-coupled FET logic) (Takada *et al.*, 1981) is classified into HSCFL (high logic-voltage-swing source-coupled FET logic) and LSCFL (low-power source-coupled FET logic), which are described in this section.

a. HSCFL (High Logic-Voltage-Swing Source-Coupled FET Logic)

HSCFL (Takada *et al.*, 1984b) consists of DFETs, diodes, and resistors, as shown in Fig. 37. This logic operates as current steering mode similarly to Si bipolar ECL (emitter-coupled logic). The FETs, T1, T2, and T3 are switching FETs, and T4, T7, and T8 are current-source FETs. T5 and T6

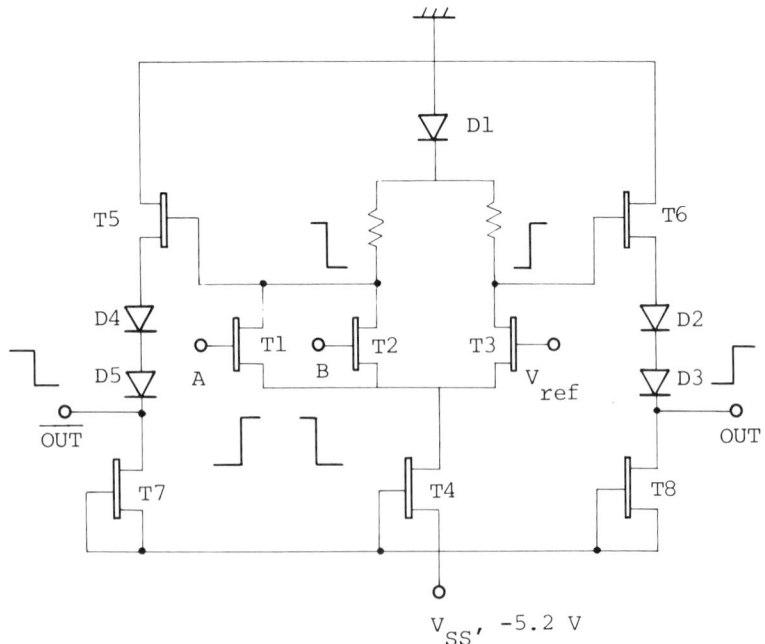

FIG. 37. Two-input OR/NOR configuration using HSCFL.

are source-follower FETs and D2, D3, D4, and D5 are the level shift diodes. D1 is specially used as a level shift diode for providing a high-speed bias condition for T5 and T6.

All of the FETs in HSCFL are operated in the drain-current saturation region. Since the gate-to-drain capacitance in this region is very small as compared with nonsaturation region as described in the section on gate capacitances, HSCFL is capable of higher-speed operation than other logics. In addition, the allowance for the threshold voltage variation is large enough to ensure a high fabrication yield. This is because the logical threshold level is determined by the DC reference voltage V_{ref}. A large logic voltage swing (1.5 to 2.0 V) is required in HSCFL to achieve high-speed operation, because it provides a high average f_T operation for implementing FETs with V_T of -0.5 to -1.0 V (Fig. 10, Takada et al., 1984b). The supply voltage V_{SS} is -5 to -8 V.

Disadvantages of HSCFL are the large power dissipation (~ 10 mW/gate) caused by the large supply voltage required, and its circuit complexity. Therefore, HSCFL is suitable for small-scale and medium-scale ICs with high-speed operation. By employing HSCFL, an ECL-compatible decision circuit was developed for optical repeaters operating at a 2.0 Gbit/sec data rate (Ohta and Takada, 1983). Since HSCFL has compatibility with ECL, it was also used for I/O circuits in an SRAM (Takahashi et al., 1985).

b. LSCFL (Low-Power Source-Coupled FET Logic)

LSCFL (Low-power source-coupled FET logic), which consists of EFETs, diodes, and resistors (Takada et al., 1985a), operates in a current steering mode. A two-input OR/NOR configuration is shown in Fig. 38.

Since LSCFL operates with dual phase signals (true and complementary signals), the transfer gain becomes twice that of HSCFL if the same FET and load resistors are used. Thus, the load resistance can be reduced to a half of that for HSCFL at the same gain. This implies a small logic voltage swing V_{LS}. Furthermore, LSCFL usually uses EFETs, which is also effective in decreasing V_{LS}.

Small V_{LS} values (0.6 to 0.8 V) provide higher speed (~ 50 psec/gate) and lower power dissipation (0.5 to 2 mW/gate) because of low supply voltage. The threshold voltage of EFET is normally designed to be $+0.1$ to 0.2 V. V_{SS} is -2.5 or -3.5 V for two- or three-level series gate configurations, respectively. The potentials at the gate electrode of current-source FETs (T1, T2, and T3) are set to be 0.6 to 0.7 V higher than V_{SS}.

The largest allowance of threshold voltage variation among GaAs logic circuits is also a great advantage of LSCFL. This is because no logical threshold voltage exists because of its differential operation. The threshold

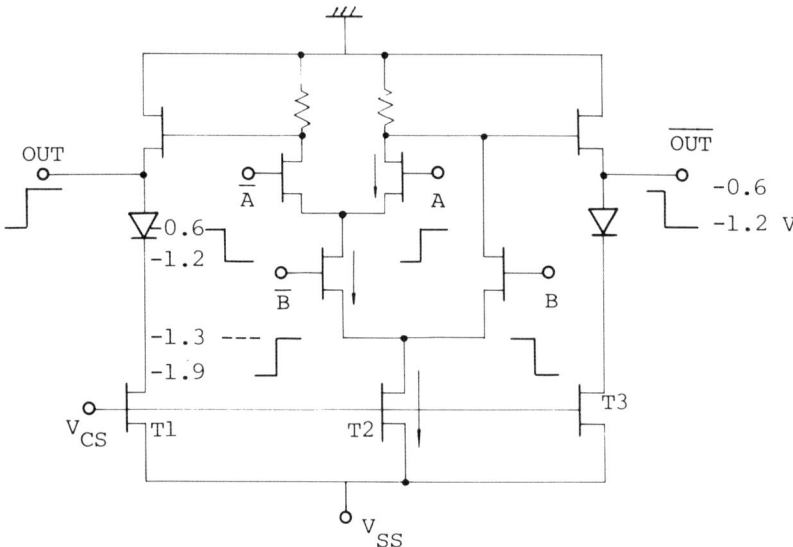

FIG. 38. Two-input OR/NOR configuration using LSCFL.

voltage margin ranges from +0.25 to −0.2 V for accurate operation. This feature is important in obtaining a high fabrication yield.

The use of the series gating technique, one of the most favorable advantages of LSCFL, reduces the propagation delay time and the power dissipation in complex-function circuits. This is effective especially in fabricating sequential logics such as flip-flops and frequency dividers as described in Section 18. An example of a three-level series gating for combinatorial logics is shown in Fig. 39 (Togashi *et al.*, 1984). When the three-input exclusive OR is configured by other logic circuits such as DCFL, BFL, or SDFL using NOR gates, three stages of passing gates are necessary, as shown in Fig. 40. On the other hand, LSCFL requires only one stage. A digital time-switch LSI opeating at 2 Gbit/sec data rate (Takada *et al.*, 1985b) and a high-speed 8-bit-ALU (arithmetic logic unit) with a 0.8 nsec critical pass delay time (Ino *et al.*, 1986) have been developed using the two- and three-level series gating LSCFL, respectively.

The one disadvantage of LSCFL is circuit configuration complexity. Because the interconnection line-area required for LSCFL is twice as large as that for other circuits (true and complementary signal lines are drawn for all gates), chip size becomes larger than that for circuits using single phase signals. Since this limits the number of gates, LSCFL is suitable for use in high-speed LSIs having less than about 3K gates.

4. GaAs LSI CIRCUIT DESIGN

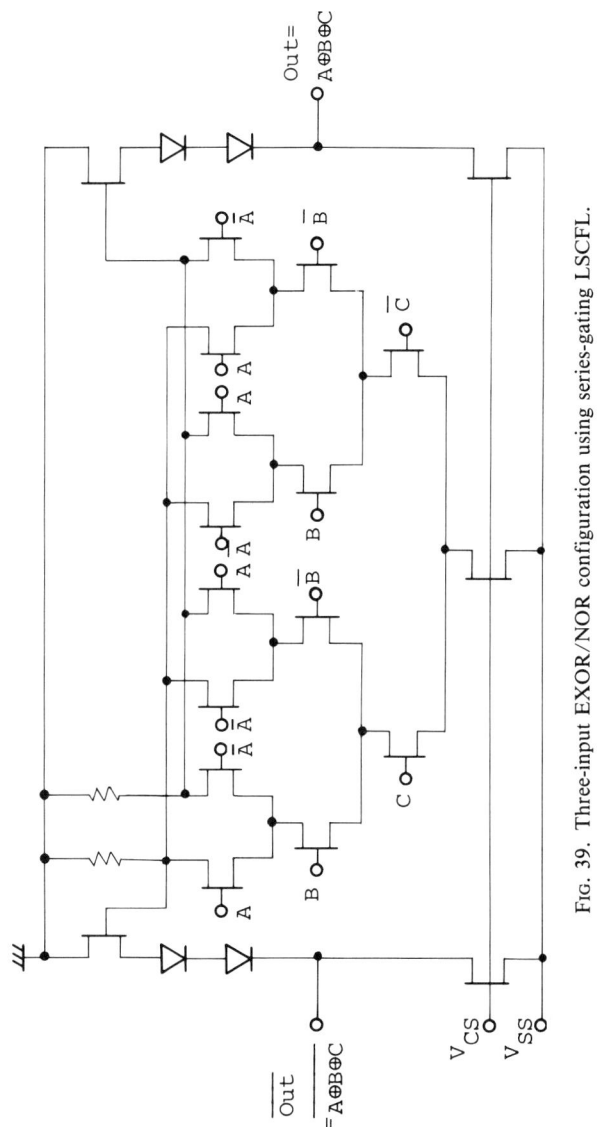

FIG. 39. Three-input EXOR/NOR configuration using series-gating LSCFL.

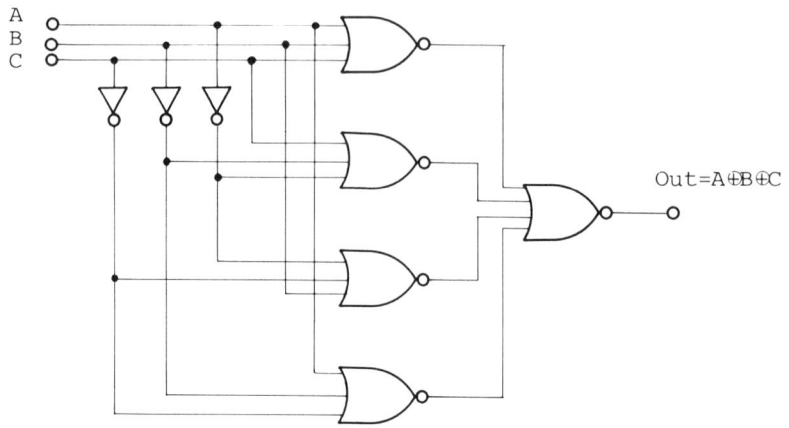

FIG. 40. Three-input EXOR configuration using NOR gates.

VI. LSI Design

The integration scale and speed of GaAs LSI are expected to be improved year by year. Especially for logic LSIs, the function generally becomes more complex as the gate number increases. The design obviously becomes more important, particularly to optimize the LSI performance and to reduce the development cost. In this part, the method of designing memory and logic LSIs will be described.

15. MEMORY

The GaAs static random access memory (SRAM) has been under active development by many companies. The speed and dissipation power of state-of-the-art Si and GaAs static and dynamic RAMs are summarized in Fig. 41. As a first attempt at a GaAs SRAM, a 16-bit test RAM having a 10 nsec access time was reported by Asai *et al.* (1981). Subsequently, 256-bit (Ohwada *et al.*, 1982) and 1-kbit (Ino *et al.*, 1982b) SRAMs were fabricated by NTT. Following these developments, a 1K SRAM using High Electron Mobility Transistors (HEMTs) (Nishiuchi *et al.*, 1984) and a 4K SRAM (Hirayama *et al.*, 1984) were reported. A 16K RAM was successfully fabricated recently by Ishii *et al.* (1984). All of these RAMs are based on the DCFL. Another approach using LPFL (low-pinchoff FET logic), an 8-bit, 0.8 nsec RAM, was demonstrated by Bert *et al.* (1982).

The design procedure and techniques for the 1-kbit GaAs SRAM developed by Ino *et al.* (1984) are described in this section. Here, 1 μm gate SAINT FETs developed by Yamasaki *et al.* (1982) were used and the FET model mentioned in Section 5 was adapted to the RAM design.

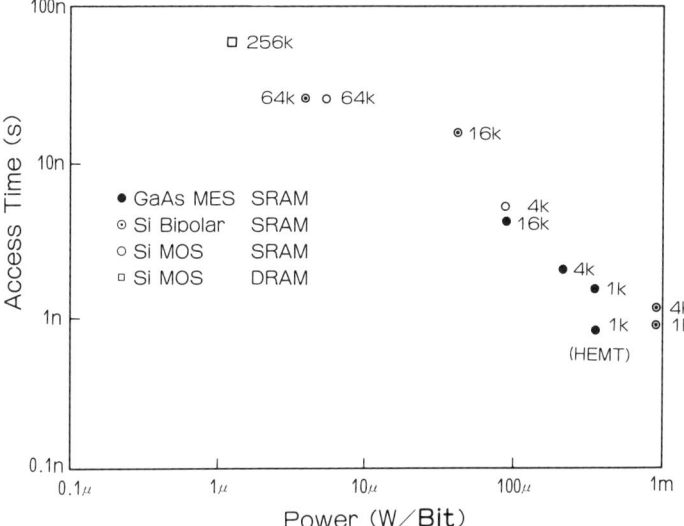

FIG. 41. State-of-the-art LSI memory development: (●) GaAs MES SRAM; (⊙) Si bipolar SRAM; (○) Si MOS SRAM; (□) Si MOS DRAM.

a. Circuit Configuration

In the first step, the design targets of speed (access time), power dissipation, chip size, and input–output level are to be determined. Since an access time below 2 nsec, a power dissipation below 500 mW and a chip size of about 3 mm × 3 mm are required, the basic circuit is determined to be DCFL. The next step in the process is the decision of the circuit configuration for each circuit block, such as a memory cell and a decoder. Here, again considering speed, power and a pattern size, a six-transistor type memory cell, a differential type sense amplifier and a five-input NOR type decoder are adopted. Following the internal circuit design, the input and output buffer circuits are to be designed according to the input and output levels, e.g. ECL or TTL.

b. Circuit Parameters

The third step is the optimization of the circuit parameters, that is, the FET threshold voltage and gate widths. Only two types of FETs, EFETs and DFETs, having the same gate lengths are used for the sake of simplicity of the fabrication process. The threshold voltages for EFETs (V_{TE}) and for DFETs (V_{TD}) are optimized to be $V_{TE} = 0.1$ V and $V_{TD} = -0.4$ V. This is accomplished from the viewpoints of switching speed and operating margin (especially the allowable range of threshold voltage fluctuations) through circuit simulation of the typical peripheral circuit, i.e., a decoder. As will be mentioned later, the statistical circuit simulation is accomplished using a threshold voltage standard deviation (σV_T) of 50 mV. In the peripheral circuits such as decoders and address buffers, the gate width ratio between the EFETs and DEFTs is set at 2:1, which facilitates a wider range of threshold voltage (V_{TD}) deviation and a smaller chip size. After the threshold voltage is determined, the gate widths of FETs in each circuit block are designed.

In the design, the gate width of a driver FET in the memory cell is initially chosen to be 9 µm, taking into account a reasonable chip size. Then, gate widths of the transfer gate FET (W_T) and the load FET (W_L) are determined to optimize speed, power, and read–write stability. A multicell circuit including four cells in two columns and two rows is analyzed to obtain the dynamic stability of the cells. The main part of the multicell circuit model in the worst case is shown in Fig. 42, in which cells 3 and 4 correspond to all the other 31 cells in the same rows as cell 1 and 2, respectively. The analytical procedure is as follows: (1) write "0" in cell 3 and "1" in cell 4; (2) write "1" in cell 1, then write "0" in cell 2; and (3) read cell 2, cell 1 then cell 2, continuously. The write stability is calculated during step 2, while read stability is calculated during step 3. Figure 43 shows the simulated read and write stability for a memory cell. In this figure, the larger W_T is, the higher

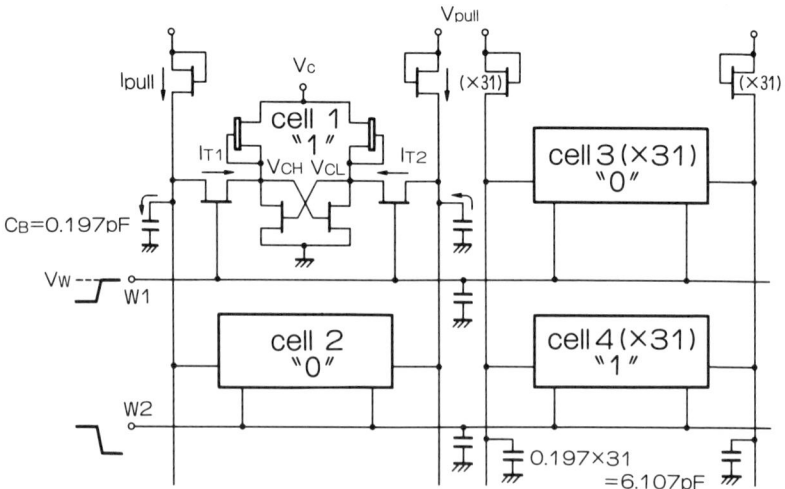

FIG. 42. Circuit model for analyzing cell stability. Although entire circuits are included in the calculation, only the main part is shown in this figure. The right-hand-side components are 31 times larger than those of the left-hand side.

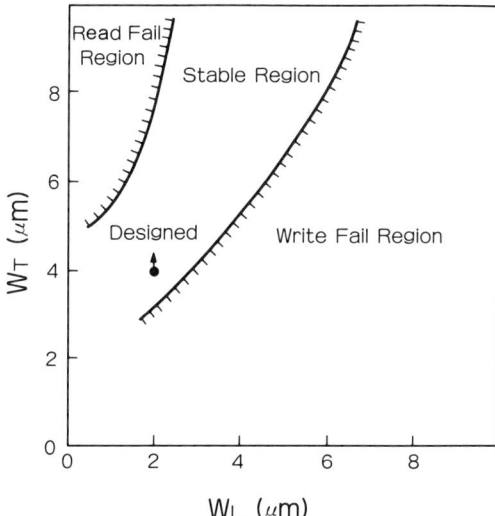

FIG. 43. The region in which write and read are stable in the W_L vs. W_T plane. W_L and W_T are gate widths for load FETs and transfer gate FETs, respectively.

is the access speed, and the larger W_L is, the larger is the power consumption. From these results, W_T and W_L are designed to be 4 μm and 2 μm, respectively.

A column sense amplifier is also constructed with an E–D-type differential inverter. An inverter sense amplifier is capable of directly driving the next-stage E–D-type gates without the use of a level shift or a multistage sense circuit. Therefore, it has a higher speed as compared with a multistage sense amplifier. Another merit of this single-stage sense circuit is its simple fabrication process, which requires neither a diode nor a resistor. The disadvantage of such a sense circuit, however, is its relatively small fluctuation tolerance for the input voltage level (bit line signal), due to the existence of a logical threshold voltage (V_{TL}). V_{TL} of the sense inverter is designed to be 0.6 V, which is nearly one-half of the bit-liner signal level.

In the buffer circuits for address, chip-select, write-enable, and data-in circuits, E–D-type gates are used that have large gate widths of between 60 and 120 μm. An E–E pushpull circuit is applied to the data-out final stage, where the output current into the 50-Ω load is designed at 12 mA. The main part of the designed GaAs 1 kbit SRAM circuit is shown in Fig. 44.

c. Threshold Voltage Margin

The threshold voltage margin is an important design factor in GaAs ICs, especially in DCFL circuits. The threshold-voltage deviation allowed for the

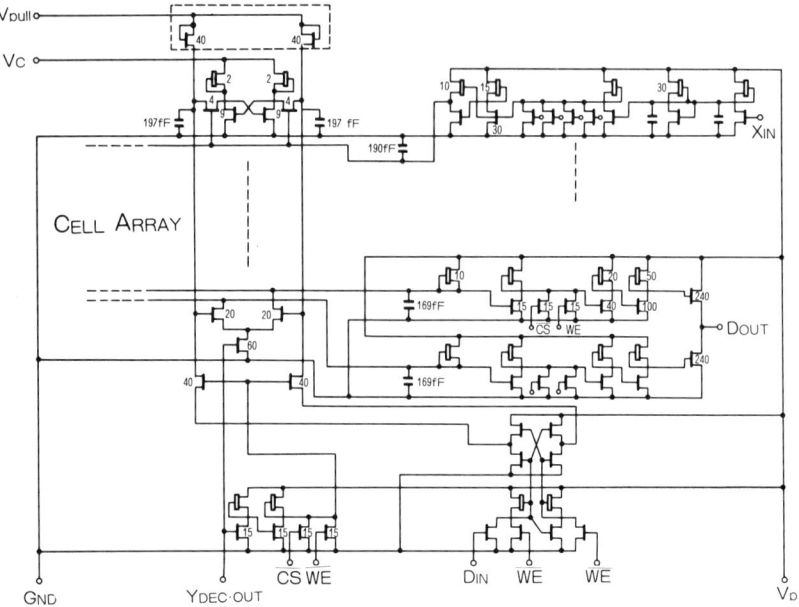

Fig. 44. 1 kbit SRAM circuit. Numbers inside FETs denote gate-widths in μm.

normal operation of the designed RAM is calculated by simulation. The result is shown in Fig. 45. The area inside the boundary line corresponds to the normal operating region having read–write stability and a date-out voltage swing of more than 0.3 V. In obtaining the calculated results shown in Fig. 45, it was assumed that both V_{TE} and V_{TD} are uniform in a unit chip, which means that $\sigma V_{TE} = \sigma V_{TD} = 0$.

The permissible region for threshold voltage deviation is also seen to be $0\text{ V} < V_{TE} < 0.25\text{ V}$ and $-0.55\text{ V} < V_{TD} < -0.15\text{ V}$, by adjusting the bit line pull-up voltage (V_{pull}) in the figure. The optimum combination is indicated by the solid circle, the same as the designed values of $V_{TE} = 0.10$ V and $V_{TD} = -0.40$ V, where V_{pull} is 0.75 V. The design takes the speed and the threshold voltage margin into consideration. Under this condition, the access time is simulated to be 1.5 nsec with a 300 mW power dissipation. Finally, V_{pull} must be supplied by a power source separated from the others in order to adjust the deviation of the threshold voltages of actual FETs.

The statistical circuit simulation is then carried out to obtain the allowable threshold voltage fluctuation in each circuit block. In the simulation, both V_{TE} and V_{TD} are assumed to be randomly distributed in a certain circuit block with no correlation existing among the same-type FETs and different-type FETs, although the threshold voltages are constant in the other blocks. The

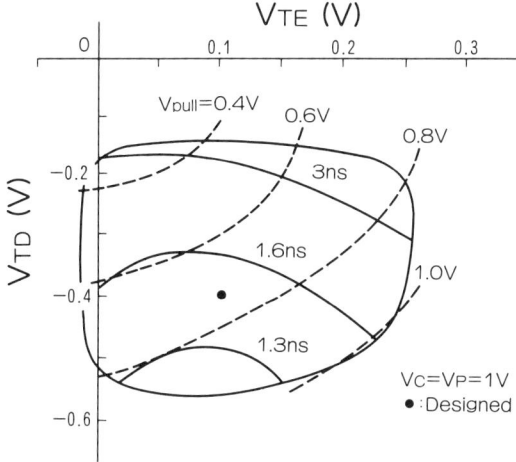

FIG. 45. Calculated results for allowable V_{TE} and V_{TD} ranges for 1K SRAM.

Monte Carlo method is applied to the simulation using the circuit simulation program ASTAP. V_{TE} and V_{TD} have a Gaussian distribution determined by random numbers in the range of V_T (nominal) $\pm 3\sigma V_T$ (standard deviation). The nominal threshold voltages are fixed at the designed values of $V_{TE} = 0.10$ V and $V_{TD} = -0.40$ V. The simulation results on σV_T are shown in Fig. 46. The allowable σV_T is more than 50 mV in a simple NOR circuit such as an address buffer, decoder or data-out buffer where a full logic swing of about 0.5 V is available. However, it is 30 mV in a sense circuit having an input voltage swing of about one-third that of a full swing. In the present fabrication technology, the best standard deviation of the threshold voltages of FETs is about 30–40 mV over a 2-inch wafer, and 20–30 mV over the area of a chip about 3 mm square. As reported by Matsuoka *et al.* (1983), however, the σV_T correlates with the distance between FETs. According to

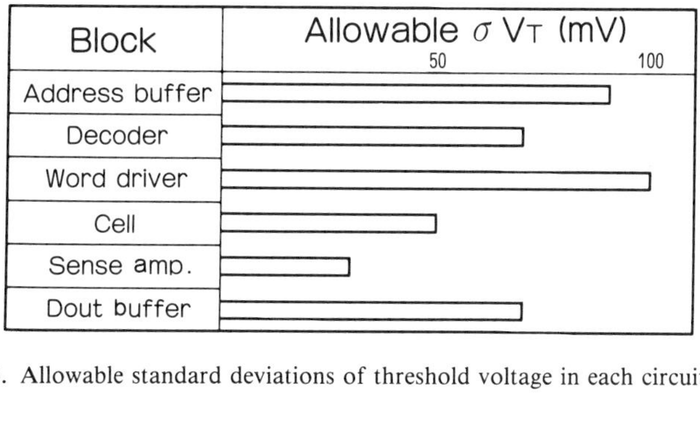

Block	Allowable σV_T (mV)
Address buffer	50 100
Decoder	
Word driver	
Cell	
Sense amp.	
Dout buffer	

FIG. 46. Allowable standard deviations of threshold voltage in each circuit block.

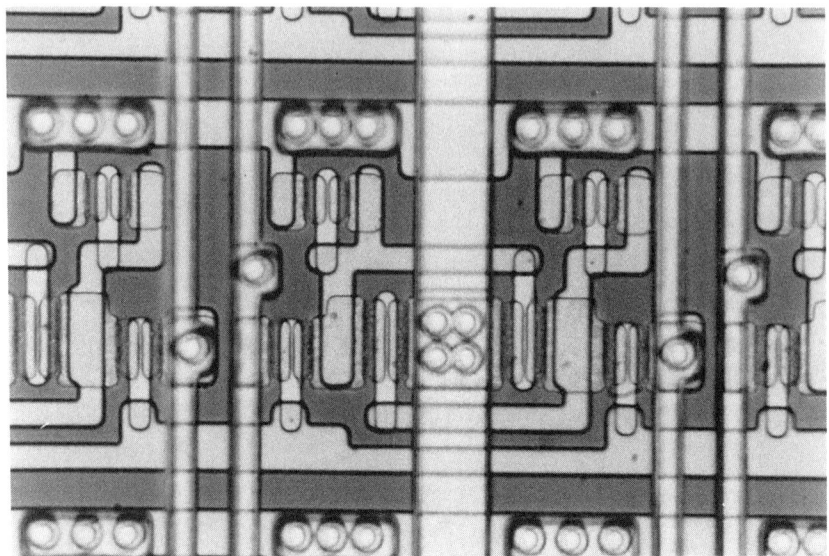

FIG. 47. Micrograph of fabricated memory cell. Cell size is 69 μm × 57 μm.

their result, σV_T between two driver FETs located only 30 μm apart in a sense circuit is less than 30 mV. Consequently, by this design, a 1.5 nsec access time with 300 mW power dissipation and a reasonable threshold voltage margin is estimated to be realistic and sufficient as a desired design target.

d. Performance

The 1K SRAM designed as stated in the previous subsections was fabricated using 1-μm-gate SAINT FETs. The fabricated memory cell pattern having a cell size of 69 μm × 57 μm is shown in Fig. 47. The wiring rule is 1.5 μm/1.5 μm for the line/space of the first metal and 3 μm/3 μm for that of the second metal. The latest developed 16K SRAM memory cell is 41 μm × 32.5 μm, which is made more compact by employing the scaled-down rule. Figure 48 is a microphotograph of a 1K SRAM chip having a size of 3.42 mm × 3.40 mm.

The minimum X-address access time was 1.5 nsec at 369 mW power consumption with a data-out signal swing of about 0.5 V into a 50 Ω load, as shown in Fig. 49. The minimum write pulse width was less than 2 nsec, which was limited by pulse generator performance. From the simulation, a 0.8 nsec 1K SRAM that uses 0.6-μm-gate FETs and the 2 μm pattern rule is estimated.

4. GaAs LSI CIRCUIT DESIGN

FIG. 48. Microphotograph of GaAs 1K RAM chip with a chip size of 3.42 mm × 3.40 mm.

16. FULL-CUSTOM LOGIC

Logic LSIs are classified into two groups: full-custom LSIs designed for use in specific systems (e.g. communications systems), and semicustom LSIs designed for wider applications. Full-custom LSIs are almost entirely manually designed to obtain the best performance. Semicustom logic LSIs

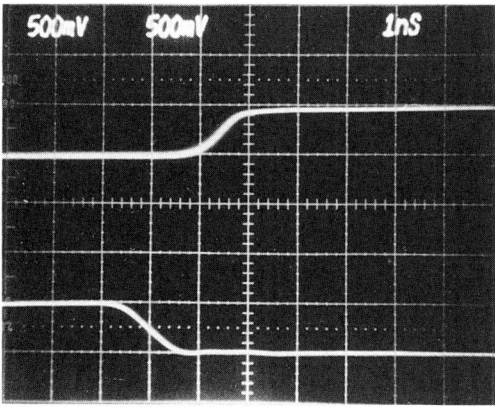

FIG. 49. Data-out (upper) and X-address input (lower) waveforms. Access time is 1.5 nsec.

including gate arrays and macro cell arrays are designed partly using CAD (computer aided design) and DA (design automation) systems and partly manually to achieve a short turnaround time. As the procedure for designing logics is essentially the same as that for designing memories, in this subsection we shall take a look at design results for both types of logics. And in the final subsection, an explanation is given for the design of a high-speed frequency divider that is a key device for communication systems.

Several kinds of GaAs full-custom logic LSIs have been developed. The 16 × 16 parallel multiplier reported by Nakayama *et al.* (1983) is the largest GaAs logic IC in terms of integration scale. A typical example of a GaAs full-custom logic LSI is a digital time-switch for application to digital switching systems. A design of a four-channel digital time switch LSI having a 2 Gbit/sec throughput and a 500 MHz channel rate developed by Takada *et al.* (1985b) is described below. The channel data rate is sufficient for application to a 100 Mbit/sec or higher high-definition TV (HDTV) data transmission system.

a. Digital Time-Switch Construction

A digital time switch exchanges data signals in a frame from input to output, as shown in Fig. 50. Address inputs control the output channel order. The block diagram of the digital time switch is shown in Fig. 51. The LSI consists of 4-bit shift registers, data latches, a counter, a control unit and I/O buffer gates. The switch LSI itself is a kind of sequential logic and is mainly constructed by various flip-flops.

The basic circuit used is an LSCFL because of its superiority in F/F (flip-flop) performance compared with other basic circuits. The input buffer

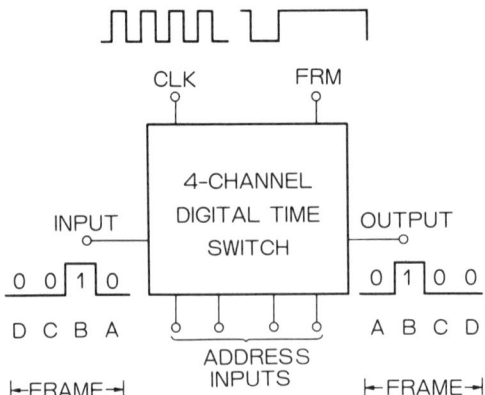

Fig. 50. Function of four-channel digital time switch.

4. GaAs LSI CIRCUIT DESIGN

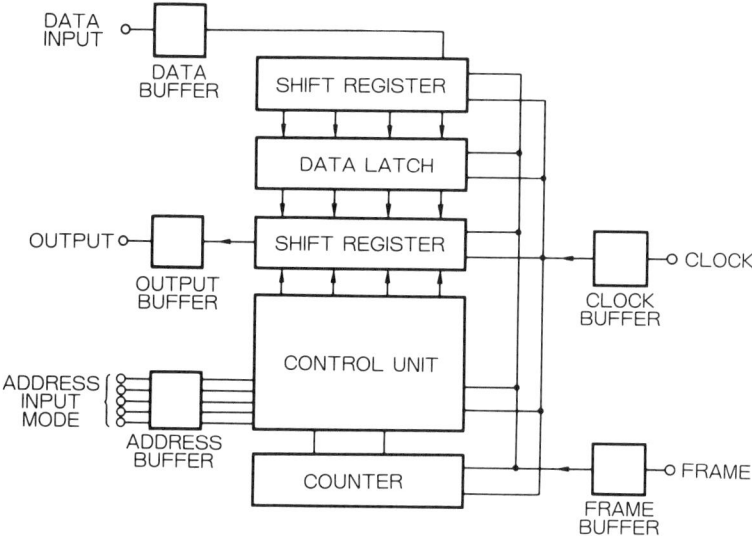

FIG. 51. Block diagram of digital time switch LSI.

is designed to convert the ECL level to the LSCFL level defined as $0 \leftrightarrow -0.45$ V. The output buffer having a 100 Ω output impedance provides a 0.45 V LSCFL level.

b. Fundamental LSCFL Design

The LSCFL has true (D, Q) as well as complementary ($\bar{\text{D}}$, $\bar{\text{Q}}$) terminals for input and output use without the need for the reference voltage (V_{ref}) usually

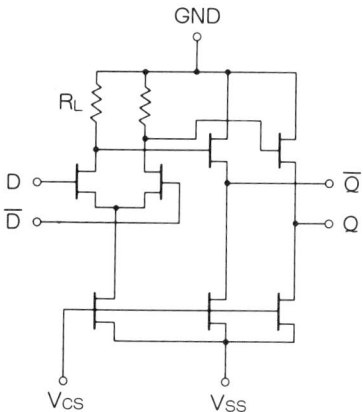

FIG. 52. Inverter configuration of LSCFL.

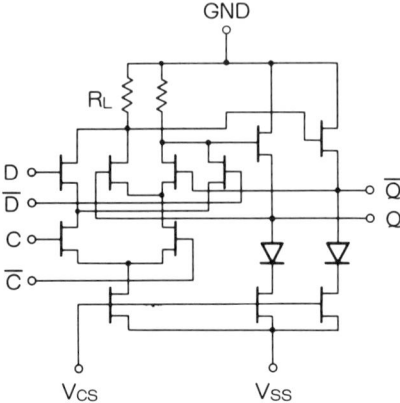

FIG. 53. Reset-set (R-S) flip-flop circuit using two-level series gate LSCFL.

employed in GaAs HSCFLs and Si ECLs. The inverter configuration of an LSCFL is shown in Fig. 52. The LSCFL has twice as high a transfer gain as the HSCFL employing a V_{ref}, when using the same load resistor, because of true and complementary operation. Therefore, the LSCFL logic-swing can be reduced by a half with half of the resistor value as compared with the HSCFL, resulting in higher-speed operation and lower V_{SS}. Furthermore, by adopting an EFET, the LSCFL enables low power operation to be achieved. The two-level series gate configuration where $V_{SS} = -2.5$ V is adopted for the construction of various F/Fs and NOR gates. A typical example of the two-level series gate LSCFL is the R.S. (reset-set) F/F shown in Fig. 53. Another merit of LSCFL is its wider threshold voltage tolerance. The LSCFL allows nearly twice as wide a V_{TE} range as the DCFL, according to the simulated results shown in Fig. 54.

c. Performance

A time switch LSI consisting of 231 equivalent gates was fabricated by 0.55 μm gate buried *p*-layer SAINT FETs (Yamasaki et al., 1984). The experimental propagation delay-time of the LSCFL inverter is 48 psec/gate with 5.4 mW/gate. Owing to the logic capability of the two-level LSCFL utilized in this LSI, the equivalent power dissipation per gate is reduced to 1.4 mW, which is small compared to that of the DCFL. The LSCFL has a large drive capability to interconnection line capacitance, because of the existence of source-followers. The measured additional delay-time per 1 mm line is 22 psec. A microphotograph of a fabricated digital time switch having a chip size of 2.0 mm × 2.4 mm is shown in Fig. 55. The fabricated LSI operated correctly for all test patterns up to and including the 2.0 Gbit/sec

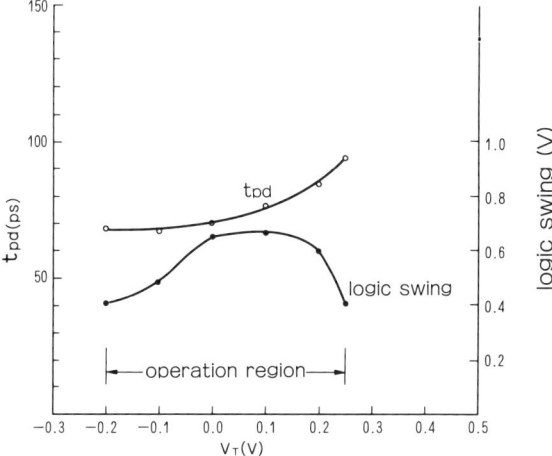

FIG. 54. Delay time t_{pd} and logic swing as a function of threshold voltage in LSCFL. The threshold voltage region for normal operation is $-0.2\text{ V} < V_T < 0.25\text{ V}$.

input data rate. The power dissipation was 640 mW at $V_{SS} = -2.5$ V. A 2 Gbit/sec data exchange of the digital time switch is shown in Fig. 56. It can be seen that the input channel order ABCD is changed into the output channel order BDAC. The output rise and fall times (20%–80%) were 90 and 80 psec, respectively, as shown in Fig. 57. This LSI effectively demonstrated a high fabrication yield of 75%.

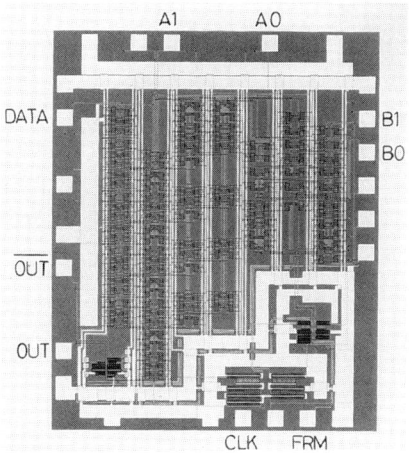

FIG. 55. Microphotograph of digital time switch LSI. Chip size is 2.0 mm × 2.4 mm.

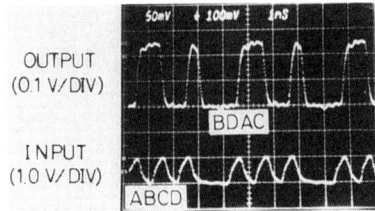

Fig. 56. 2 Gbit/sec data exchange operation in time-switch LSI. Input channel data (A, B, C, D) are changed into output channel data sequences (B, D, A, C).

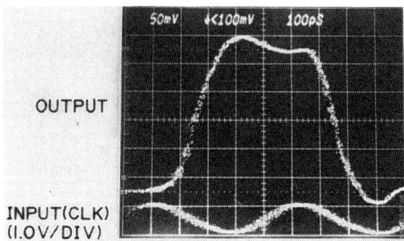

Fig. 57. Waveforms of output and input pulses in time switch LSI. Rise time is 90 psec and fall time is 80 psec.

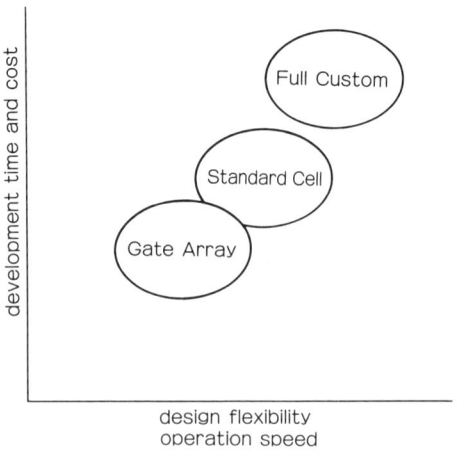

Fig. 58. Comparison of custom logic LSIs in terms of development time and cost, design flexibility and speed.

17. GATE ARRAY

A gate array is an LSI in which the layout of transistors and passive elements is fixed but the wiring is specialized so that a customer's requirements can be satisfied. A gate array is based upon a master-slice system that employs a substrate already prepared except for the final interconnection process, called a "master". The advantages of a gate array are its short turnaround time and its short design time. Another advantage is its good reproducibility of LSI performance, since masters which consist of only uniform FETs can be selected and stocked. This means the most effective utilization of present GaAs fabrication technology.

Three kinds of custom LSI approaches will now be compared from the viewpoint of development time and cost, design flexibility and operation speed, as shown in Fig. 58. A standard cell is similar to a gate array in terms of design method, because both are constructed by interfunctional cell routing. However, since all masks (process steps) are needed for the standard cell, it is considered to be closer to a full-custom IC. Although it has a relatively low speed, a gate array is suitable for small-scale production with variations. In this subsection, the current GaAs gate-array designs are described.

Here, a gate array composed of only simple gates such as NOR or NAND gates is called a simple gate array. A typical example of GaAs simple gate arrays is the 1K gate array reported by Ikawa *et al.* (1984) using a DCFL three-input NOR gate as a basic cell. The basic cell layout and corresponding circuit representation are shown in Fig. 59. It has 14 columns, and each column has 75 basic cells and 1050 gates in total. There are 13 interconnection tracks between the columns. This gate array showed an unloaded gate delay time of 100 psec and a loaded delay time of 350 psec under the conditions of $L = 3$ mm and fanout number $= 3$.

Recently, higher speeds of 42 psec unloaded and 215 psec loaded gate delays have been achieved in a 2K gate array by Toyoda *et al.* (1985). It has an organization similar to the aforementioned 1K gate array. Simple gate arrays have also been developed by many companies. A heterojunction bipolar-transistor gate array was also fabricated by Yuan (1982).

A second kind of gate array is called a macro-cell array. The main difference between the macro-cell array and a simple gate array is in its basic cell configuration. The basic cell in the former consists of more elements to form several kinds of function cells. Tektronix has developed a GaAs macro-cell array called the Quick Chip (Pengue *et al.*, 1984). Each cell in the chip contains 20 depletion mode FETs, 8 diodes and 12 resistors. A D (delayed) latch can be formed in just one cell, and a master–slave flip-flop in just two. Simple NOR and NAND gates are also included in the macro-cell library. The

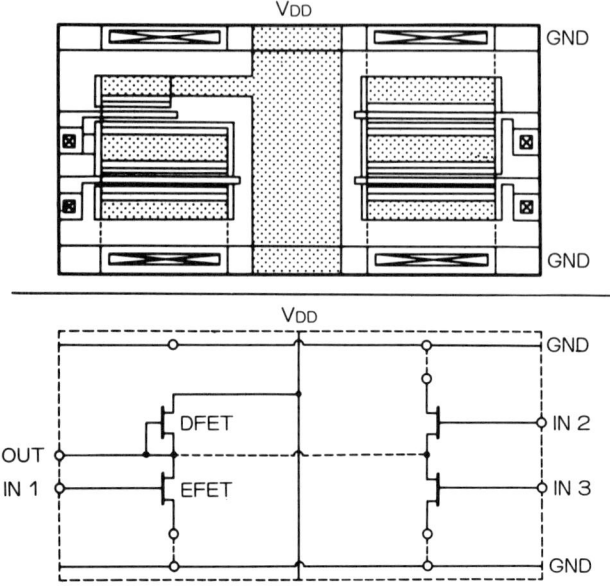

FIG. 59. Basic cell layout and corresponding circuit representation for a 1K-gate DCFL gate array.

array uses a BFL to build logic gates. The typical delay time for the BFL gate is 190 psec with F.I./F.O. (fanin/fanout) = 3/3 and an airbridge interconnect length of 0.5 mm. The macro-cell array features a speed and design turn-around time intermediate between that of a full-custom logic and that of a simple gate array.

18. FREQUENCY DIVIDER

A frequency divider is one of the most important digital circuits. As with ring oscillators, this circuit is often used to test new processes and circuit technology and to evaluate a figure of merit of the circuit at an early stage of development. This is because the figure of merit of an LSI can be estimated directly from the tested divider performance. In addition, the divider itself is widely applicable to such areas as phase-lock loop systems and synthesizers.

In this subsection, a description is given of circuit configurations for static frequency dividers, along with a comparison of their performance. Dynamic-type dividers, which feature the highest speed, are also discussed. Finally, the section will present a look at variable modulus divider circuits, which are applied to communication systems.

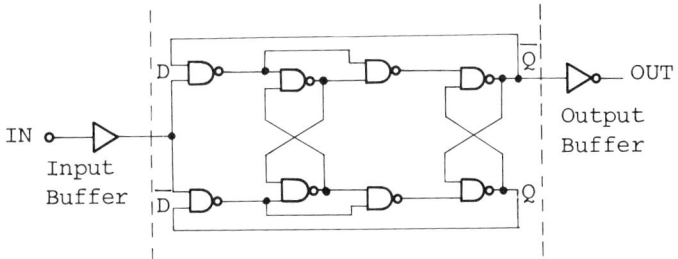

FIG. 60. Divide-by-two circuit using NOR gates.

a. Static Divider

The fundamental configuration of a divide-by-two static divider is shown in Fig. 60. This divider (T-type flip-flop, T-F/F) is constructed by connecting the Q ($\bar{\text{Q}}$) output to the $\bar{\text{D}}$ (D) input in a D-type master–slave flip-flop (D-F/F) and needs eight NOR-gates. The maximum operating frequency (toggling frequency, F_{tog}) in this circuit is determined by $1/(4\tau_{pd})$, where τ_{pd} is the propagation delay time of a NOR gate. Although all kinds of basic logic circuits can use this configuration, BFL, HSCFL, and LSCFL usually use different configurations to increase the F_{tog} using series gating techniques (NOR/NAND gates). In these configurations, F_{tog} is increased to $1/(2\tau_{pd})$. Figures 61 and 62 show such improved configurations for BFL (Liechti *et al.*, 1982) and LSCFL (Takada *et al.*, 1985a).

The toggle frequencies of state-of-the-art static dividers using GaAs MESFETs are plotted against power dissipation per first-stage T-F/F in Fig. 63. Also plotted for reference are the performances of dividers using

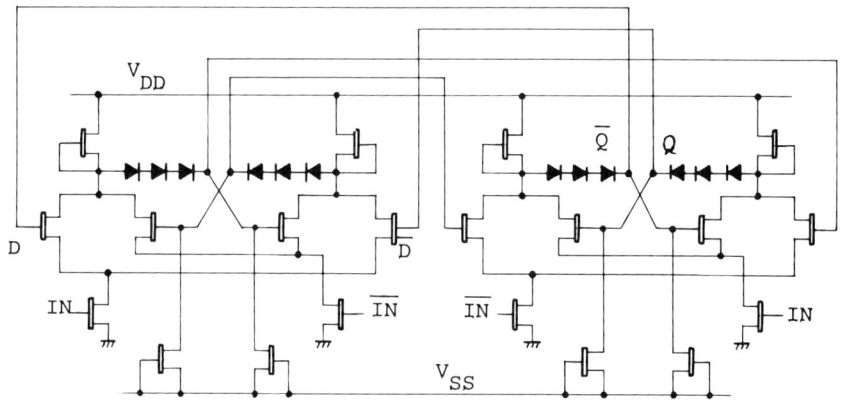

FIG. 61. T-type flip-flop configuration using series-gating BFL.

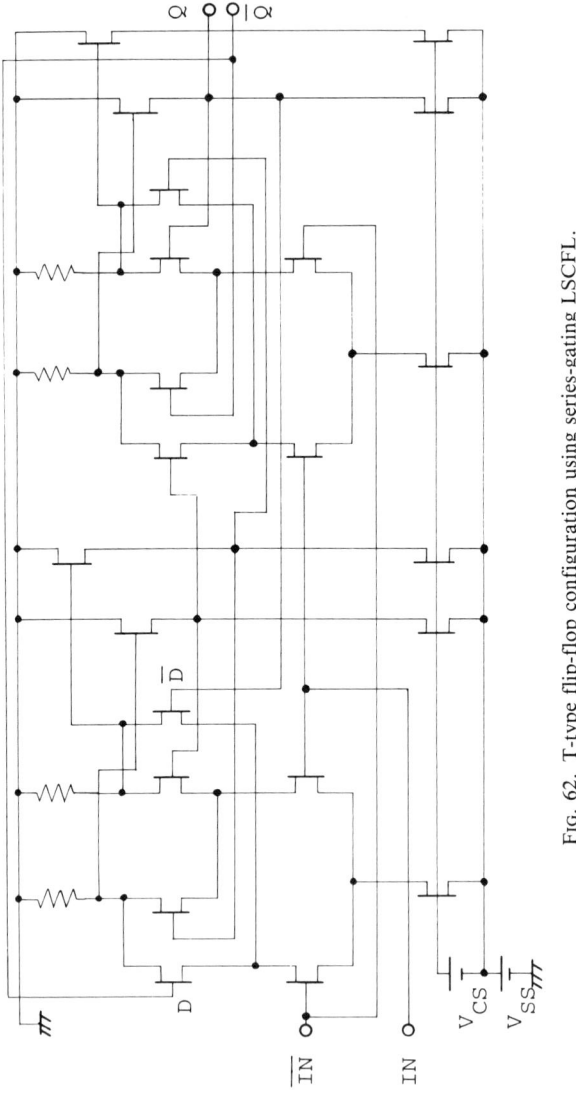

Fig. 62. T-type flip-flop configuration using series-gating LSCFL.

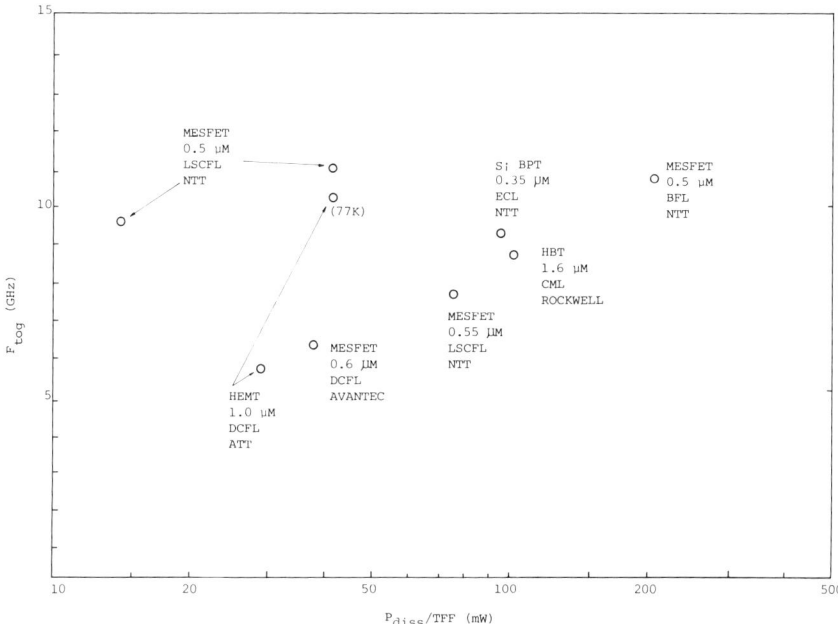

FIG. 63. Toggle frequency vs. dissipation power for state-of-the-art frequency dividers.

other devices, such as Si bipolar (Suzuki *et al.*, 1985) and GaAs–GaAlAs hetero-bipolar transistors (Asbeck *et al.*, 1984) and HEMTs (Pei *et al.*, 1984).

As shown in this figure, DCFL (Andrade and Anderson, 1985) and LSCFL provide low power dissipation, while BFL allows high speed at large power dissipation. The best figure of merit is achieved by the divide-by-four divider with LSCFL configuration (Takada *et al.*, 1986). This was fabricated using 0.5-μm-gate depletion-type FETs with V_T of -0.2 V and an airbridge interconnection line technique to obtain high speed, although LSCFL circuit normally uses enhancement-type FETs. The voltage waveforms at the input and the output are shown for low-power operation (9.7 GHz, 52 mW of total power) and for high-speed operation (11 GHz, 149 mW) in Figs. 64a and 64b.

b. Dynamic Divider

A dynamic divider using GaAs MESFETs was first proposed by M. Rocchi and B. Gabillard (1983). The configuration of this divider is shown in Fig. 65. A dynamic divider basically consists of an inverter and transfer gates. In the figure, BFL is implemented as an inverter. The F_{tog} is determined by $1/(2\tau_{pd})$ when true and complementary input signals are used.

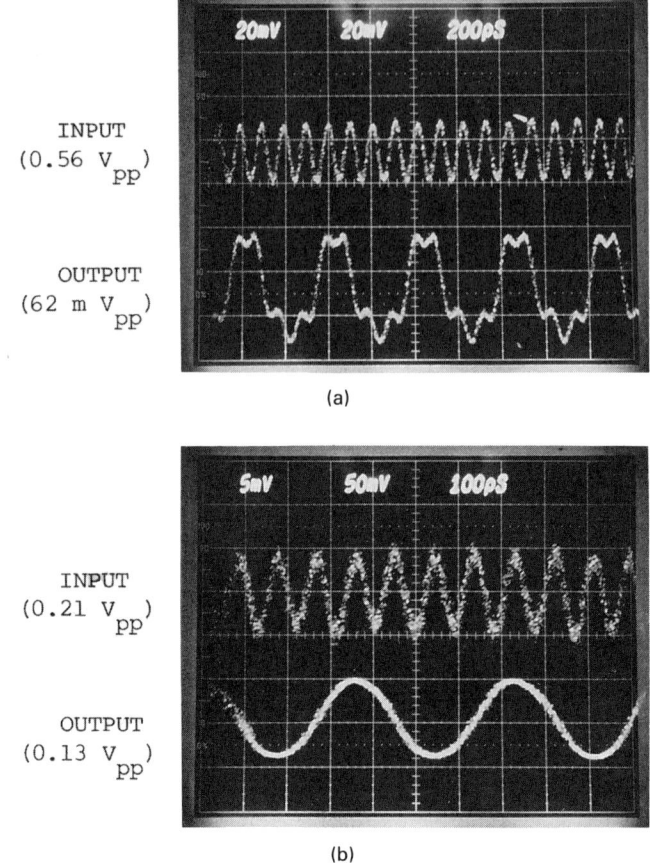

FIG. 64. Input and output waveforms in LSCFL 1/4 divider (a) at low power (9.7 GHz, 52 mW) and (b) for high-speed (11 GHz, 149 mW).

Since the fanout is smaller in the circuit shown in Fig. 65 than in circuits using the series-gating technique as described above, and since in addition, F_{tog} increases by adjusting phases of the two input signals, a higher F_{tog} can be obtained. F_{tog} values of 10.2 and 13.0 GHz were individually reported, both using 0.7 µm gate-length FETs, by Rocchi and Gabillard (1983) and Osafune et al. (1986), respectively.

However, this type of divider does not operate in the low frequency region (< 500 MHz) because of the discharge of the noninverting stage. Thus, applications of these dividers are limited to special uses requiring no DC operations, such as in stabilized frequency sources and RF receivers. In addition, since the implementation of dynamic circuits together with static

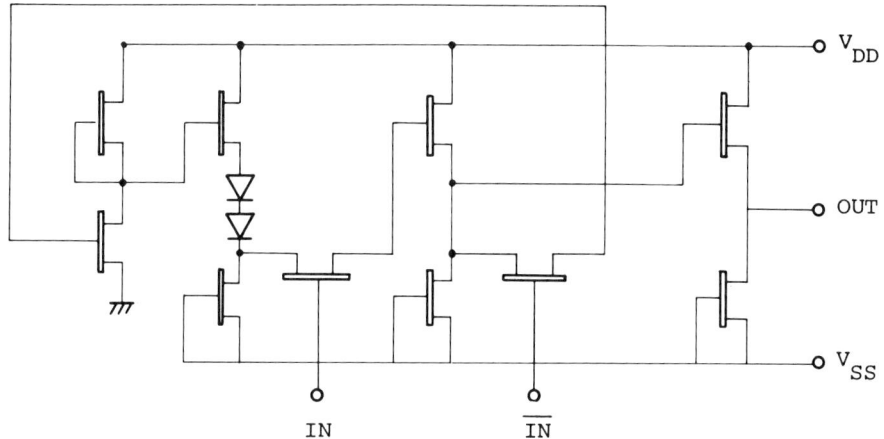

FIG. 65. Dynamic frequency divider configuration using BFL.

circuits in LSIs is not easy, the characteristics of a dynamic divider do not provide a direct indication to evaluate LSI performance.

c. *Modulus Divider*

Modulus dividers are finding numerous applications in prescalers for counters, frequency synthesizers, and digital phase-locked loops. In particular, frequency synthesizers operating with low current have recently become increasingly important for constructing battery-operated VHF band mobile communication systems. The modulus dividers are the key circuits in these systems. Since these dividers usually require low current characteristics, circuit configurations that realize low current operation, such as DCFL, LSCFL, and SDFL, are employed.

Figure 66 shows a block diagram of the LSCFL 128/129 dual modulus divider (Takada *et al.*, 1985c), which was developed for use in the synthesizers of automobile telephone systems. As is clear, the divider consists of NOR gates, D-type flip-flops (D-F/F), T-F/Fs, and an OR gate. When the mode control terminal (MC) is maintained at a high level or a low level, divide-by-128 or divide-by-129 is selected. The NOR1 and NOR2 gates in Fig. 66 are designed to be included into the D-F/F1 using five-level series gating, as shown in Fig. 67. This was designed to decrease current and to increase speed. The dissipation current of the divider at 1 GHz is as low as 5 mA, which is six times less than that for the commercially available Si bipolar dividers (Akazawa *et al.*, 1983) having the same functions. Figure 68 shows a microphotograph of the developed GaAs 128/129 dual modulus divider.

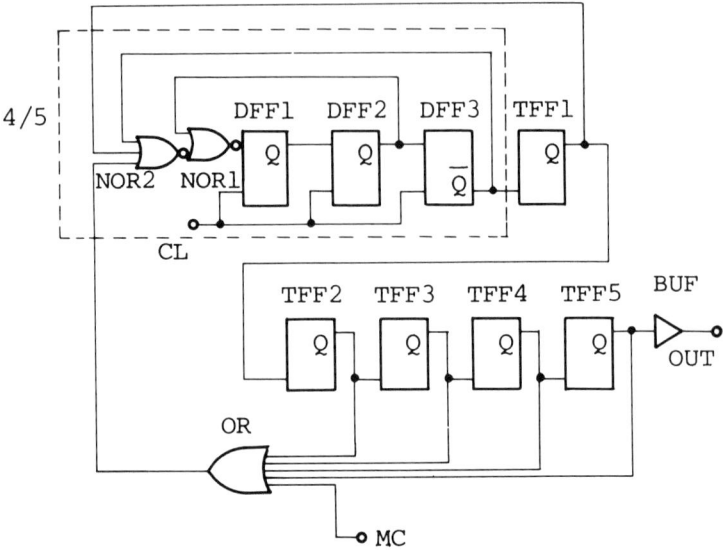

FIG. 66. Block diagram of GaAs 128/129 dual modulus frequency divider.

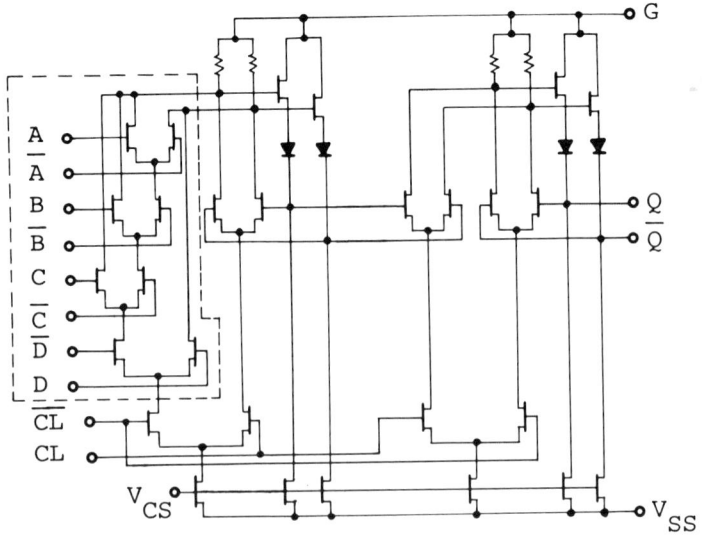

FIG. 67. First stage D-type flip-flop configuration using five-level series-gating LSCFL.

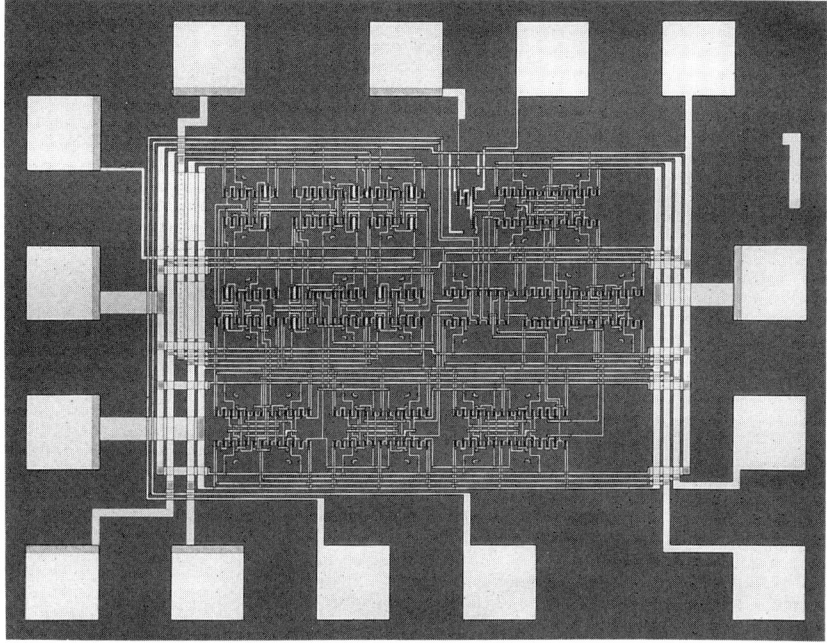

FIG. 68. Microphotograph of GaAs 128/129 dual modulus frequency divider. Chip size is 1.0 mm × 1.3 mm.

VII. Testing Technology

19. HIGH-FREQUENCY PROBE CARD

IC performance is initially tested under the "on-wafer" condition using probe cards. Functional testing is generally carried out only at low frequency (<100 MHz) because ordinary measurement systems using probe cards do not have a sufficient electrical capability for measuring high-frequency performance. After being functionally tested in a wafer form, ICs are divided into individual chips. Those ICs that have passed the functional tests are put into packages or chip carriers. Finally, high-frequency tests are performed with a measurement setup that is carefully designed to provide high-frequency performance.

It consumes considerable time and, therefore, increases the cost to test the functions and the high-frequency performance by using different setups. In order to solve this problem, high-frequency "on-wafer" probe-card testing systems, which have recently been developed, can carry out not only functional tests but also high-frequency tests.

FIG. 69. Coaxial-type high-frequency on-wafer probe card.

A photograph of a coaxial-type probe card is shown in Fig. 69. This probe card consists of three types of probes, as shown in Fig. 70. Structure (a) is used as an input signal probe, in which the center pins and ground metal of the two cables are electrically connected at the cable edges (coupled coaxial probe, CCP) (Hirayama et al., 1985). The first and second coaxial cables are used for the input signal and an input signal monitor, respectively. The second cable is connected to the input terminal of an oscilloscope whose input impedance is 50 Ω. Through this structure, both monitoring and the input signal termination can be suitably performed. A ground probe is also located adjacent to the CCP, to provide higher frequency capability. Since the probe-wire lengths of these probes are minimized as much as possible (<2.0 mm), small input wave reflections are assured. The insertion loss from port A to port B is shown as a function of frequency in Fig. 71. A loss of less than 1.0 dB, which is sufficiently small for digital pulse measurements, is attained at frequencies less than 7.5 GHz.

The coaxial cable is also used to transmit the output signal. Moreover, the ground probe is located in the vicinity of this coaxial cable as well.

The bare wires shown as (c) in Fig. 70 are used for DC bias probes and signal probes, which require no high-frequency performance. These probes are electrically connected to the ground using chip capacitors to ensure a small inductance of the voltage supplies.

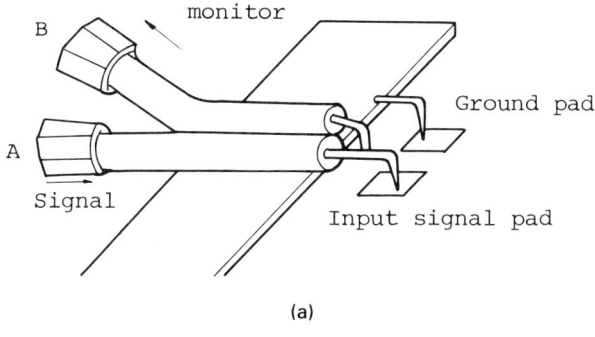

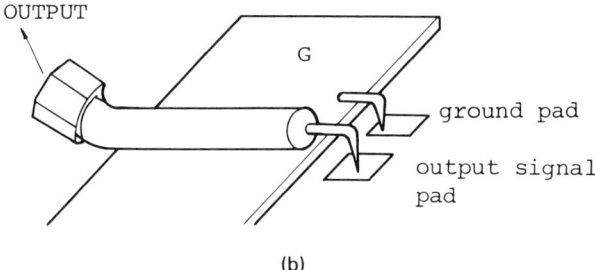

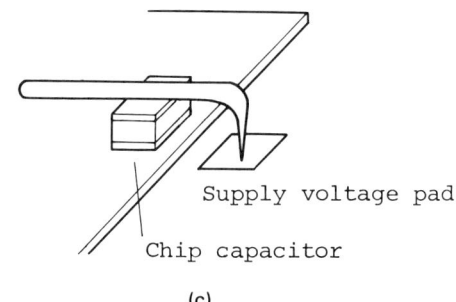

FIG. 70. Cut-out views of test probes used in on-wafer probe-card: (a) is the input signal probe (coupled coaxial probe, CCP), (b) is the output signal probe, and (c) is the DC probe.

The other type of probe, which has been developed by Strid and Gleason (1982), utilizes a structure consisting of coplanar lines on an alumina substrate. This probe provides higher frequency performance (DC to 12.8 GHz) than that of the coaxial-type probe card. However, it is difficult to fabricate large numbers of probes on one card. This type of probe is,

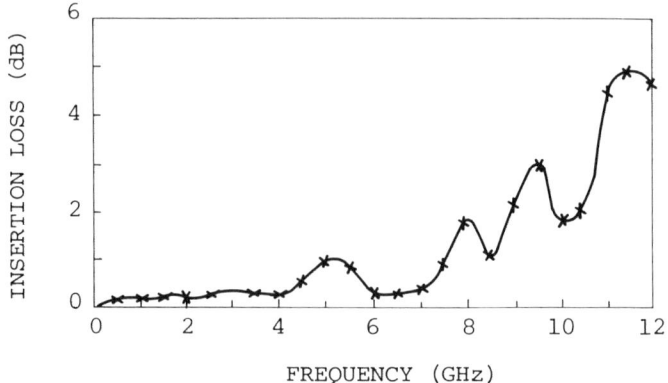

FIG. 71. Insertion loss in coupled coaxial cable as a function of frequency.

therefore, suitable for use in analog IC evaluations such as S parameter measurement, rather than for digital IC measurements.

20. PACKAGE

Assembling bare chips on a circuit board is preferable for obtaining the highest possible circuit speed. However, the enclosure of IC chips must satisfy the requirement for easy handling as well as for hermetic sealing to ensure reliability.

Dual in-line packages and flat packages are commonly used for Si integrated circuits. For GaAs high-speed ICs such as standard logic SSI families, specially designed flat packages or chip carriers are used (Wilson et al., 1982; Gheeuralo et al., 1985). In this section, the design of packages for high-speed IC will be described in detail. The typical structure of flat packages and an equivalent circuit are shown in Figs. 72 and 73. To obtain high frequency performance, the following five requirements are essential.

(1) The length of metal leads (ℓ_1) must be as short as possible.
(2) The characteristic impedances of metal leads (Z_{01}, Z_{02}) must be nearly equal to 50 Ω.
(3) The space/width (S/W) of signal lines must be large enough to reduce crosstalk (less than 20 dB).
(4) The package cavity size must be nearly equal to the chip size for minimizing the bonding wire length ($\sim \ell_{01}$).
(5) Package dielectric and conductor losses must be as small as possible.

Figure 74 shows simulation results of pulse waveforms at an input terminal of a GaAs MESFET IC chip (B point in Fig. 73) as a function of ℓ_1. Z_{11} and

4. GaAs LSI CIRCUIT DESIGN

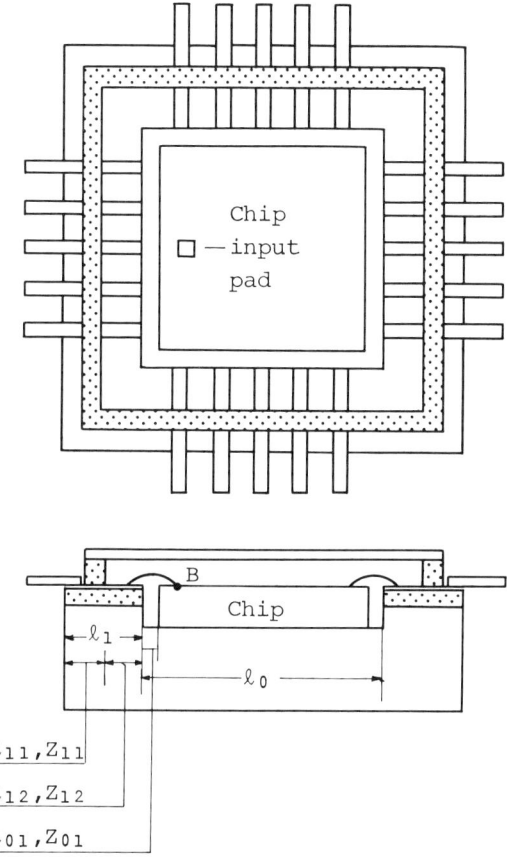

FIG. 72. Typical structure of flat package.

Z_{12} are assumed to be both 50 Ω, and the bonding-wire length (ℓ_{01}) is ignored assuming that the cavity size is ideally selected. The internal impedance of the signal source is assumed to be 10 Ω, and the external signal cable length (ℓ_2) is 0 mm for a simplified calculation. An ideal 50 Ω terminator is connected in front of the package (A point). Rise and fall times of 50 psec (0–100%) for an input signal pulse are selected considering a logic operation of up to about 10 GHz.

It is found that ringing increases as ℓ_1 increases. This ringing possibly causes erroneous switching when it overshoots the logic threshold voltage of an IC. If the IC response time is slow enough, this erroneous switching does not occur even at a large ringing magnitude. However, consideration must be given to the ringing noise, especially for high-speed circuits such as GaAs

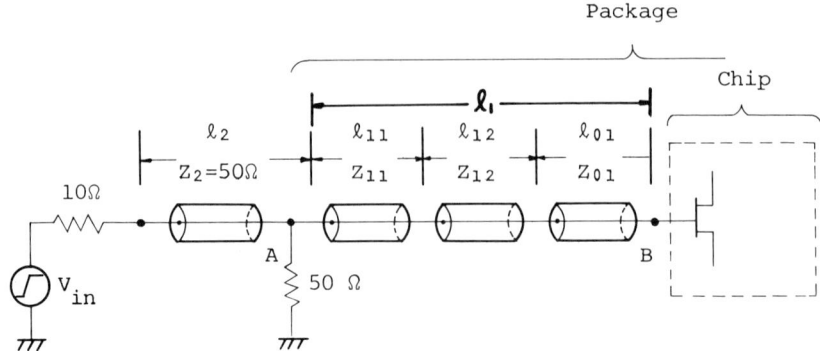

FIG. 73. Equivalent circuit of flat package. ℓ_{01}, ℓ_{12}, and ℓ_{11} are lengths of package sections, and Z_{01}, Z_{12}, and Z_{11} are their characteristic impedances. ℓ_2 and Z_2 are the length and characteristic impedance of the external input signal cable.

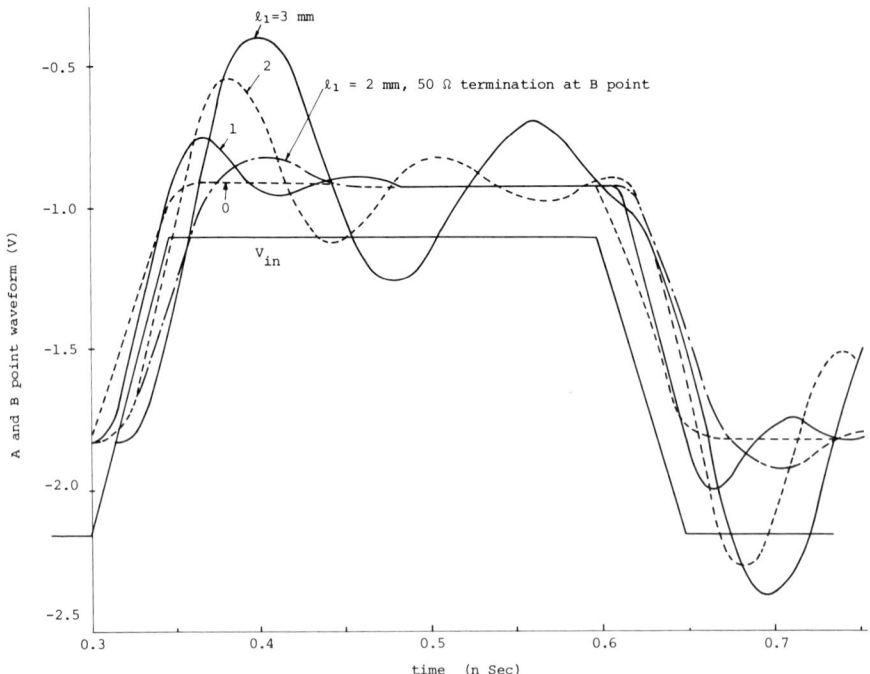

FIG. 74. Pulse waveforms at A and B points in the circuit shown in Fig. 73 with ℓ_1 as a parameter.

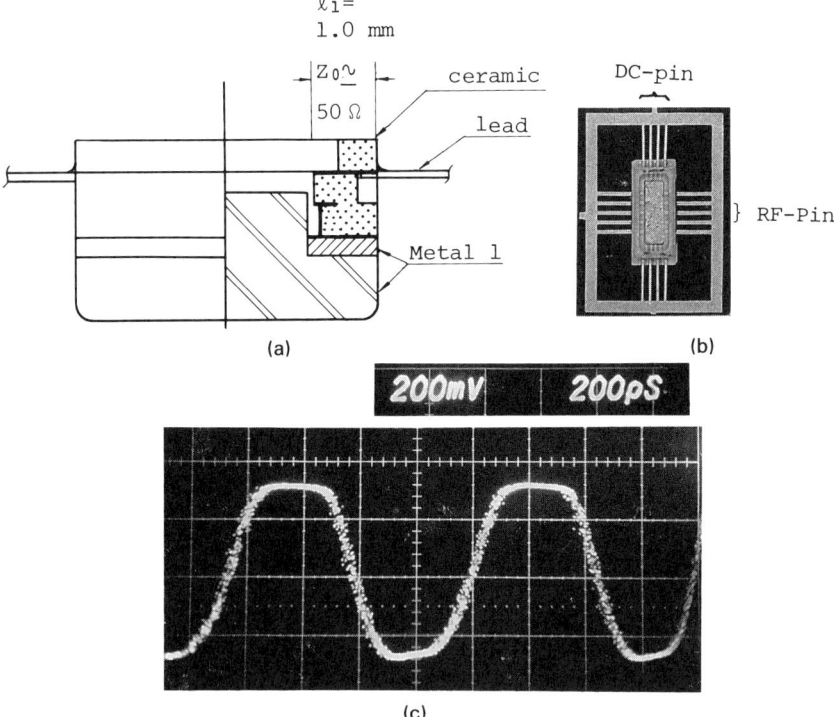

FIG. 75. (a) Structure, (b) microphotograph of the high-speed IC package, and (c) output waveform from the packaged IC.

ICs, since erroneous switching can easily occur. The best method for decreasing the ringing is to make 50 Ω termination possible at the input pad of the IC chip (B point).

The pulse waveform for this case is also shown in Fig. 74 for ℓ_1 of 2 mm, where ringing is very small. However, termination in IC chips is not suitable when multi-fanout circuits are constructed with these IC chips. In this case, only a single termination is required among all of the IC input terminals. As a result, internal terminations of ICs are generally avoided. It is found from Fig. 74 that an ℓ_1 value of less than 1 mm is necesary to obtain ringing with less than a 20% logic voltage-swing.

The structure and microphotograph of a package that was developed on the basis of this design concept are shown in Fig. 75a and 75b (Takada and Ida, 1983). The DC pins and RF pins are located separately, and the characteristic impedance of the ℓ_1 region is designed to be 50 Ω. This high-speed IC package has been successfully employed for a GaAs decision IC for use in

1.6 Gbit/sec optical repeaters (Ohta and Takada, 1983). The output waveform of the IC packaged in the developed package is shown in Fig. 75c, in which a clean output waveform without ringing is obtained.

VIII. Summary

The design of digital GaAs LSIs has been discussed in this chapter. This design includes device analysis and modeling, the actual LSI design and various testing methods. Furthermore, emphasis has been placed on the circuit design. Use of CAD programs for pattern layout (e.g., automated routing) was not mentioned here, because the CAD technique for Si LSIs basically can be applied to GaAs LSIs as well. It is our hope that this chapter helps readers to better understood the GaAs LSI design process.

REFERENCES

Akazawa, Y., Kikuchi, H., Iwata, A., Matsuura, T., and Takahashi, T. (1983). *IEEE J. of Solid-state Circuits*, **SC-18**, 115.

Andrade, T., and Anderson, J. R. (1985). *IEEE Electron Device Letters* **ED1-6**, 83.

Asai, K., Ino, M., Kurumada, K., Kawasaki, Y., and Ohmori, M. (1981). *Int. Symp. GaAs and Related Compounds*, p. 533.

Asbeck, P. M., Miller, D. L., Anderson, R. J., Deming, Chen, R. N., Liechti, C. A., and Eisen, F. H. (1984). GaAs *IC Symp.*, p. 133.

Bert, G., Morin, J. P., Nuzillat, G., and Arnodo, C. (1982). *IEEE Trans. Microwave Theory and Techniques* **MTT-30**, 1014.

Eden, R. C. (1981). U.S. Patent 4,300,064 (1981): Schottky Diode FET Logic.

Eden, R. C. (1984). GaAs *IC Symp.*, p. 11.

Gheeuralo, Wala, and Thsharr (1985). *Microelectron. J.* **16**(2), 30.

Gibbons, J. F., Johnson, W. S., and Mylroie, S. W. (1975). *In* "Projected Range Statistics," Semiconductors and Related Materials, 2nd Ed., Halstead Press.

Grey, P. E. (1963). *In* "Physical Electronics and Circuit Models of Transistors." Wiley, New York.

Higashisaka, A., and Hasegawa, F. (1980). *Electron. Lett.* **16**, 412.

Hirayama, M., Ino, M., Matsuoka, Y., and Suzuki, M. (1984). **ISSCC**, 46.

Hirayama, M., Takada, T., and Osafune, K. (1985). *National Convention Record on IECE Japan*, p. 393.

Hockney, R. W., Warriner, R. A., and Reiser, M. (1974). *Electron. Lett.* **10**, 484.

Ikawa, Y., Toyoda, N., Mochisuki, M., Terada, T., Kanazawa, K., Hirose, M., Mizoguchi, T., and Hojo, A. (1984). **ISSCC**, 40.

Ino, M., and Ohmori, M. (1980). *IEEE Trans. Microwave Theory and Techniques* **MTT-28**, 456.

Ino, M., Kurumada, K., and Ohmori, M. (1981). *IEEE Electron Dev. Lett.* **EDL-2**, 144.

Ino, M., Hirayama, M., Kurumada, K., and Ohmori, M. (1982a). *IEEE Trans. Electron Devices* **ED-29**, 1130.

Ino, M., Hirayama, M., Ohwada, K., and Kurumada, K. (1982b). GaAs *IC Symp.*, p. 2.

Ino, M., Togashi, M., Hirayama, M., Kurumada, K., and Ohmori, M. (1984). *IEEE Trans. Electron Devices* **ED-31**, 1139.

Ino, M., Takada, T., Sutoh, H., Kato, N., and Ida, M. (1986). SSDM, 371.
Ishii, Y., Ino, M., Idda, M., Hirayama, M., and Ohmori, M. (1984). GaAs *IC Symp.*, p. 121.
Lee, C. P., Zucca, R., and Welch, B. M. (1981). GaAs *IC Symp.*, p. 13.
Lee, F. S., Kaelin, G. R., Welch, B. M., Zucca, R., Shen, E., Asbeck, P., Lee, C. P., Kirkpatrick, C. G., Long, S. I., and Eden, R. C. (1982). *IEEE J. Solid State Circuits* **SC-17,** 638.
Lee, S. J., Vahrenkam, R. P., Kaelin, G. R., Hou, L. D., Zucca, H. R., and Kirkpatrick, C. G. (1983). GaAs *IC Symp.*, p. 74.
Lehovec, K., and Zuleeg, R. (1970). *Solid-State Electron* **13,** 1415.
Liechti, C. A., Baldwin, G. L., Gowen, E., Joly, R., Namjoo, M., and Rodell, A. F. (1982). *IEEE Trans. Electron Devices* **ED-29,** 1094.
Matsui, N., and Moriya, K. (1983). *National Convention Record on IECE, Japan*, p. S4-2.
Matsuoka, Y., Yamane, Y., and Hirayama, M. (1983). Presented at *Japan Applied Physics Society Ann. Meet.*
Nakayama, Y., Suyama, K., Shimizu, H., Yokoyama, N., Shibatomi, A., and Ishikawa, H. (1983). *ISSCC*, p. 48.
Namordi, M. R., and White, W. A. (1982). *IEEE Electron Device Lett.* **EDL-3,** 264.
Nishiuchi, K., Kobayashi, N., Kuroda, S., Notomi, S., Mimura, T., Abe, M., and Kobayashi, M. (1984). *ISSCC*, p. 48.
Ohta, N., and Takada, T. (1983). *Electron. Lett.* **19,** 983.
Ohwada, K., Ino, M., Mizutani, T., and Asai, K. (1982). *Electron. Lett.* **18,** 299.
Osafune, K., Ohwada, K., and Sugeta, T. (1986). *ISSCC*, 198.
Pei, S. S., Shah, N. J., Hendel, R. H., Tu, C. W., and Dingle, R. (1984). GaAs *IC Symp.*, p. 129.
Pengue, L., McCormack, G., Strid, E., Smith, D., and Bowman, T. (1984). GaAs *IC Symp.*, p. 27.
Reiser, M. (1973). *IEEE Trans. Electron Devices* **ED-20,** 35.
Rocchi, M., and Gabillard, B. (1983). *IEEE J. Solid State Circuits* **SC-18,** 3.
Schneider, M. V. (1969). *Bell. Syst. Tech. J.* **48,** 1421.
Seki, S., and Hasegawa, H. (1982). GaAs *IC Symp.*, p. 119.
Strid, E. W., and Gleason, K. R. (1982). *IEEE Trans. Electron Devices* **ED-29,** 1065.
Suzuki, M., Hagimoto, K., Ichino, H., and Konaka, S. (1985). *IEEE Electron Device Lett.* **EDL-6,** 181.
Takada, T., and Ida, M. (1983). *National Convention Record on Semiconductors and Materials of IECE Japan*, p. 219.
Takada, T., Idda, M., and Sudo, T. (1981). *National Convention Record on Semiconductors and Materials of IECE Japan*, p. 123.
Takada, T., Yokoyama, K., Ida, M., and Sudo, T. (1982). *IEEE Trans. Microwave Theory and Techniques* **MTT-30,** 719.
Takada, T., Togashi, M., and Hirota, T. (1984a). *Paper Tech. Group on Solid State Devices, IECE, Japan* **SSD83,** 124.
Takada, T., Togashi, M., Kato, N., and Idda, M. (1984b). *Extended Abstracts of the 16th (1984 International) Conference on Solid State Devices and Materials, Kobe, Japan*, p. 403.
Takada, T., Togashi, M., Ida, M., Yamasaki, K., and Hoshikawa, K. (1985a). *Paper Tech. Group on Solid State Devices, IECE, Japan* **SSD84-116,** 97.
Takada, T., Shimazu, Y., Yamasaki, Y., Togashi, M., Hoshikawa, K., and Idda, M. (1985b). *Proc. IEEE Microwave and Millimeter-Wave Monolithic Circuits Symp.*, p. 22.
Takada, T., Saito, S., Kato, N., and Idda, M. (1985c). *Electron. Lett.* **21,** 731.
Takada, T., Kato, N., and Ida, M. (1986). *IEEE Electron Device Lett.*, **7,** p. 47.
Takahashi, K., Maeda, T., Katano, F., Furutsuka, T., and Higashisaka, A. (1985). *ISSCC*, p. 68.

Togashi, M., Takada, T., and Ida, M. (1984). *National Convention Record on Communications of IECE Japan*, p. 1-80.
Tomizawa, M., Yoshii, A., and Yokoyama, K. (1983). *Trans. IECE Japan* **C-114,** 1072.
Toyoda, N., Uchitomi, N., Kitaura, Y., Mochizuki, M., Kanazawa, K., Terada, T., Ikawa, Y., and Hojo, A. (1985). *ISSCC,* p. 206.
Van Tuyl, R. L., Liechti, C. A., Lee, R. E., Gowen, E. (1977). *IEEE J. Solid State Circuits* **SC-12,** 485.
Willing, H. A., Rauscher, C., and Santis, P. (1978). *IEEE Trans. Microwave Theory Tech.* **MTT-26,** 1017.
Wilson, D., Frick, N., Kwiat, J., Lo, S., Churchill, J., and Barrera, J. (1982). GaAs *IC Symp.*, p. 13.
Yamaguchi, K., Asai, S., and Kodera, H. (1986a). *IEEE Trans. Electron Devices* **ED-23,** 1283.
Yamagishi, K., and Kodera, H. (1976b). *IEEE Trans. Electron Devices* **ED-23,** 545.
Yamamoto, R., Higashisaka, A., Asai, S., Tsuji, T., Takayama, Y., and Yano, S. (1983). *ISSCC,* p. 40.
Yamasaki, K., Asai, K., and Kurumada, K. (1982). *IEEE Trans. Electron Devices* **ED-19,** 1972.
Yamasaki, K., Kato, N., and Hirayama, M. (1984). *Electron. Lett.* **20,** 1029.
Yokoyama, K., Tomizawa, M., Takada, T., and Yoshii, A. (1983). *IEEE Trans. Electron Devices* **ED-30,** 719.
Yuan, H. (1982). GaAs *IC Symp.*, p. 100.
Yuan, H. T., Lin, Y., and Chiang, S. (1982). *IEEE Trans. Electron Devices* **ED-29,** 4, 639.

CHAPTER 5

GaAs LSI Fabrication and Performance

M. Hirayama

NTT LSI LABORATORIES
3-1, MORINOSATO WAKAMIYA, ATSUGI-SHI
KANAGAWA 243-01, JAPAN

M. Ohmori

ELECTRONIC MATERIALS AND COMPONENTS LABORATORIES
NIPPON MINING COMPANY
TODA, SAITAMA 335, JAPAN

K. Yamasaki

NTT RESEARCH AND DEVELOPMENT HEADQUARTERS
YAMATO SEIMEI BLDNG. UCHISAIWAICHO, CHIYODA-KU
TOKYO 100, JAPAN

   I. INTRODUCTION 234
      1. *History of GaAs LSI Development* 234
      2. *Heterojunction Devices* 237
      3. *GaAs IC Industrial Growth* 237
  II. CHARACTERIZATION OF SEMI-INSULATING GaAs WAFERS FOR
     IC APPLICATION 238
      4. *GaAs Wafers for IC Application* 238
      5. *Wafer Selection* 238
      6. *Substrate Evaluation Using MESFET* 242
 III. TEG SELECTION FOR LSI FABRICATION PROCESS AND CHIP/
     WAFER LAYOUT 245
      7. *Fabrication Process TEG for LSIs* 245
      8. *Chip/Wafer Layout* 249
  IV. FET FABRICATION AND PERFORMANCE FOR LSIs 251
      9. *MESFET Structures for LSIs* 251
     10. *Active Layer Formation* 253
     11. *Gate Formation* 255
     12. *Source/Drain Formation* 260
     13. *Device Isolation* 262
     14. *Fabrication and Performance of Self-Aligned FET* . . . 263
     15. *Submicron Gate Length FETs* 272
   V. CIRCUIT ELEMENTS AND INTERCONNECTION 276
     16. *Diodes, Resistors, and Capacitors* 276
     17. *Interconnection* 279

VI.	ASSEMBLY TECHNIQUES AND PACKAGING	283
	18. Dicing and Bonding	283
	19. Package	284
VII.	LSI FABRICATION AND PERFORMANCE	287
	20. Uniformity and Controllability of Threshold Voltage	287
	21. Memory LSIs	293
	22. Logic LSIs	296
	23. Preliminary Yield and Reliability	298
VIII.	SUMMARY AND FUTURE TRENDS	300
	REFERENCES	301

I. Introduction

Since the first GaAs digital IC was reported by Van Tuyl and Liechti (1974), who demonstrated the potential for a very high speed propagation delay-time per gate—below 100 psec—to be achieved, twelve years have passed. In recent years, GaAs metal semiconductor field effect transistors (MESFET) digital large-scale integration (LSI) has reached a maximum integration scale of 16-kbit static RAM, which consists of about 102,000 MESFETs (Ohmori, 1984; Ishii et al., 1984a), and a minimum propagation delay time per gate below 10 psec (Yamasaki et al., 1984). This technology has resulted in factory production of commercial SSI and MSI circuits, in the USA and later in Europe and Japan. The rapid progress of LSI technology has been due to improvements in material growth technique, as well as in processing technology and circuit design.

In this chapter, technologies leading to the present state of GaAs digital LSI will be reviewed and discussed, mainly focusing on materials, fabrication processes and LSI performance.

1. HISTORY OF GaAs LSI DEVELOPMENT

GaAs LSI development can be divided into four generations, according to characteristic technologies developed as shown in Fig. 1 (Ohmori, 1984). The first generation is from 1974 to 1977. GaAs MESFET IC complexity increased from the elementary stage of logic gates to SSIs and MSIs. Circuits were realized using epitaxial layers with mesa device isolation. The next generation is from 1978 to 1980. The fabrication process was greatly improved using selective ion implantation in a semi-insulating substrate. The planar and uniform process developments resulted in MSIs and the first GaAs LSIs. The third generation, 1980 to 1984, is referred to as the LSI generation. Direct coupled FET logic (DCFL) with low power dissipation was used, and several kinds of LSI-directed self-aligned FET structures were developed. The fourth generation started in 1985 with the attainment of GaAs IC commercial production and has been continued along with research

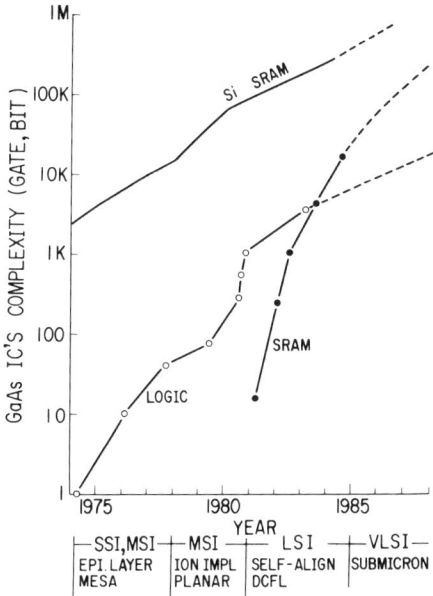

FIG. 1. GaAs IC complexity vs. year. [From Ohmori (1984).]

and development of submicron gate and VLSI technologies. A more detailed description of this development is given in the following.

In 1974, Van Tuyl and C. Liechti of Hewlett Packard (HP) reported the first GaAs FET logic of buffered FET logic (BFL). It consisted of 1 μm gate-length MESFETs and Schottky barrier diodes fabricated on an n-type liquid phase epitaxial layer on a Cr-doped semi-insulating substrate and showed a minimum propagation delay time as fast as 60 psec. HP researchers then increased IC complexity to flip-flops, frequency dividers and multiplexers and reached about 60-gate 8-bit multiplexer/data generators in 1977 (Van Tuyl et al., 1977). The divide-by-two frequency dividers operated up to 4.5 GHz. The delay time of a BFL NOR gate was 86 psec; however, power dissipation per gate was as high as 40 mW.

In 1977, Fujitsu reported greatly reduced power dissipation of 264 μW/gate with a delay time of 280 psec, by applying normally-off mode logic, DCFL (Ishikawa et al., 1977), which had been used in GaAs junction FET (JFET) logic circuits by McDonnell-Douglas (Nottoff et al., 1975).

The second generation was initiated and led by Rockwell researchers who developed a planar and uniform process utilizing localized ion implantation into a semi-insulating substrate (Welch and Eden, 1977). n-type channel and n^+ low resistance layers for FETs or diodes were formed by Se, S or, later,

Si ion implantations. The threshold voltage standard deviation obtained was 94 mV over a 14 mm square area, which was a remarkably improved value at that time. They also invented the Schottky diode FET logic (SDFL), in which a buffered FET logic (BFL) input circuit was replaced from FET to diode. In addition, dimensions of MESFETs and diodes for the level shift circuit were reduced in design, and power dissipation was reduced to near 1 mW/gate (Eden et al., 1978). Parallel multipliers of 3×3, 5×5, and 8×8 bits were fabricated using this technology in 1980 (Lee et al., 1980). The 8×8 bit parallel multiplier was the first LSI in GaAs; it contained 1008 gates and presented 5.25 nsec multiple time with an equivalent gate delay time of 150 psec and power dissipation of 0.6 to 2 mW (Lee et al., 1982). The speed was about four times, and power dissipation was one-fourth, that of corresponding silicon circuits at that time.

BFL gate delay time was reduced to 34 psec with 41 mW power dissipation using a gate length of 0.5 μm, which was delineated by electron beam direct writing by Hughes (Greiling et al., 1978). DCFL gate delay time was also dramatically reduced to 77 psec (977 μW) and 30 psec (1.9 mW) with 0.8 and 0.6 μm gate length, respectively, by applying electron beam direct writing by NTT (Mizutani et al., 1980).

In the third generation, LSI circuits using DCFL gate were developed predominantly in Japan. Features of the DCFL gate were its low power dissipation of less than 1 mW per gate and simple gate construction of E/D pair FETs. However, highly uniform material and process technology were required for enhancement-mode FETs with high transconductance. Several kinds of self-aligned MESFET structures were developed.

In 1983, Fujitsu fabricated 3000 gate 16×16 bit parallel multipliers (Nakayama et al., 1983) and 1-kbit static RAMs (Yokoyama et al., 1983) using refractory metal gate self-aligned MESFET. After reporting the first GaAs 16-bit static RAMs in 1981 (Asai et al., 1981), NTT fabricated a 1-kbit static RAM in 1982 (Ino et al., 1982), a 4-kbit static RAM (Hirayama et al., 1984), and the largest integration scale 102,000 device 16-kbit static RAM in 1984 (Ohmori, 1984; Ishii et al., 1984a). This was accomplished through application of tri-level resist mask self-aligned FET technology, namely self-aligned ion implantation for n^+-layer technology (SAINT) (Yamasaki et al., 1982a). From 1983 to 1984, other laboratories in Japan such as Oki, Hitachi, Toshiba, NEC, and Mitsubishi also fabricated DCFL gate arrays and static memories using their own self-aligned FET technologies.

As IC complexity increases, realization of complete-operation LSI chips largely depends on crystal uniformity and process yield. Through statistical analysis, deviation of the FET threshold voltage was found to be mainly caused by a crystal dislocation effect (Miyazawa et al., 1983) and gate length scattering (Ohmori, 1983). Accordingly, by applying an indium-doped dislocation-free LEC crystal and fine lithography with a 10-to-1 projection aligner, the threshold voltage standard deviation for SAINT FETs was dramatically decreased to 20 mV over a 2-inch wafer. Using this process technology, a 16-kbit static RAM with 4.1 nsec access time and a fully operational 4-kbit static RAM were realized as decribed in Section 21.

2. Heterojunction Devices

Other noteworthy developments in the third generation include a high electron mobility transistor (HEMT) in 1980 (Mimura et al., 1980) and fabrication of a heterojunction bipolar transistor (HBT) in 1982 (Yuan, 1982). Detailed descriptions are given in Vol. 30, Chapters 3 and 4.

The propagation delay time per gate in HEMT has been reduced from the first-reported 17.1 psec at 77K in 1981 (Mimura et al., 1981) to 5.8 psec at 77K in 1986 (Shah et al., 1986). HEMT IC has also been developed for a 4-kbit static RAM (Kuroda et al., 1984) and a 10 GHz frequency divider (Hendel et al., 1984).

Very recently, a MIS-like FET using an undoped AlGaAs/undoped GaAs two-dimensional electron gas active layer began to be investigated.

The HBT was theoretically analyzed by Kroemer in 1957, and fabricated after the development of thin epitaxial layer technology was developed in 1983. In 1982, a prototype IIL (integrated injection logic) gate array with an integrated gate number of 114 was fabricated through the application of an n-AlGaAs/p-GaAs/n-GaAs heterostructure made by molecular beam epitaxy. Delay time per gate has been reduced from 185 psec for ECL (emitter coupled logic) in 1983 to 29.3 psec for NTL (non-threshold logic) in 1984 (Asbeck et al., 1984a). The maximum operating toggle frequency for a HBT frequency divider is 8.6 GHz and the maximum integration complexity is 1K gate array.

3. GaAs IC Industrial Growth

The commercial production and sale of GaAs digital ICs began in 1984. The history of GaAs IC development entered into the fourth generation. Typically, standard SSIs and MSIs including gate arrays are now on the sales products. In a 1984 IEDM paper, Liechti noted that "approximately 50 companies in North America, 13 in Japan and 10 in Western Europe are now developing and/or producing GaAs ICs in small quantities". North American production of GaAs devices, according to the reference by Liechti, are projected to grow 45% annually from $90 million (in 1984) to $830 million (in 1990), and in 1990, the digital and analog/microwave IC segments are estimated to have equal market shares of approximately $280 million.

The GaAs IC production of the Western world including Japan is projected to reach $1,250 million in 1990.

Large projects for research and development of GaAs LSIs are now underway in the USA, Japan, and Western Europe. One of the DARPA projects involves utilizing a pilot line with a throughput capability of one hundred 3-inch wafers per week and fabricating static RAMs of 16-kbit to 64-kbit and gate arrays of 6K to 10K gates by 1986 (Iversen, 1983). Another recent project seeks to increase the current limit of the RISC computer by employing a GaAs. DARPA's target is to have a 10,000-gate 32-bit GaAs chip by 1988.

II. Characterization of Semi-insulating GaAs Wafers for IC Application

4. GaAs Wafers for IC Application

Rapid progress in GaAs IC integration complexity and performance has been primarily attributed to advances in direct ion implantation technology and in quality of bulk GaAs substrates. The ion implantation technique has a very high throughput for preparing uniform and reproducible active layers on a wafer. However, characteristics of ion-implanted layers, such as carrier concentration, carrier profile, and electron mobility, are strongly dependent on the nature of the semi-insulating bulk substrate used.

Necessary specifications of a semi-insulating GaAs substrate for ion implantation in LSI fabrication are (1) high resistivity larger than 10^7 Ω-cm; (2) no thermal conversion after activation annealing at near 800°C; (3) high activation efficiency over 70%; (4) high electron mobility over 3700 cm^2/V-cm at 10^{17} cm^{-3} carrier concentration; (5) high reproducibility with threshold voltage in a range of less than 100 mV for wafer-to-wafer and lot-to-lot; (6) high uniformity with threshold voltage standard deviation less than 50 mV over a wafer; and (7) round 3-inch diameter size. High uniformity was achieved in a crystal of very low crystal defects such as dislocations, pits, and slips in a wafer.

5. Wafer Selection

A typical procedure for selecting GaAs wafers as substrates for ion implantation is described in Table I. Investigation was made of GaAs bulk wafers; Cr-doped (0.45–2 wt ppm) horizontal Bridgeman (HB)-grown crystals (Mizutani et al., 1982) and lightly Cr-doped (0.01–0.4 wt ppm) and undoped liquid-encapsulated Czochralski (LEC)-grown crystals (Nanishi et al., 1981; Yamazaki et al., 1981).

TABLE I

GaAs Wafer Selection Procedure for LSIs

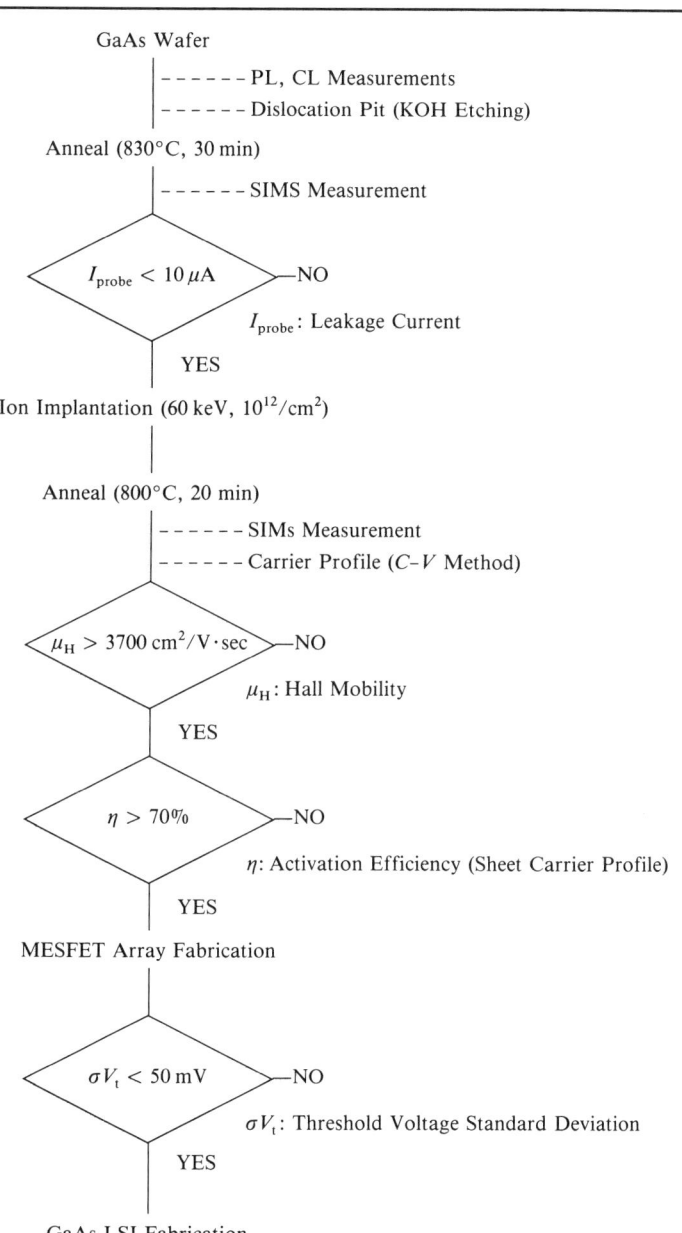

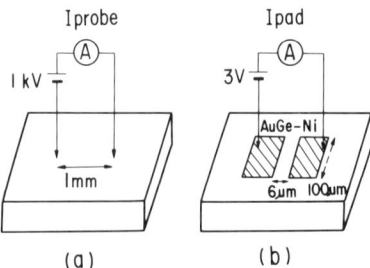

FIG. 2. Leakage current measurement methods: (a) with two probes separated by 1 mm (I_{probe}) and (b) with two pads, each of 100 μm width, separated by 6 μm (I_{pad}). [From Mizutani et al. (1982).]

First, thermal stability is to be examined. The wafer is encapsulated with 1500-Å-thick plasma-enhanced CVD SiN film and annealed at 800–900°C for 30–60 min (typically 830°C for 30 min) in flowing N_2 gas. The wafer is then evaluated by leakage current measurements employing two different methods shown in Figs. 2a and 2b. The two-probe method (a) measures leakage current (I_{probe}) under room light illumination at room temperature by applying 1 kV DC voltage between two tungsten probes separated by 1 mm. This method is very simple and convenient. The other method measures leakage current (I_{pad}) between two 100 μm-wide AuGe/Ni pads separated by 6 μm at 3 V applied voltage. The latter is devised to evaluate leakage current of MESFETs in an integrated circuit.

The I_{pad} versus I_{probe} relation obtained for HB-grown wafers annealed at 800°C for 30 min is shown in Fig. 3; a close correlation is found. In order to keep leakage currents from FETs in fabricated LSIs very low, I_{pad} must be less than 1 μA and the corresponding I_{probe} less than 10 μA. Thus, the selection rule for good substrate as to thermal stability is determined.

The thermal conversion—i.e., the conductive n-layer formation—at the interface is caused by Cr redistribution during annealing, because residual shallow donors cannot be compensated because of Cr depletion (Huber et al., 1979).

I_{probe}, leakage current, is distributed inversely to Cr concentration in an HB-grown ⟨100⟩ wafer, which is used as an LSI substrate. Since the crystallographic ⟨100⟩ plane is inclined by 54.7° from the growth direction ⟨111⟩, the ⟨100⟩ plane does not coincide with the solid–liquid interface during growth and Cr concentration is not uniform in a ⟨100⟩ wafer.

For an undoped semi-insulating LEC GaAs, I_{probe} variation corresponds inversely to dislocation density over a wafer (Miyazawa et al., 1982).

Wafers sliced from ingots that passed the leakage current test are ion-implanted with Si at 60 keV with doses of 10^{12}–10^{13} cm^{-2} and activation

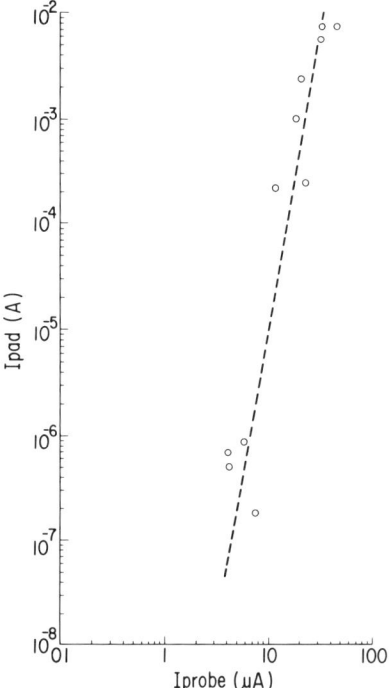

FIG. 3. I_{pad} and I_{probe} correlation for Cr-doped HB-grown GaAs. [From Mizutani et al. (1982).]

annealing is followed with a 1500-Å-thick plasma–CVD SiN cap at 800°C for 20 min in N_2. Hall measurements are carried out by the Van der Pauw method, and electron mobility (μ_H) and sheet carrier concentration (N_s) are measured; from the data the activation efficiency (η) is obtained.

N_s is well represented by the following equation in the low-dose range of 1–4×10^{12} cm^{-2}: $N_s = \eta\phi - N_{res}$, where ϕ is an ion dose and N_{res} is the residual acceptor. When lightly Cr-doped and undoped LEC crystals are used as substrates, a typical value of η is 60–70% and N_{res} is about 6×10^{11} cm^{-2} (Yamazaki et al., 1981).

μ_H versus Cr concentration in LEC wafers with N_s ranging 4.0 to 4.7×10^{11} cm^{-2} is shown in Fig. 4. μ_H increases from 3000 to 4600 cm^2/V-sec as Cr concentration is reduced from 0.5 to 0.06 wt ppm. Further Cr reduction decreases μ_H to 3600 cm^2/V-sec at the undoped level. The activation efficiency showed the same tendency. It is considered that point defect-related centers are acting as scattering centers and activation killer centers in very low Cr-doping or undoped LEC wafers. However, no clear conclusion

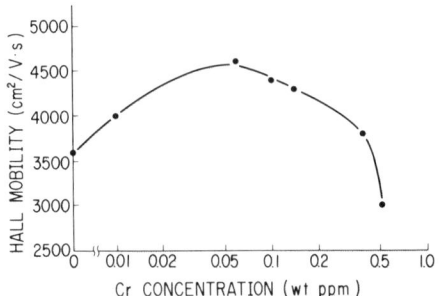

FIG. 4. Hall mobility dependence on Cr concentration in LEC GaAs implanted with Si ions. Sheet carrier concentration is $4.0-4.7 \times 10^{11}$ cm^{-2}.

to explain these decreases can be drawn from the experimental results (Nanishi *et al.*, 1981).

From these results, it was concluded that lightly Cr-doped (0.05–0.1 wt ppm) LEC crystals show high mobility and the highest activation efficiency. Therefore, they were used for LSI fabrication in the present authors' laboratory in the early stage of development from 1981 to 1983. Wafers cut from the front, middle, and tail section in an ingot are tested for the purpose of selecting GaAs ingots.

6. Substrate Evaluation Using MESFET

After wafer selection is finished, the uniformity of a bulk GaAs is directly evaluated by fabricating a MESFET array over a whole wafer using ion

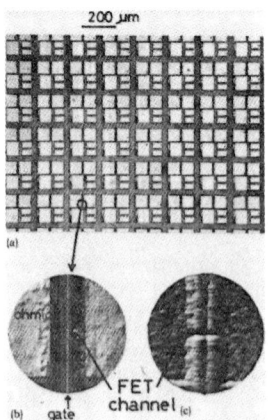

FIG. 5. (a) Photomicrograph of FET array fabricated on GaAs substrate; (b) an enlarged FET channel; (c) an etched FET. [From Miyazawa *et al.* (1983).]

Fig. 6. Ungated drain-to-source current distribution for Si ion–implanted Cr-doped LEC GaAs; darker area corresponds to higher current. Area is 1.4 cm × 1.8 cm. [From Nanishi et al. (1981).]

implantation. Automatic measurements of DC parameters of FETs are also performed using a computer-controlled on-wafer probing system (Ishii et al., 1984b). For macro-scale homogeneity characterization, MESFETs with 1 μm gate length, 20 μm gate width, and 6 μm source-to-drain distance are fabricated with intervals of 1.5 mm horizontally and 1.0 mm vertically. For micro-scale characterization, FETs with 1 μm gate length, 5 μm gate width and 6 μm source–drain distance were fabricated with 200 μm intervals as shown in Fig. 5 (Miyazawa et al., 1983).

Source and drain ohmic contacts are formed by alloying AuGe/Ni, and a Schottky gate is formed by evaporating Ti/Pt/Au. Distributions of ungated drain–source current (I_{ung}), gated drain–source current (I_{ds}), and threshold voltage (V_T) are measured over a wafer and displayed by maps in which the magnitudes are represented by the gray scale.

Typical maps of ungated drain–source current are shown in Figs. 6 and 7. In Fig. 6, 6300 small FETs covered a 1.4×1.8 cm^2 area near the central part of a 3-in. diameter Cr-doped (0.16 wt ppm) LEC GaAs, into which 1.4×10^{12} cm^{-2} Si was implanted at 60 keV. Several concentric circles centered at the upper left part of the figure can be distinguished. These circles coincide with the growth striations revealed through chemical etching. It is considered that strong micro-scale fluctuations in Cr concentration are

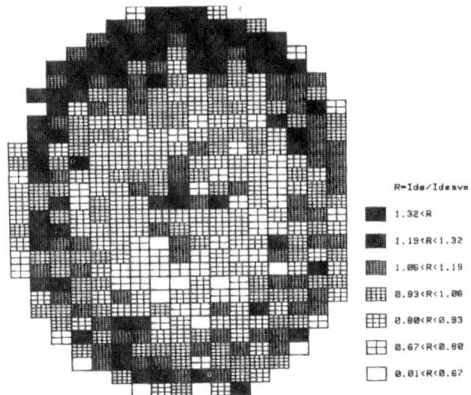

FIG. 7. Ungated drain-to-source current distribution for Si-ion–implanted undoped LEC GaAs; darker area corresponds to higher current. [From Nanishi et al. (1982).]

introduced by thermal fluctuation during growth in semi-insulating GaAs. This result suggests that much better micro-scale homogeneity could be expected in undoped semi-insulating GaAs.

Figure 7 shows an I_{ung} distribution measured in a 404-FET array fabricated on an entire ⟨100⟩ wafer cut from the directly-synthesized undoped semi-insulating LEC GaAs grown in a PBN crucible. The active layer was formed by implanting 2.0×10^{12} cm^{-2} Si ions at 60 keV. I_{ung} is high at the center and in the periphery, and low I_{ds} was observed in the middle concentric area. The I_{ung} distribution shows clear correlation with the W-shaped distribution of etch pit density (EPD) on the adjacent ⟨100⟩ wafer revealed by KOH etching at 450°C for 7 min (Fig. 1 in Nanishi et al., 1982). The observed W-shaped distribution is as high as 1×10^5 cm^{-2} at the center and in the periphery and as low as 1×10^4 cm^{-2} at the middle.

Accordingly, high I_{ung} and low I_{ung} correspond to high EPD and low EPD, respectively. In undoped or very low Cr-doped LEC GaAs, a dislocation distribution in a wafer predominantly determines uniformity of electrical properties. It is emphasized that elimination of such dislocations is very important to obtain uniformity of distribution of FET performance.

The influence of a dislocation pit on FET performance was more directly investigated. After V_T of each FET was measured in an array, the wafer with FETs was chemically etched with molten KOH at 450°C for a few minutes to reveal the dislocation pit. The measured V_T was plotted against the distance between the FET channel and a nearest neighbor dislocation pit for 60 FETs randomly, as shown in Fig. 8. The average V_T is -0.012 V and its standard deviation is 97 mV. It can be clearly seen that V_T values for a FET located

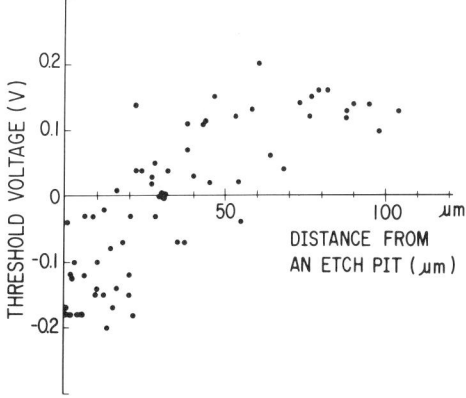

Fig. 8. FET threshold voltage vs. distance between FET channel and a nearest neighbor dislocation, measured randomly for 60 FETs along the ⟨110⟩ direction on a wafer. [From Miyazawa *et al.* (1983).]

within 20–30 μm from the nearest dislocation pit are lower as compared to those of FETs located more than 30 μm from the nearest dislocation (Miyazawa *et al.*, 1983). The difference between maximum and minimum V_T is about 300 mV. It can be seen that in a *denuded zone* within the critical distance of 20–30 μm from a *dislocation pit*, the activation efficiency of implanted Si ions is higher than that far from dislocation; this allows higher drain–source current, resulting in a shift of V_T to a lower value. The conclusion derived from the above discussion is that dislocation-free GaAs crystals have been strongly required to decrease the standard deviation of V_T and have been realized (Yamazaki *et al.*, 1984).

The threshold voltage uniformity of self-aligned MESFETs, SAINT, is tested to evaluate the feasibility of LSI fabrication processes. SAINT FETs are fabricated on the same wafer on which actual GaAs LSIs are fabricated. These results will be described in Section 20 as a test element group (TEG) evaluation.

III. TEG Selection for LSI Fabrication Process and Chip/Wafer Layout

7. FABRICATION PROCESS TEG FOR LSIs

One of the difficulties in developing an LSI fabrication process is the diagnosis of problems that arise especially during the initial stage of the process establishment. However, if the appropriate test element group (TEG) has been prepared, these problems can be estimated and understood by electronic measurements for TEG.

TABLE II

PROCESS EVALUATION AND CIRCUIT TEGs FOR GaAs LSI TECHNOLOGY DEVELOPMENT

TEGs	Element	Parameter/Object	Measurement
A: Process TEG			
1. Basic recognition TEG	Device isolation	Current leakage	I at 3 V
	TLM pattern	Ohmic contact	R_c, sheet R (for n^+ layer)
	Long gate FET	Mobility	μ, N_D profile
2. Circuit element TEG	E/D FETs	W_g, gate direction	V_t, I_{on}, g_m, g_d, n-value, V_{bi}, C_g
	Diode		V_f
	Resistor		R
	Capacitor		C_g
3. Interconnection TEG	TEG1	First/second level	Sheet R
	TEG2	Via contact	R
	TEG3	Short/open	R
B: Circuit TEG			
1. Fundamental circuit TEG	Two-stage inverter		V_H, V_L
	Ring oscillator	W_g, FI/FO	t_{pd}, P_{dis}
2. Extended circuit TEG	Example; decoder		Operation

5. GaAs LSI FABRICATION AND PERFORMANCE   247

The most important factor in determining a TEG is the selection of elements from its large variety. The use of many elements is desired for diagnosis but requires a wide chip area. On the other hand, the use of few elements requires a small chip area but is insufficient for problem analysis. Furthermore, the selection depends on the state of fabrication process technology development. For the first evaluation of a new process sequence, many TEGs are needed on a LSI photolithography mask, so that one-fourth to one-third of the masks' area is initially required. Finally, the photolithography mask set does not require any TEGs, and only a few circuit elements are needed for LSI fabrication (Zucca *et al.*, 1980).

The first GaAs 1-kbit RAM using the SAINT process was fabricated using a TEG similar to that shown in Table II. This TEG contains almost all of the elements for self-aligned FET and two-level metal layer interconnection LSI processes. Also shown in the table are the circuit TEG and the process TEG, which has been shown to be sufficient for LSI fabrication. The circuit TEG clarifies the circuit operation margin or delay time and the functions. As shown in the table, the process TEG is divided into three types. A basic recognition TEG should be principally confirmed by a preliminary test run for LSI fabrication. In addition, the circuit element and interconnection TEGs are essential parts of the process TEG. Furthermore, this interconnection TEG becomes even more important to the circuit element TEG.

Concerning interconnection, a difficulty in achieving completely operational LSI chips is the occurrence of shorts or disconnections in interconnection lines. The importance of these problems is very easily understood considering the interconnection area, which is larger than the circuit element area in fabricated LSI chips. The ratio of these areas is especially high, up to 0.8 or 0.9, in logic LSIs.

The configuration used in the circuit TEG is the same as that for an LSI chip. Transfer characteristic measurements using a two-stage inverter clarify the signal swing voltages in DC. The dynamic operation speed is measured using a ring oscillator having many varieties with different fan-in and fan-out numbers. The modified ring oscillator characteristics are also important for new versions of the process because of the unknown interconnection and stray capacitances. Furthermore, circuit designers sometimes use an extended circuit TEG to test the most important circuit functions. However, this TEG is a general part of an LSI circuit. As a result, it is not useful because the operation of the critical circuits determines the overall operation of the LSI.

Measurement conditions and definitions are shown in Table III. In the table, some of the definitions are determined for GaAs MESFETs, Schottky diodes, and DCFL circuits. Furthermore, two definitions for threshold voltage are listed in the table. Threshold voltage V_T, which is calculated from

TABLE III
MEASUREMENT AND CODE DEFINITIONS FOR GaAs LSI TEG EVALUATION

Measurement Item	Code	Definition	Element
Threshold voltage	V_t	$I_d = K(V_{gs} - R_s I_d - V_t)^2$	FET
	V_{tc}	V_t (at $I_d = 5\,\mu\text{A}/10\,\mu\text{m}\,W_g$) at $V_{ds} = 1\,\text{V}$	FET
On-current	I_{on}	I_d at $V_{gs} = 0.6\,\text{V}$, $V_{ds} = 1\,\text{V}$	FET
Saturation current	I_{dss}	I_d at $V_{gs} = 0.0\,\text{V}$, $V_{ds} = 1\,\text{V}$	DFET
Transfer-conductance	g_m	ΔI_d ($V_{gs} = 0.6$–$0.5\,\text{V}$)/ΔV_{gs} at $V_{ds} = 1\,\text{V}$	FET
Drain conductance	g_d	ΔI_d ($V_{ds} = 1.0$–$0.9\,\text{V}$)/ΔV_{ds} at $V_{gs} = 0.6$ or $0.0\,\text{V}$	FET
Load resistance	R	V_{ds}/I_d at $V_{ds} = 1$ or $5\,\text{V}$	Resistor
Built-in potential	V_{bi}	$1/C^2 - V^2$ relation or $\log I_{gs} - V_{gs}$ relation	Long gate FET
Forward rising voltage	V_f	V_{gs} at $10\,\mu\text{A}$ or $100\,\mu\text{A}$	Diode
Ideal factor	n	$\log I_{gs} - V_{gs}$ relation	FET
Gate capacitance	C_{gs0}	C_{gs} at $V_{gs} = 0\,\text{V}$	DFET
Logic high level	V_H	$V_{in} - V_{out}$ relation	Inverter
Logic low level	V_L	$V_{in} - V_{out}$ relation	Inverter
Transfer gain	G_T	$V_{in} - V_{out}$ relation at $V_{in} = V_{out}$	Inverter
Evaluation factor	N_G	$\log I_d - V_{ds}$ relation	FET

I-V characteristics, is defined by the equation in which drain current is proportional to the square of $V_{GS} - V_T$. This definition requires the measurement of I-V characteristics and the calculation of V_T. The other definition of threshold voltage V_{TC} is simply given as the voltage at a constant current of $5\,\mu\text{A}/10\,\mu\text{m}$ gate width. This constant level was determined from the measurement results for SAINT FETs with threshold voltages of -0.2 to $+0.2$ V. This definition enables the easy determination of the threshold voltage and requires no calculation. Based on a comparison between V_T and V_{TC}, the difference was found to be less than ± 30 mV.

Next, the subthreshold characteristics are measured to evaluate short-channel effects using the parameters of drain conductance g_d and subthreshold current factor N_G. The n^+ self-aligned FET has embedded n^+ layers neighboring a metal gate to reduce the series resistance. Since the aspect ratio of channel layer thickness to length is selected to be small enough to eliminate short-channel effects, the subthreshold current flows under the channel layer between the embedded n^+ layers. In this case, g_d is usually less than one-tenth of g_m, and N_G is less than 2.

A resistor is used for E/R direct coupled FET logic (DCFL) and source-coupled FET logic (SCFL) circuits as a load element. In addition, the level shift in the SCFL circuit can be estimated from the forward rising voltage measurement of a diode. Moreover, the gate capacitance for a $10\,\mu\text{m}$ gate-width MESFET is measurable in a range of 5–10 fF using a high-accuracy capacitance meter.

8. Chip/Wafer Layout

An example of a TEG layout on a chip is shown in Fig. 9. The pads for the probe contact or wire bonding are $80\,\mu\text{m}$ square in size and are arrayed with a period of $160\,\mu\text{m}$. The dependence of FET characteristics on crystal orientation for the FET gate directions is measured in $\langle 011 \rangle$, $\langle 01\bar{1} \rangle$, and $\langle 010 \rangle$. Next, the device isolation is measured by using a current leakage test pattern that has two ohmic square patterns separated by several microns. Three kinds of ring oscillators are used to measure time delays depending on the gate width and fan-out number.

Half of the TEG area is occupied by the interconnection TEG. Three of these TEGs are the resistance measurement TEG for a gate and for the first- and second-level interconnections. A TEG is used to measure the total resistance of the metal through 100 sequentially connected via holes. From this measurement, the contact resistance between the first- and second-level metals is calculated by subtracting the first- and second-level metal resistances from the total resistance.

The crossover interconnection TEG is considered to be the most significant. In this TEG, the first-level metal is configured in the form of a zigzag

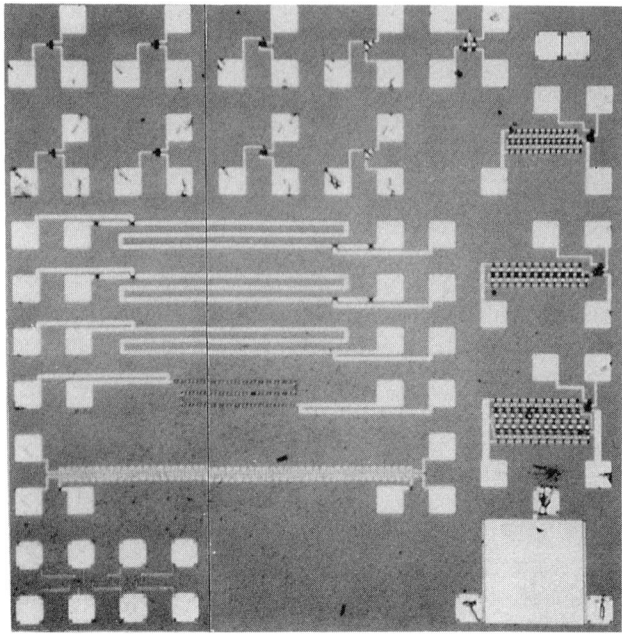

FIG. 9. Fabrication process evaluation TEG layout. FETs are in the upper part, interconnection in the center, and C-V pattern in the lower part.

FIG. 10. Chip layout pattern for GaAs 1-kbit RAM and process evaluation TEG on a 2-in. LEC wafer.

pattern. The second-level interconnection lines are placed in a set of two parallel lines on top of the first-level pattern. The quality of the interconnection is then fully tested for disconnections, poor step coverage, and shorts between the second level lines.

Ohmic contact and n^+ layer resistances are also measured by the transmission line method (TLM). A long-gate FET is used to estimate the electron drift mobility and to measure the carrier profile by the C–V method.

An example of a chip layout on a wafer is shown in Fig. 10. Here, GaAs 1-kbit RAMs using the 3 μm pattern rule and TEGs are arrayed on a 2-in. diameter 450-μm-thick wafer. One repeated unit area contains three chips having 1-kbit RAMs and a TEG area. The area ratio of RAM to TEG is three to one. In a TEG area, 10–20 E- and D-type FETs are arrayed to evaluate FET threshold voltage deviation on a chip and on a full wafer.

IV. FET Fabrication and Performance for LSIs

9. MESFET STRUCTURES FOR LSI

Since 1974, many GaAs MESFET structures have been proposed and developed for attaining high-speed integrated circuits. The cross-sections of some typical devices are illustrated in Fig. 11.

The first GaAs IC was fabricated on an epitaxial layer with mesa-defined isolated regions, as shown in Fig. 11a (Van Tuyl and Liechti, 1974). This mesa-epitaxial FET encountered difficulties in achieving uniformity and reproducibility in the threshold voltage required by ICs. This is because the thickness and doping density of epitaxial active layers are difficult to control precisely.

This problem was successfully resolved by using planar FET technology (Eden and Welch, 1977). In this technology, the active layers are formed by selective ion implantation directly into GaAs semi-insulating substrates (Fig. 11b). The unimplanted substrate around the devices functions to electrically isolate each device. Moreover, the doping profile in an active layer is optimized for each type of device (e.g., MESFETs, diodes, resistors, etc.) by multiple localized implantation.

Enhancement-mode FET (EFET) circuits, also called direct-coupled FET logic (DCFL) circuits, are preferred to depletion-mode FET (DFET) circuits for LSI applications. This is because of low power dissipation and circuit simplicity. An enhancement-mode (normally-off mode) FET is fabricated in such a way that the thickness of the FET channel is less than that of the depletion region with zero gate bias. This device requires forward gate bias to enable current flow in the channel. In an EFET with a conventional structure, the surface depletion layer between source and gate electrodes causes large source resistance because the active n layer is thin.

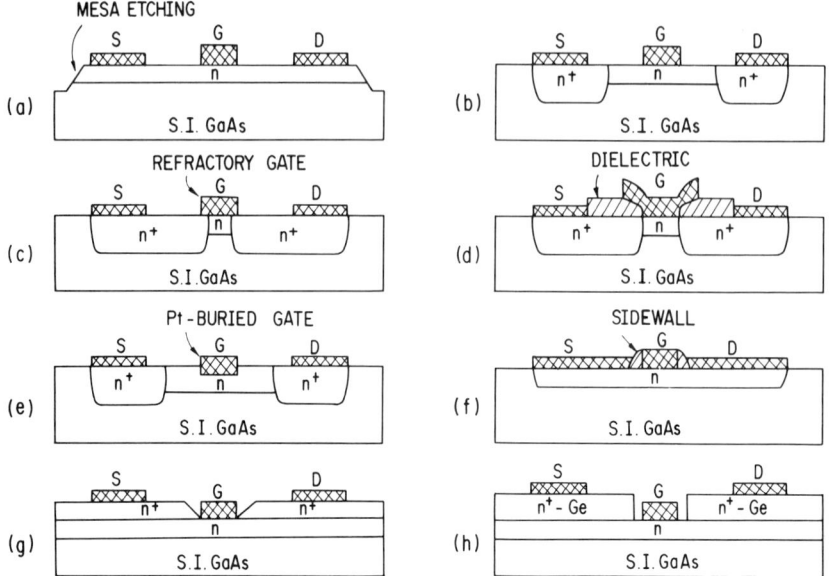

FIG. 11. Cross-sections of various MESFET structures developed for GaAs ICs: (a) mesa-epitaxial MESFET; (b) planar MESFET with selective ion implantation; (c) refractory gate n^+-self-aligned MESFET; (d) self-aligned implantation for n^+-layer technology (SAINT) FET; (e) Pt-buried gate MESFET; (f) sidewall-assisted self-aligned technology (SWAT) FET; (g) selectively grown n^+ source/drain MESFET; (h) n^+-Ge/GaAs source/drain MESFET.

Since a surface passivation technology is not well established for GaAs, various self-aligned structures, which can eliminate the source-to-gate space region, have been developed to decrease source resistance. The source and drain n^+ layers are formed close to the gate by using a refractory metal or its alloy for the gate that acts as a mask for n^+ implantation (Fig. 11c) (Yokoyama et al., 1981). In Fig. 11d, which illustrates another self-aligned FET fabricated by self-aligned implantation for n^+-layer technology (SAINT) (Yamasaki et al., 1982a), n^+ layers are embedded by using a T-shaped multilayer resist for a dummy gate acting as the implantation mask. In this structure, the n^+–gate spacing is controlled by undercutting the resist to decrease parasitic gate capacitances. The SAINT fabrication process and FET characteristics will be described in detail in Section 14.

In another self-aligned FET, a Pt gate is buried by using a Pt–GaAs reaction (Toyoda et al., 1981). As a result, the channel layer is effectively recessed, as shown in Fig. 11e. In a SWAT (side-wall-assisted self-aligned technology) FET, gate and ohmic electrodes are closely separated by thin dielectric films formed on the side-walls of a gate electrode (Fig. 11f) (Higashisaka et al., 1983).

Substrate leakage current between closely spaced source and drain n^+ layers is an essential problem of n^+ self-aligned MESFETs with submicron gate lengths (short-channel effects). To suppress this effect, n^+-GaAs layers are formed on the channel layer by selective epitaxial technology, as shown in Fig. 11g (Imanura et al., 1984). Finally, a unique structure fabricated by using Ge/GaAs heterojunction nonalloyed ohmic contacts is shown in Fig. 11h (Yamane et al., 1985).

10. ACTIVE LAYER FORMATION

An active layer is formed by either ion implantation (bare and through-implantations) or epitaxy (vapor-phase epitaxy [VPE], molecular-beam epitaxy [MBE] and metallorganic chemical vapor deposition [MOCVD]). These technologies are compared in Table IV in terms of uniformity, reproducibility, productivity, and device performance.

At present, ion implantation technology has kept pace with requirements for LSI application by reducing the dislocation density in the GaAs substrate. Specifically, a standard deviation of the threshold voltage smaller than 20 mV has been successfully achieved by using low-dislocation wafers (Ohmori, 1984).

The activation mechanism of implanted dopants, typically Si^+, is very complicated because Si becomes both a donor (Si_{Ga}) and an acceptor (Si_{As}). The activation efficiency depends on crystal characteristics and process

TABLE IV

COMPARISON OF VARIOUS TECHNIQUES FOR ACTIVE-LAYER FORMATION

	Epitaxy			Ion Implantation	
	VPE	MBE	MOCVD	Direct Implantation	Through Implantation
Thickness					
Uniformity	Poor	Good	Good	Good	Good
Reproducibility	Poor	Good	Good	Good	Good
Scale-down	Poor	Very good	Very good	Fair	Good
Doping density					
Uniformity	Poor	Good	Good	Good	Good
Reproducibility	Poor	Fair	Fair	Good	Good
High density	Good	Good	Good	Fair	Fair
Mobility	Good	Good	Good	Fair	Fair
Productivity	Poor	Poor	Fair	Very good	Very good
Device isolation	Poor	Poor	Poor	Fair	Fair

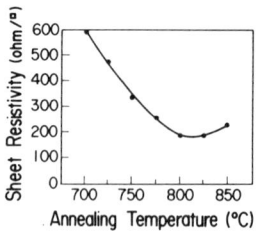

FIG. 12. Sheet resistance of n-layer fabricated by Si ion implantation (200 keV through 0.15 μm SiN film with 4×10^{13} cm^{-2}) vs. activation annealing temperature. [From Kato et al. (1985a).]

parameters, such as implanted dose, annealing temperature and time, and cap materials and atmosphere during annealing. Therefore, the crystal and process parameters should be controlled for better reproducibility.

An example of the relationship between sheet resistivity and annealing temperature is shown in Fig. 12 (Kato et al., 1985a). In this study, Si$^+$ was implanted through a plasma-assisted chemical vapor deposited (PCVD) SiN film (0.15 μm thickness) at 200 keV acceleration energy with a dose of 4×10^{13} cm^{-2}. The implanted layer is usually used for source and drain n^+ regions. The wafers encapsulated with the SiN films were annealed at temperatures ranging from 700 to 850°C for 10 min in N$_2$ atmosphere. As shown in the figure, when the annealing temperature is elevated from 700 to 800°C, the sheet resistance decrease by a factor of three. On the other hand, the sheet resistance slightly increases with temperature above 800°C. As a result, the optimum temperature providing the minimum sheet resistance, that is, the maximum activation efficiency, is 800°C. The optimum temperature becomes lower with a decrease in the implantation dose.

For FET performance improvement, that is, an increase in transconductance and suppression of short channel effects, one of the most effective methods is to reduce the active-layer thickness and to increase the carrier density. The thickness is reduced by decreasing the acceleration energy. For example, an increase in transconductance by shallow implantation is clearly shown in Fig. 13, in which the transconductance versus threshold voltage curves are compared for different acceleration energies (30 and 67 keV) of Si implantation. Here, the activation annealing was performed at 800°C, and the characteristics were measured in buried p-layer SAINT FETs with 0.5 μm gate lengths, which will be described in Section 15. From the plot, the transconductance improvement factor is 1.3 for 230 mS/mm at threshold voltage of 0 V.

Another technique for forming shallow implanted layers is through-implantation (Nishi et al., 1983). WSi gate self-aligned FETs were fabricated

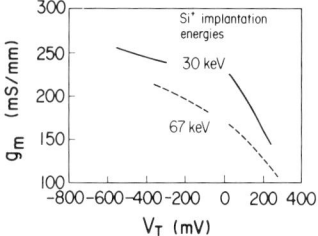

FIG. 13. Comparison of transconductance (g_m) vs. threshold voltage (V_T) for two types of n-channel layers, fabricated by Si ion implantation at 30 and 67 keV.

on a thin channel layer formed by Si ion implantation at 59 keV through a thin (55 nm) AlN film deposited by reactive sputtering. Then, activation annealing was performed at 850°C for 15 min with AlN film as a protective cap. The transconductance of the through-implanted FETs is increased by 30 to 50% as compared with that of conventional self-aligned FETs.

It is also important to reduce thermal diffusion during the high-temperature annealing process. Diffusion can be reduced by employing low-temperature annealing or short-time annealing (rapid thermal annealing), or by using good cap materials that prevent Ga and As atoms from outdiffusing.

Modern epitaxy techniques, such as MBE and MOCVD, are very promising for attaining a high doping density and a thin active layer. Moreover, the optimum doping profile for high-speed operation is obtained by using the epitaxy. However, current problems with these epitaxial methods are poor reproducibility and poor productivity.

11. Gate Formation

Various materials are used for formation of Schottky barriers on n-GaAs, and properties of such typical materials are summarized in Table V. The required characteristics for LSI applications are large barrier height, low electric resistivity, good adhesion to the GaAs surface, thermal stability, and fabricability (deposition and patterning).

The gate electrodes are patterned by lift-off or etching techniques, as schematically shown in Fig. 14. In order to obtain smooth edges, spacer and T-shaped resist-assisted lift-off techniques have been developed, as shown in Figs. 14b and 14c, respectively. These techniques have now taken the place of the conventional direct lift-off technique, shown in Fig. 14a. In the new techniques, the overhang structures are formed by gas plasma or reactive ion etching. In addition, these techniques can also be applied to interconnection line formation. On the other hand, since refractory metals and their alloys, which are used in self-aligned FETs, are usually deposited by sputtering, the

TABLE V
COMPARISON OF VARIOUS GATE MATERIALS

Material	Barrier Height (V)	Resistivity ($\mu\Omega$-cm)	Thermal Stability	Deposition	Pattern Formation	Comments
Al	0.74	2.7	400°C	Evaporation	Lift-off Wet etching	
Ti	0.73	47	450°C	Evaporation	Lift-off	Good adhesion Used in Ti/Pt/Au
Mo	0.68	5.3	450°C (Peel-off)	Evaporation Sputtering	Lift-off Plasma etching	
Pt	0.84	10.4	250°C (Compound formation)	Evaporation	Lift-off	Used for buried gate
TiW	0.78	—	800°C	Sputtering	Plasma etching	Used for refractory gate metal in n^+ self-aligned FET
WSi	0.76	—	850°C	Co-sputtering	RIE	
WAl	0.71	60	900°C	Co-sputtering	Plasma etching	
WN	0.84	70	800°C	Reactive sputtering	RIE	
TaSi	0.78	100	800°C	Sputtering	Plasma etching	
SiGeB	0.9–1.0	10^6–10^7	450°C	LP-CVD	RIE	Used in Al/SiGeB

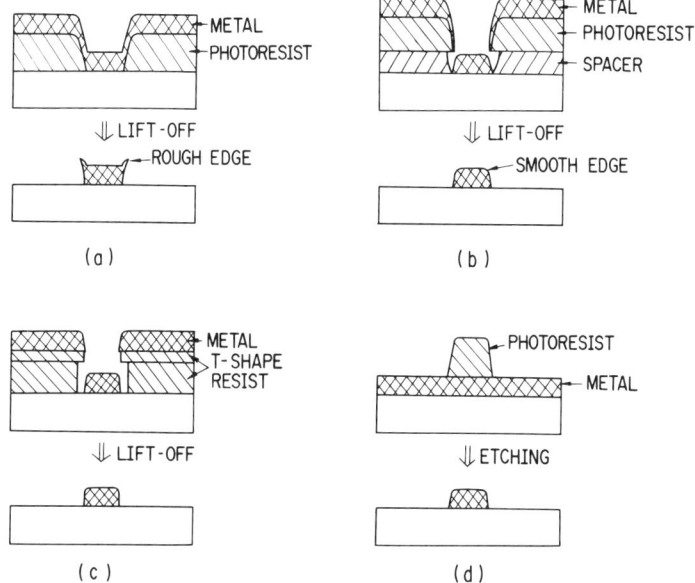

FIG. 14. Comparison of fabrication technologies for gate-metal patterning: (a) conventional direct lift-off using photoresist; (b) spacer lift-off; (c) lift-off using T-shaped resist; (d) etching.

lift-off techique cannot be applied. Therefore, these materials must be patterned by dry etching (Fig. 14d).

Surface cleaning prior to gate-metal deposition improves Schottky contact characteristics. A dry process is desirable, since the gate regions to be cleaned are very narrow (about 1 μm). Hydrogen plasma treatment in a reactive ion etcher after etching SiN on gate regions in gas plasma ($CF_4 + O_2$) removes the surface contaminations and improves the electrical characteristics (Muraguchi et al., 1984). Uniformity of ideality factor and barrier height is significantly improved, as shown in Fig. 15, which shows scattering of ideality factor versus barrier height before (a) and after (b) H_2-plasma treatment. Since the cleaning was carried out at low power and for a short period (20 sec), no etching or damage was observed in the GaAs surface.

Multilayer metallization systems, such as Mo/Au and Ti/Pt/Au, are used to reduce the gate electrode resistance. As shown in Table III, refractory gate materials have larger resistivity than conventional metal systems by a factor of about ten. Unfortunately, this large gate resistance is disadvantageous for microwave circuit applications.

The barrier height of n-GaAs Schottky junctions is about 0.7–0.8 V and depends only slightly on the work function of the material. This independence is due to the pinning of the surface Fermi level by interface defects

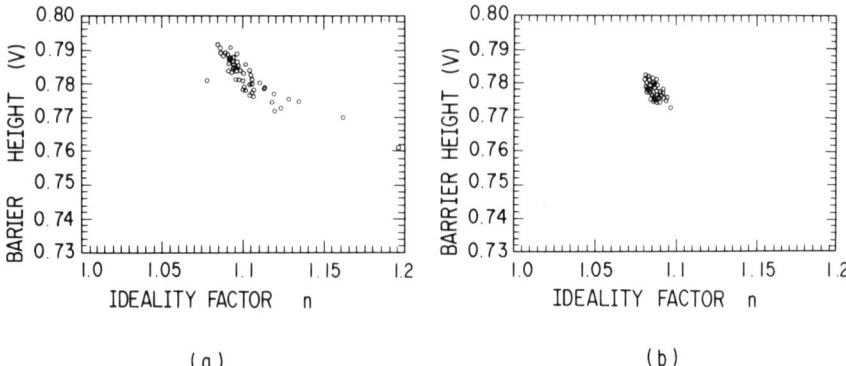

Fig. 15. Effects of surface cleaning by hydrogen reactive ion etching on Schottky characteristics (barrier height vs. ideality factor) for (a) an untreated sample; (b) a hydrogen plasma-treated sample. Both samples were annealed at 325°C after Schottky metal deposition.

formed during metal deposition (Spicer *et al.*, 1979). The barrier height limits the forward gate bias and the logic swing, especially in DCFL circuits.

In order to improve the operational margin, a large barrier height is desirable. Larger built-in potential is provided by a p^+–n junction (approximately 1.1 V), which is used as a gate in a JFET shown in Fig. 16b. A JFET is superior to a MESFET for enhancement-mode FETs. However, a JFET is more difficult to fabricate because an additional p^+-doping process is needed, and precise junction-depth control is necessary to obtain a uniform threshold voltage. In addition, submicron-gate JFETs are also difficult to fabricate because of p^+ lateral diffusion.

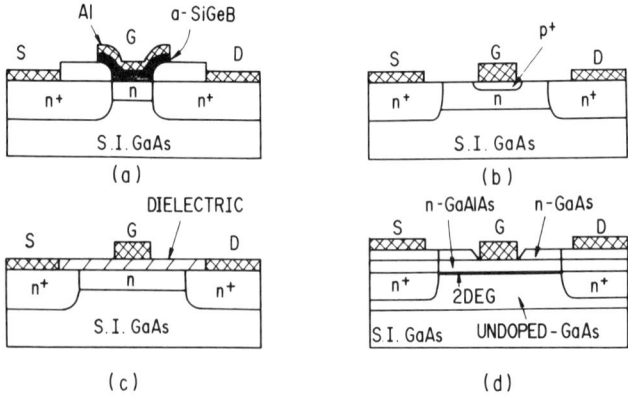

Fig. 16. Cross-sections of FETs with larger barrier heights than that of an ordinary MESFET: (a) MASFET using metallic-amorphous SiGeB gates; (b) junction FET (JFET); (c) insulated Gate FET (IGFET); (d) high electron mobility transistor (HEMT).

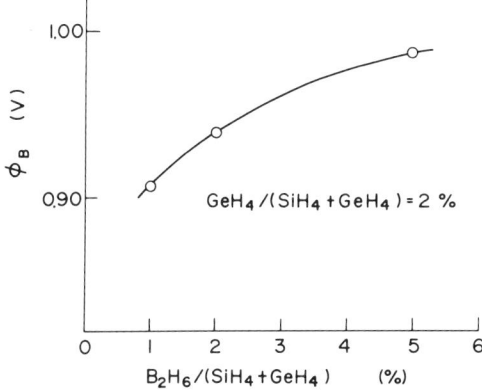

FIG. 17. Barrier height of a-SiGeB/GaAs contact vs. deposition condition (equal to reactive gas flow rate).

A large barrier height of 1.0 V is easily achieved by using an amorphous SiGeB/GaAs junction (Suzuki et al., 1983). The a-SiGeB film is deposited by thermal decomposition of a SiH_4–GeH_4–B_2H_6 mixture in a low-pressure furnace at 450°C. The barrier height is controlled by the B_2H_6 flow rate, as shown in Fig. 17. A metallic amorphous silicon gate FET (MASFET) has been fabricated using the SAINT process (Fig. 16b). The gate electrode is made of Al/a-SiGeB. The aluminum patterned by lift-off reduces the gate electrode resistance and also acts as an etching mask for defining the amorphous film by $CF_4 + O_2$ gas plasma. The high-level voltage of the DCFL MASFET inverter, determined by the forward characteristics of the next-stage Schottky gate, is 0.94 V. This is apparently larger than that of the MESFET inverter (normally 0.7 V). Moreover, the feasibility of fabrication-yield improvements as well as the enlargement of the logic swing have also been confirmed (Suzuki et al., 1984).

Theoretically, insulated gate FET (IGFET) technology would completely eliminate the logic swing limitation. Unfortunately, the lack of a suitable insulator on GaAs associated with a large interface state density has resulted in several unsuccessful attempts. Presently, InP is being used for IGFETs, and this material may be more promising than GaAs.

A high electron mobility transistor (HEMT) uses a heterojunction of GaAlAs/undoped GaAs, as shown in Fig. 16d (Mimura et al., 1980). GaAlAs with a larger bandgap takes the place of an insulator in an inversion-mode IGFET. Thus, HEMT is very attractive for very high-speed LSIs because of the high mobility of two-dimensional electron gas accumulated in the potential well at the interface, especially at low temperatures.

12. Source/Drain Formation

Alloyed AuGe-based contacts are widely used for making ohmic contacts (source/drain electrodes) on n-type GaAs. This alloying is performed by heating the sample to a typical temperature of about 450°C for a short period (about 1 min). During the alloying procedure, Au and Ge are indiffusing and Ga is outdiffusing. As a result, an n^+ heavily doped layer of about 10^{19} cm^{-3} is created by Ge substituting for Ga sites.

The conduction band diagram of an alloyed ohmic contact is schematically shown in Fig. 18a (Dingfew and Heime, 1982). When the donor density N_D in the n-GaAs substrate is lower than about 10^{18} cm^{-3}, the specific contact resistance r_c is limited by the n^+-n junction, so r_c is proportional to N_D^{-1}. Based on previous experimentation, the minimum values satisfy the relationship $r_c = 2 \times 10^{11} N_D^{-1}$ Ω cm^2 (N_D: cm^{-3}). In the higher doping region, on the other hand, the specific contact resistance is limited by metal-n^+ junction tunneling, so it is independent of N_D.

For planar devices, the specific contact resistance is determined by the transmission line method (TLM), as shown in Fig. 19. The measured resistance R is determined between the two square pads of length L_C and width W. These two pads are separated by distance L on an active layer of sheet resistance $R_\square$. Thus, R is given by,

$$R = (R_\square/W)(L + 2L_T \coth(L_C/L_T)),$$

where L_T, the transfer length, is related to r_c by

$$L_T = \sqrt{r_c/R_\square}.$$

The slope of the R versus L plot yields $R_\square$, and the intercept with the ordinate enables r_c to be found.

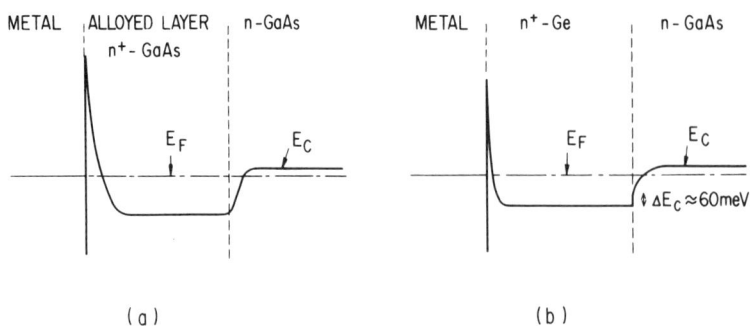

FIG. 18. Conduction band diagram of ohmic contacts: (a) alloyed contact; (b) nonalloyed Ge/GaAs contact. [From Dingfew and Heime (1982).]

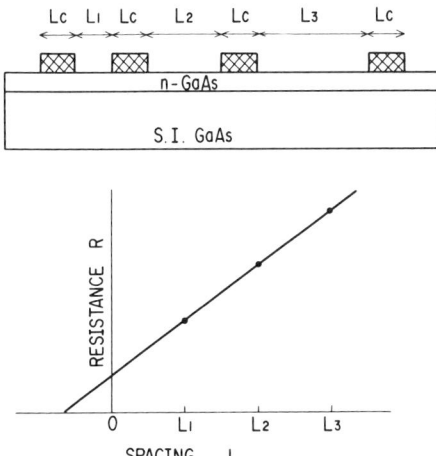

FIG. 19. Transmission line method (TLM) for determining specific contact resistance of ohmic contacts.

When the electrode length L_c is much longer than the transfer length, the measured resistance is independent of the electrode length since $\coth(L_C/L_T)$ is about unity. However, when the electrode length decreases to that of the transfer length, the measured resistance increases. This measured resistance corresponds to the source/drain series resistance. A comparison of dependence of resistance on electrode length for both theory and experiment is shown in Fig. 20 (Yamane et al., 1984). The active layer measured was fabricated by Si ion implantation at 120 keV through 0.15 μm SiN film with a dose of 1.6×10^{14} cm^{-2}. This layer was used as source and drain n^+ regions. From the best fit between the experimental result and the

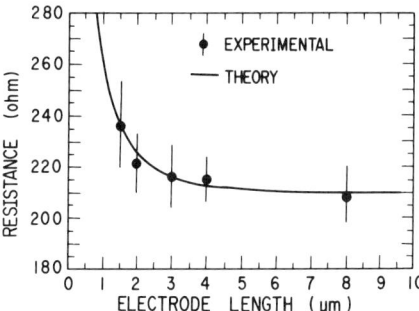

FIG. 20. Relationship between ohmic resistance and electrode length: (●) experimental; (solid line) theory.

theoretical curve of the resistance R versus electrode length L_c, the transfer length L_T is determined to be 2.4 µm. This obtained value is similar to the source or drain electrode length used in digital LSIs. Therefore, it is necessary to decrease the transfer length, that is the specific contact resistance r_c, in order to reduce the physical electrode size required for higher density integration without increasing the total resistance.

The thermal stability of an ohmic contact is important, since it limits high-temperature processing following the formation of an ohmic contact; it is also important obtaining long life reliability. The processing temperature after ohmic metallization should be below the alloying temperature. Moreover, AuGe/Pt alloyed contacts deteriorate rapidly when interconnection lines (Ti/Pt/Au) are overlaid, because Pt diffuses into GaAs during thermal aging even at 250°C (Lee et al., 1981). To suppress this degradation, it was beneficial to use AuGe/Ni instead of AuGe/Pt or to increase the Ti thickness of the interconnection.

In order to reduce the specific contact resistance and also to improve the thermal stability, nonalloyed ohmic contacts have been developed. One of the candidates for this is Ge/GaAs heterojunction contacts (Stall et al., 1979). The band diagram for Ge/GaAs is shown in Fig. 18b. The electron affinity difference between Ge and GaAs is very small (about 60 meV) and it is not expected to be a significant barrier to electron flow. Since the doping level of Ge can be made heavier than that of GaAs by a factor of about ten, the metal/n^+-Ge tunneling contact resistance is expected to be lower than the metal/n^+-GaAs contact resistance. In Stall's experiment, an extremely low value of specific contact resistance, below 1×10^{-7} Ω cm^2, was achieved for Ge (1.4×10^{20} cm^{-3})/GaAs (1.5×10^{18} cm^{-2}) contacts grown by MBE. Recently, a FET using the nonalloyed Ge/GaAs contact has been successfully fabricated, as shown in Fig. 11h. The fabrication process and characteristics of this unique FET will be described in Section 15.

13. Device Isolation

Electrical isolation between closely spaced devices is required for successful design and fabrication of high density integration circuits. Isolation by mesa etching of an epitaxial n layer was applied to the first generation of GaAs SSIs and MSIs (Fig. 21a). However, the mesa isolation approach is not suitable for LSIs, because the nonplanar approach limits the maximum density and yield. The planar technology has been developed by selective ion implantation for active layer formation. In this technology, a semi-insulating substrate itself acts to electrically isolate devices (Fig. 21b).

However, the so-called sidegating effect—that is, the interaction between adjacent MESFETs through a semi-insulating substrate—is observed

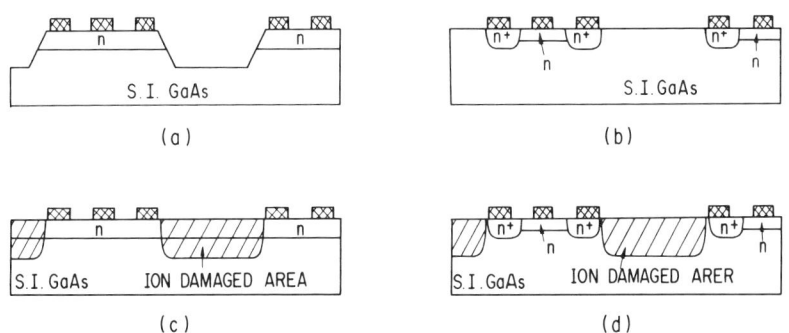

FIG. 21. Comparison of various types of device isolation technology: (a) mesa-etching of active layer; (b) selective ion implantation directly to semi-insulating substrate; (c) ion bombardment (proton implantation) into active layer; (d) selective ion implantation followed by ion bombardment.

especially in DFET circuits operating at a relatively large supply voltage. The drain current of a FET at $V_{DS} = 2.0$ V begins to decrease significantly when -2.8 V is applied to a sidegate placed 26 μm apart from the FET (Lee et al., 1982). The threshold voltage at which the sidegating effect becomes observable coincides with the trap-filling voltage in an n–i–n structure where the i layer corresponds to the separation between the FET and the sidegate. The space-charge-limited current flows in the surface region in processed semi-insulating GaAs wafers. This is possibly because of deep-trap outdiffusion during post-implantation annealing. Proton implantation improves the electrical isolation as well as the sidegating effects (Fig. 21d). The sidegating threshold increases up to -7 V with 4×10^{14} cm^{-2} (at 150 keV) proton implantation. Proton bombardment induces lattice disorder in the GaAs surface and creates very high-density deep traps. Therefore, this bombardment increases the trap-filling voltage. Proton implantation technology has been also applied to electrically isolate devices fabricated by epitaxy or nonselective ion implantation, as shown in Fig. 21c (D'Avanzo, 1982).

14. Fabrication and Performance of Self-Aligned FET

As described in the previous section, self-aligned GaAs MESFETs are required for high-speed LSIs especially constructed with enhancement-mode FETs. In this section, the fabrication technologies of refractory gate FETs and SAINT FETs, as well as the FET performance, are described.

a. Fabrication Technology for Self-Aligned FETs

The fabrication process steps for refractory gate self-aligned MESFETs are schematically shown in Fig. 22. An *n*-type active layer is formed by using

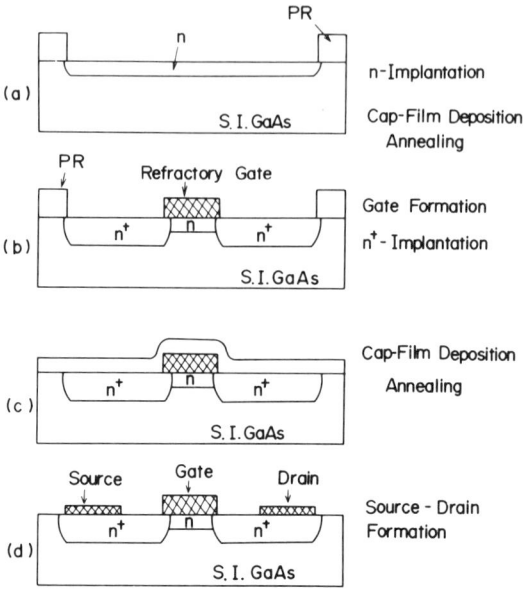

FIG. 22. Fabrication process steps of a refractory gate n^+ self-aligned MESFET: (a) n implantation, cap film deposition and annealing; (b) gate formation and n^+ implantation; (c) cap film deposition and annealing; (d) source–drain formation.

selective ion implantation technology followed by activation anealing with a protective cap film. Next, a refractory metal or its alloy (WSi, WAl, WN, etc.) is deposited by sputtering and patterned by dry etching to form a gate electrode. Then, high-dose ions are implanted by using the gate electrode as an implantation mask. After deposition of a cap layer, such as SiO_2, the sample is annealed at about 800°C to activate the dopants. As a result, n^+ layers are formed outside the gate region in a self-aligned manner. The refractory gate materials can stand and the Schottky gate characteristics do not degrade after the annealing. Finally, source and drain electrodes are formed with AuGe-based alloyed contacts.

The fabrication steps of the SAINT are shown in Fig. 23.

First, Si ions are selectively implanted at 60 keV into a semi-insulating GaAs substrate to form an n-active layer (Fig. 23a).

Next, the GaAs surface is covered with a SiN film (0.15 μm in thickness) deposited by plasma-enhanced CVD. Then, a multilayer resist is formed on the SiN film (Fig. 23b). This resist consists of a bottom resist (FPM [polytetrafluoropropyl methacrylate] or AZ 1470 photoresist), an intermediate sputtered-SiO_2 layer (0.3 μm in thickness), and a top photoresist (AZ 1470) for photolithography. When electron-beam direct drawing is employed

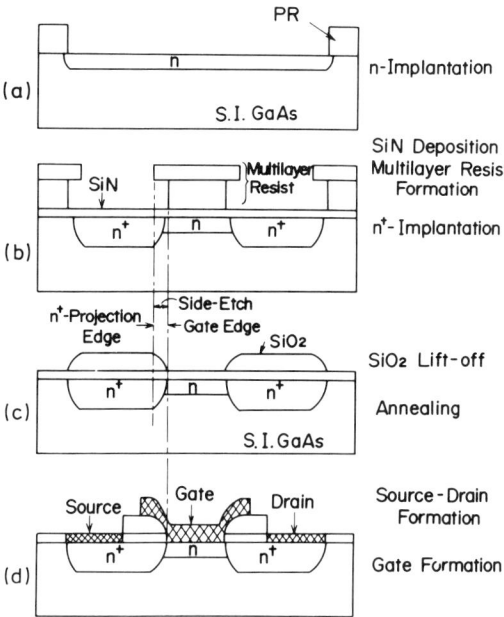

FIG. 23. Fabrication process steps of a SAINT FET: (a) n implantation; (b) SiN deposition, multilayer resist formation and n^+ implantation; (c) SiO_2 lift-off and annealing; (d) source-drain formation and gate formation.

for patterning, a thin metal film (Mo, 100 Å) is inserted between the intermediate SiO_2 and the top electron-beam resist (CMS), as shown in the top of Fig. 24 (Kato et al., 1983). This thin metal film successfully avoids charging in a semi-insulating GaAs substrate during electron-beam exposure and reduces positioning errors (Fig. 24).

In general the top resist pattern is duplicated into the intermediate SiO_2 layer by reactive ion etching (RIE). Then, the bottom resist is defined by RIE in the O_2 discharge. This bottom layer is undercut with respect to the upper SiO_2 pattern. The lateral etching rate, the rate of undercutting, is much smaller than the vertical etching rate, typically by a factor of seven. The amount of undercut is controlled within a deviation of 0.02 μm by adjusting O_2 pressure, RF-power, and etching time in RIE. A SEM photograph of a multilayer resist scheme is shown in Fig. 25. High-dose Si ion implantation is performed at 200 keV (dose of 4×10^{13} cm^{-2}) through the PCVD-SiN film for n^+-layer formation, where the intermediate layer acts as an ion stopping mask.

Then, the second SiO_2 layer (0.3 μm) is deposited by sputtering and lifted off by using the multilayer resist (Fig. 23c). As a result, sputtered-SiO_2 films

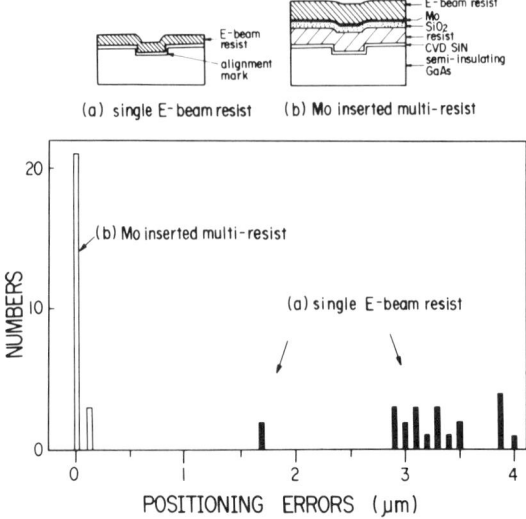

FIG. 24. Comparison of electron-beam positional errors between (a) a single electron-beam resist and (b) a multilayer resist having an inserted Mo layer. [From Kato et al. (1983).]

FIG. 25. SEM photograph of a T-shaped resist formed by reactive ion etching.

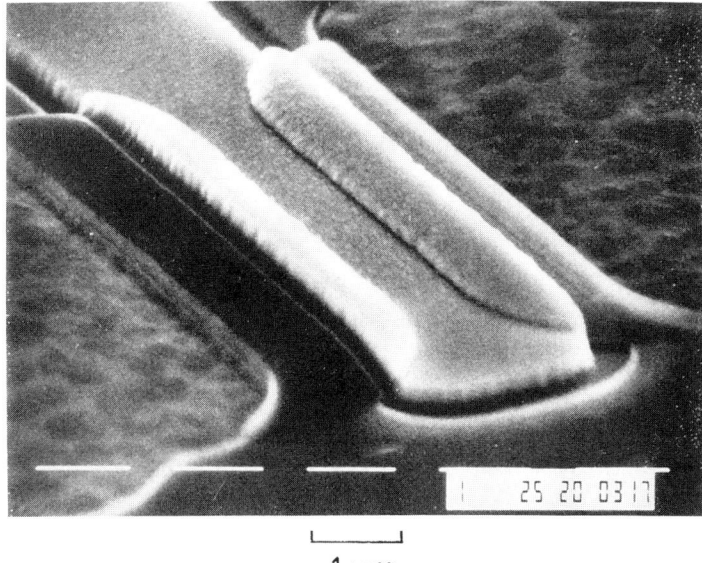

FIG. 26. SEM photograph of a SAINT-fabricated FET.

remain only above the n^+-implanted layers but extend somewhat beyond them. The electrical activation of the n and n^+ implants are made by thermal annealing at 800°C, and the ground PCVD-SiN film acts as a protective cap.

The source and drain ohmic electrodes are formed by AuGe base alloyed contacts. Finally, the gate metallization is accomplished by photoresist patterning, SiN plasma etching, and evaporating conventional metal (Ti/Pt/Au) localized by lift-off with the resist (Fig. 23d). The gate metal overlaps the SiO_2 and SiN dielectric film. The Schottky-gate length is defined by the spacing between the edges of the two sputtered-SiO_2 islands because the etching rate for SiO_2 is much smaller than that for SiN. A SEM photograph of a SAINT FET is shown in Fig. 26.

The undercut of the bottom resist defines the spacing between the projection edge of the n^+ implantation and the gate-contact edge (see Fig. 21). This is accomplished with the aid of the second SiO_2 layer formed by the lift-off. With this technology, the position of the gate-contact edge can be controlled with respect to the edge of n^+ region by taking into account the lateral spreading of the n^+ implants caused by ion recoil and thermal diffusion.

The advantages of the SAINT process and the resultant FET characteristics are summarized as follows:

(1) The source and drain parasitic series resistances are drastically reduced by embedding the self-aligned n^+ layers.

(2) Optimization of the spacing between the n^+ layer and the gate is attained from the trade-off between the external source resistance and parasitic gate capacitance.

(3) Submicron gate lengths shorter than the initial resist pattern length defined by lithography are achieved by the undercutting of the multilayer resist.

(4) Conventional gate metals with low resistivities, or unique gate materials such as a-SiGeB with large barrier heights, can be used.

(5) The threshold voltage could be adjusted by gate recessing after monitoring the ungated drain current.

(6) A decrease in gate resistance even with extremely short gate lengths is achieved by using the T-shaped gate electrode in the cross-section.

The parasitic capacitance caused by the overlapping of the gate electrode and the dielectric film is at most 10–20% of the intrinsic gate capacitance for a 1 μm gate device. Recently, the overlapping region has been successfully eliminated by flat-gate technology using ion-beam milling (Enoki et al., 1985).

The flat-gate SAINT process is almost the same as the ordinary one described above, except for gate formation. The process steps are shown in Fig. 27. The p layer is buried to suppress the substrate current, which will be described in detail in Section 15. After opening the gate contact region, Mo/Au is deposited over the entire surface by sputtering. Then, the Au surface is planarized by ion-beam milling with a large ion-beam incident angle, until Au remains in the gate dip region only. This planarization

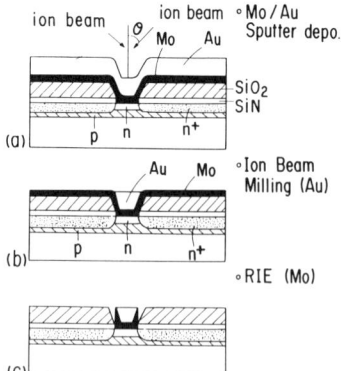

FIG. 27. Gate formation process steps for a flat-gate SAINT FET: (a) Mo/Au sputter deposition; (b) ion-beam milling (Au); (c) RIE (Mo).

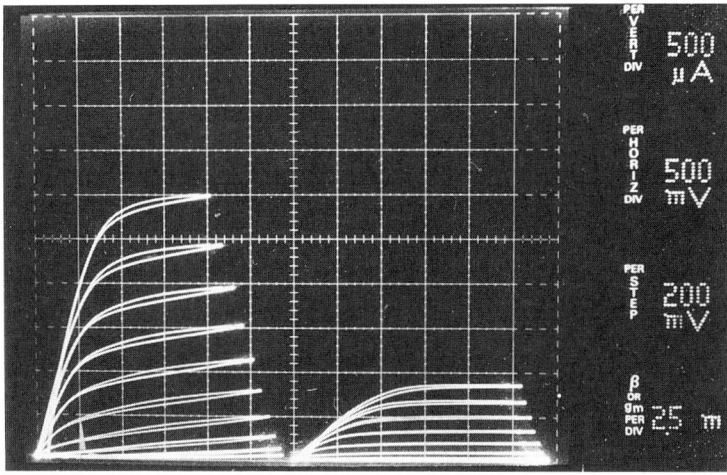

FIG. 28. Comparison of current–voltage characteristics for (a) a SAINT FET and (b) a conventional MESFET without an n^+ layer. The dimensions are 1 μm for gate length and 10 μm for gate width. Horizontal axis: 0.5 V/div; vertical axis: 0.5 mA/div, gate voltage: 0.2 V/step from −1.0 to 0.8 V.

method accounts for the cosine dependence of the Au etching rate on the ion-beam incident angle. Finally, the Au pattern is duplicated in the Mo layer by reactive ion etching. As a result, the gate electrode is embedded only in the gate opening region.

b. FET Performance

Typical drain current versus voltage characteristics for a SAINT FET and a conventional planar MESFET without n^+ implantation are compared in Fig. 28. The conventional FET was fabricated by the same process as the SAINT FET, except for self-aligned n^+ implantation. The gate length, gate width, and source–drain distance are 1, 10, and 5 μm, respectively. The drain conductance in the linear region, the saturation drain current, and the transconductance are all dramatically improved. The maximum transconductance of the SAINT FET is 2.8 mS (280 mS/mm), which is about three times the value (0.9 mS) of the conventional FET shown here.

The optimization of n^+-gate spacing from the trade-off of source resistance and gate capacitance is of great importance in determining switching speed. Using a two-dimensional simulation, the influence of the n^+-to-gate spacing on gate capacitance was investigated, and the result is shown in Fig. 29 (Yamasaki *et al.*, 1982b). In this figure, the vertical scale denotes the sheet density of carriers that are swung or swept out from the gate channel

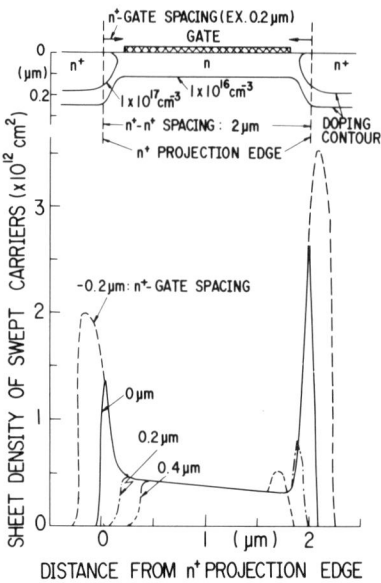

FIG. 29. Effect of n^+-layer and gate spacing on the capacitance component around the gate. Sheet densities of carriers swung or swept out from the channel layer at the on–off interstate transition are compared for various n^+-gate spacings. [From Yamasaki et al. (1982b).]

layer at the on–off interstate transition. Compared with the case of nonspacing, a total movement charge for a 0.2 μm spacing FET with 2 μm source/drain n^+ spacing decreases by a factor of 1.6. The gate capacitance increase with n^+ embedding is effectively restrained by the n^+-gate spacing. An experimental approach was taken to evaluate the influence of n^+-gate spacing on source resistance (Yamasaki et al., 1982c). The source resistances were measured with variation of the n^+-gate spacing, and the result is shown in Fig. 30. The significant improvement in the source resistance caused by the self-aligned n^+ layer ceases at an n^+-gate spacing of about 0.3 μm. This 0.3 μm length might correspond to a lateral spreading of the n^+ implants. From the experiment, the optimum n^+-gate spacing was found to be about 0.3 μm.

The threshold voltage of a short-channel FET depends on the gate orientation with respect to the crystallographic orientation. Two kinds of SAINT FETs, which are at right angle to each other ($\langle 011 \rangle$ and $\langle 01\bar{1} \rangle$), were fabricated on a $\langle 100 \rangle$ wafer under the same implantation conditions, as shown in the inset of Fig. 31. For gate lengths below 2 μm, a FET with a gate parallel to the $\langle 011 \rangle$ direction has a lower threshold voltage than the $\langle 01\bar{1} \rangle$

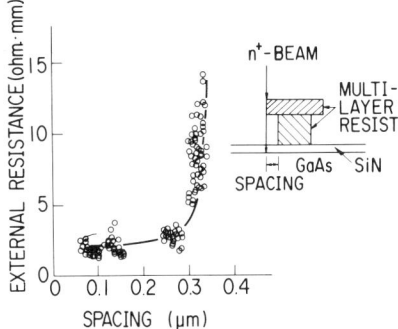

FIG. 30. External source resistance as a function of n^+-gate spacing for enhancement-mode SAINT FETs. [From Yamasaki et al. (1982c).]

oriented FET, as shown in Fig. 31 (Yamasaki et al., 1982d). Such orientation dependence of the threshold voltage shift is commonly observed in GaAs MESFETs including SAINT FETs, refractory gate FETs, and non-self-aligned FETs. However, the gate direction that gives the smallest threshold shift is different among these FETs.

An origin of such orientation dependence has been found to be the piezoelectric charge induced by the elastic stress near windows in dielectric overlays (Asbeck et al., 1984b). The differences in the orientation dependence found by various experiments can be explained by differences in the stress sign of the dielectric overlays. For example, SiO_2 film is in compressive stress and SiN film is in tensile stress on GaAs (Ohnishi et al., 1985).

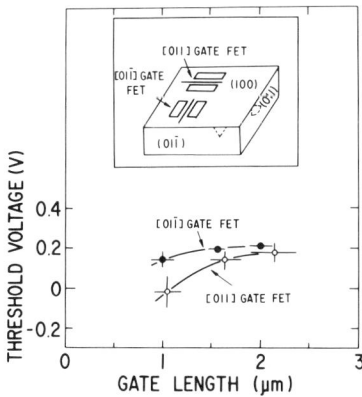

FIG. 31. Orientation dependence of threshold voltage shift on reducing gate length for SAINT FETs. [From Yamasaki et al. (1982d).]

15. Submicron Gate Length FETs

The most effective method for achieving faster operation is gate-length shortening. However, short-channel effects, such as threshold voltage shift, poor current saturation and subthreshold current flow, appear in the submicron gate-length region. In this section, two device structures suppressing short-channel effects, a buried p-layer SAINT FET and an n^+-Ge–GaAs source/drain MESFET, are described.

a. Buried p-Layer SAINT

It is believed that the main causes for short-channel effects in n^+ self-aligned MESFETs are substrate leakage current and lateral spreading of the n^+ layers. The lateral spreading due to thermal diffusion during activation annealing can be suppressed by using capless annealing without surface stress (Sadler and Eastman, 1983) or rapid thermal annealing (Ohnishi *et al.*, 1984). For SAINT FETs, the lateral spreading is cancelled by separating the gate from the n^+ projection edges at a distance equal to the spreading (Kato *et al.*, 1984). The substrate current, which flows in a semi-insulating substrate between the source and drain n^+ layers, is suppressed by thinning these layers (Yamasaki *et al.*, 1982e).

A more direct and effective method is to bury a barrier layer between the active layer and the substrate. In this method, a p layer, formed by Be implantation, acts as the barrier (Yamasaki *et al.*, 1984). A cross-sectional view of a buried p-layer SAINT (BP-SAINT) FET and its doping profiles are shown in Fig. 32. In this structure, the p layer causes neither instability nor a parasitic-capacitance increase, since the effect of deep levels in the buried p layer is small: shallow acceptor levels predominate and the p layer is comletely depleted by a built-in potential against the upper active layer.

To suppress the substrate current, the Debye length of the p layer should be shorter than about one-tenth of the gate length. Another restriction is that the depletion layer created by the built-in potential against the active layer should extend across the entire p layer. These conditions determine the thickness and impurity density of the p layer. Little difficulty is caused by the fabrication process since only one simple process step (Be implantation after Si implantation) is added. Furthermore, the activation efficiency of implanted Be ions is approximately 100% and thermal diffusion is negligibly small. As a result, there are no additional problems with threshold voltage control.

The threshold voltage shift and increase in the subthreshold parameter N_G are successfully alleviated by burying the p layer, as shown in Fig. 33. The subthreshold parameter N_G is defined by the dependence of subthreshold current I_{DS} on gate voltage V_{GS}: $I_{DS} \propto \exp(qV_{GS}/N_G kT)$.

5. GaAs LSI FABRICATION AND PERFORMANCE

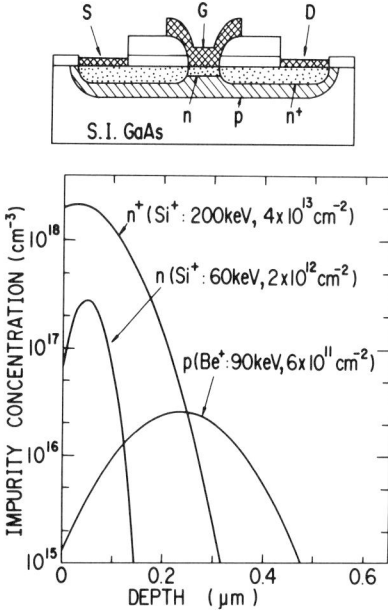

FIG. 32. Structure and implanted ion profiles of a buried p-layer SAINT FET.

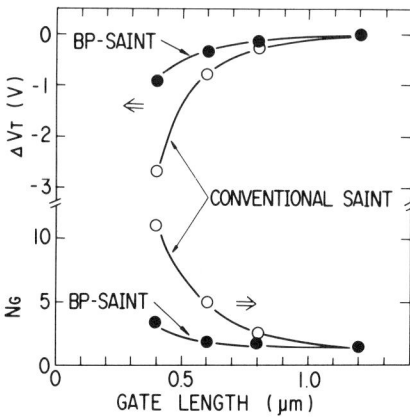

FIG. 33. Comparisons of threshold voltage shift and subthreshold parameter N_G dependencies on gate length between (○) conventional SAINT FETs and (●) buried p-layer SAINT FETs.

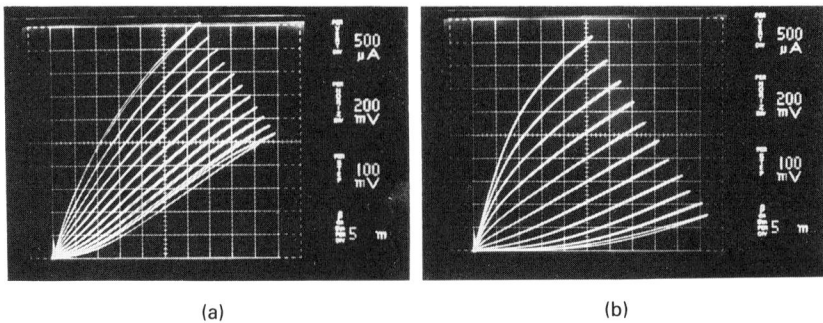

FIG. 34. Current–voltage characteristics of (a) a conventional SAINT FET and (b) a buried p-layer SAINT FET. Gate length and width are 0.4 and 40 μm, respectively. Horizontal axis: 0.2 V/div; vertical axis: 0.5 mA/div; gate voltage: 0.1 V/step from −0.3 V to 0.7 V.

The current–voltage characteristics of a BP-SAINT FET are compared with those of a conventional SAINT FET in Fig. 34. The gate lengths of both FETs are 0.4 μm. The triode-like characteristics and poor current cutoff performance of the conventional SAINT FET are remarkably improved with the buried p layer. The threshold voltages of the FETs with and without a p layer are −0.5 and −2.21 V, respectively. It should be noted that the transconductance of the BP-SAINT FET is increased by a factor of 1.6 even at high gate voltage. This improvement can account for the effective thinning of the n channel due to the depletion layer spreading into the n layer. This spreading is caused by the p–n junction between the buried p layer and the n-channel layer.

25-stage ring oscillators with an E/D DCFL configuration were fabricated using the BP-SAINT process. The propagation delay time and power dissipation measured at room temperature are compared for different SAINT structures in Fig. 35. The gate lengths are 0.4 μm. The delay time of BP-SAINT FETs at a given power is reduced by a factor of 1.6 compared with that of conventional SAINT FETs. The improvement is due to the large transconductance described above. The results indicate that the completely depleted p layer does not increase parasitic capacitance. The minimum delay time observed is 9.9 psec/gate. Even faster operation is made possible by applying the scale-down law: gate-length shortening, channel thinning, and higher doping.

b. Ge/GaAs Source/Drain MESFET

To suppress the substrate leakage current, the source and drain n^+ layers must be made shallow, but they cause an increase in source and drain series

5. GaAs LSI FABRICATION AND PERFORMANCE 275

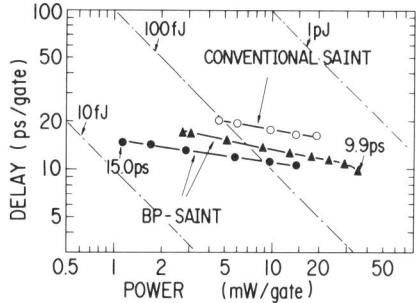

FIG. 35. Comparison of propagation delay time vs. power dissipation per gate for 0.4 μm-gate-length ring oscillators constructed with (○) conventional SAINT FETs and (●, ▲) buried p-layer SAINT FETs.

resistances. Since the intrinsic channel resistance is decreased by shortening the gate length, the source/drain external resistances should be further decreased to improve the total electrical performance in submicron-gate MESFETs.

As described in Section 12, an n^+-Ge layer epitaxially grown on GaAs forms ohmic contact with low contact resistance. Use of n^+-Ge layers in source and drain regions resolves the above problems (Yamane et al., 1985). In this structure, the substrate leakage current is stopped, because the n^+ layers are placed on the channel layer. In addition, the sheet resistance of n^+-Ge, with a carrier concentration 100 times that of GaAs, is reduced to about one-tenth that of GaAs. The gate electrode is also formed close to n^+-Ge layers by self-aligned technology.

The fabrication process steps are illustrated in Fig. 36. An epitaxial layer, constructed on n^+-Ge (0.15 μm in thickness, 5×10^{19} cm^{-3} in doping density), n-GaAs (0.105 μm, 7×10^{17} cm^{-3}), p-GaAs (0.03 μm,

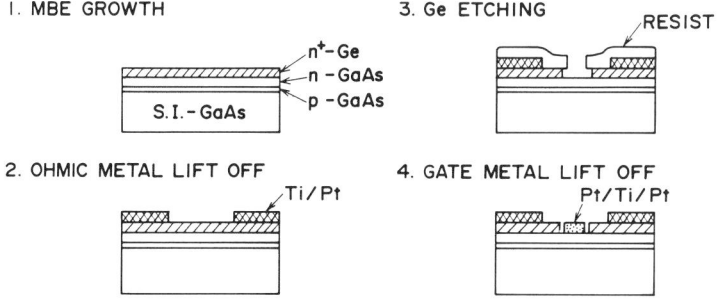

FIG. 36. Fabrication process for a Ge/GaAs source/drain MESFET: (a) MBE growth; (b) ohmic metal lift-off; (c) Ge etching; (d) gate metal lift-off.

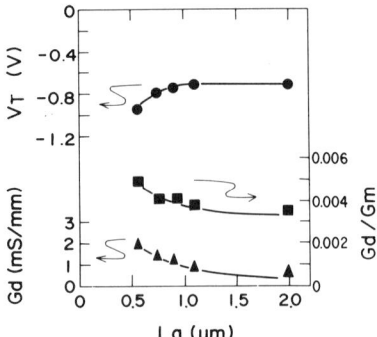

FIG. 37. (●) Threshold voltage V_T, (▲) drain conductance g_d, and (■) the ratio of g_d to the transconductance of Ge/GaAs source/drain MESFETs as a function of gate length.

1×10^{18} cm^{-3}), and undoped GaAs (0.5 μm), is grown by MBE on a semi-insulating GaAs substrate. Then, the source/drain electrode patterns are defined by photolithography. The Ti/Pt metal is evaporated and lifted off. Without any alloying process, the source and drain exhibit ohmic characteristics because of the high doping level of Ge. After gate patterning by electron-beam lithography, only the Ge layer is removed. This is done by means of RIE with 0.2 μm side-etching in CF$_4$. Then, the gate metals (Pt/Ti/Pt) are evaporated and lifted off. The Ge layer etched with an undercut ensures separation between the gate and the n^+ Ge layer.

The gate-length dependence of threshold voltage, drain conductance in the saturation region, and the drain-conductance/transconductance ratio (g_d/g_m) are shown in Fig. 37. These three parameters can be used to describe the short-channel effects. The drain conductance is reduced by a factor of about ten as compared with ordinary n^+ self-aligned MESFETs. The threshold voltage difference between 0.58 μm and 0.90 μm gate-length FETs is 0.17 V, and transconductance of a 0.58 μm gate-length FET is 360 mS/mm.

V. Circuit Elements and Interconnection

16. Diodes, Resistors, and Capacitors

The most important LSI element is a transistor. The DCFL and BFL circuits in GaAs LSIs need no elements except for E- and D-type FETs. Other circuits (e.g., SCFL, SDFL, LPFL, and CCL) require resistors, diodes, or capacitors for their circuit elements.

In an LSI fabrication process, all of the elements should be fabricated by the same process as much as possible. Since transistors and interconnections

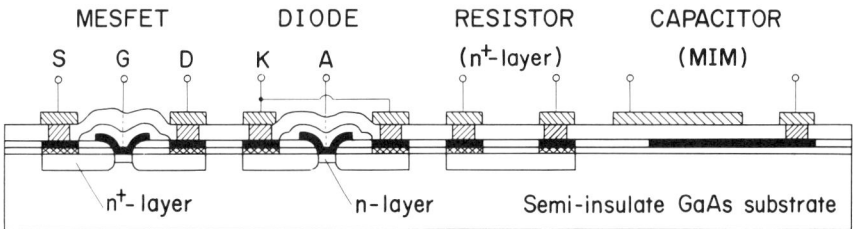

FIG. 38. Planar integrated GaAs elements with a SAINT FET, diode, resistor, and capacitor.

are the basic elements of LSIs, other elements should be fabricated based on these two important ones.

An example of an LSI cross-section fabricated using a completely compatible process for the FET, diode, resistor, and capacitor is shown in Fig. 38. In addition, a SAINT FET is also shown in the figure. The diode used for a level shift circuit or a coupling element has the same structure as the FET except that the source and drain electrodes are shorted using the first-level interconnection. The resistor uses an n^+ layer for a source/drain and the contact electrode structure is the same for the FET and diode. Furthermore, the capacitor is a metal insulator metal (MIM) structure, which is identical to the two-level metal interconnection.

The characteristics of each element are determined by the pattern layout. The diode current is determined by the fringe length in this structure, and it flows only at the fringe of the anode metal because the anode center has a large series resistance in comparison with the fringe area. It is a little difficult to accurately estimate the amount of current flowing in this structure by calculation alone using a one-dimensional diode model. If an n^+ layer is embedded under the n layer, the current flows vertically through the entire n layer. In this case, the diode current is proportional to the contact area of the anode metal.

However, the technology for embedding an n^+ layer is very complex for both GaAs and Si LSI processes. Even in matured Si bipolar processes, an embedded collector n^+ layer is formed by thermal diffusion into the p-type substrate before an n layer is grown by the CVD method. The n^+-layer embedding always requires additional processes for LSI fabrication. Since this extra processing generally results in additional processing time and lower chip yield, it is best avoided. In the GaAs LSI process, the situation is the same as for Si. As a result, the completely compatible process used for elements as shown in Fig. 38 is the best choice.

A diode is also used as a coupling element between fundamental logic gates because of its large capacitance as compared with an MIM capacitance having the same area. However, diode capacitance has nonlinear

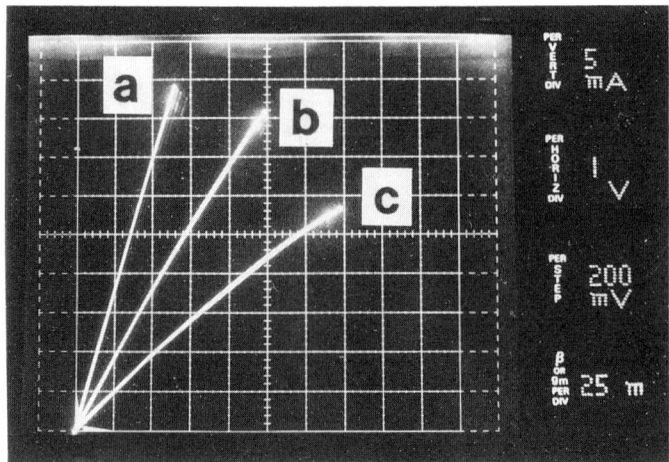

FIG. 39. Measurement results for resistors of (a) 50 Ω, (b) 100 Ω, and (c) 200 Ω designed using an ion implanted n^+ layer.

characteristics with respect to applied voltage. If the circuit requires constant capacitance, a MIM capacitor must be applied.

Resistors are very important elements in SCFL or E/R DCFL circuits. In addition, these circuits utilize a resistor as a load. The simplest resistor is made using an n or n^+ layer in a semiconductor. Examples of I–V curves of n^+ layer resistances are shown in Fig. 39. Each resistor is made by Si implantation with a 2.5×10^{13} dosage at 50 keV. The resistances obtained are 15% larger than the designed values.

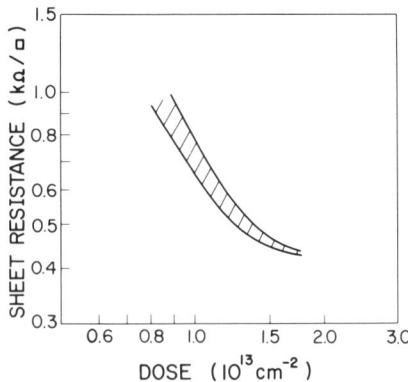

FIG. 40. Controllable sheet resistances for GaAs resistors using an ion implanted n^+ layer.

Sheet resistance of a Si-implanted layer is plotted as a function of implantation dose in Fig. 40. The implanted energy is 67 keV. The cross-hatched region shows the reproducible area from wafer to wafer or run to run. The high sheet resistance area for low dosage shows a wider zone because low dosages reduce the carrier concentration in the active layer, which is easily depleted by the charge in the GaAs surface state. On the other hand, the high-dosage area has a small zone because a high carrier density suppresses the surface state effectiveness. Furthermore, the absolute value of sheet resistance is saturated at low sheet resistances. As a result, activation saturation occurs at a carrier density of 1×10^{18} cm^{-3} for conventional thermal annealing.

In some LSI circuits, more than 1 kΩ of sheet resistance is required. For this purpose, a p layer gives a resistance ten times larger than that of an n layer because the hole mobility is one-tenth to one-twentieth of the electron mobility. Thus, p-layer sheet resistance is 10 to 20 times larger than that of an n layer. However, it should be noted that ohmic contact metal applied to the p layer is incompatible with an n layer.

A MIM capacitor shown in Fig. 38 utilizes the same insulation layer as that used for interconnection. The insulator is usually 0.3- to 0.7-μm-thick SiO$_2$ or SiN films. The fabrication processes of MIM capacitors are similar to Si LSI processes except for the deposition temperature of the layer. Since the insulator SiO$_2$ or SiN is deposited after Schottky barrier formation, the deposition temperature should be lower than 300°C, except for the processes using refractory gate metals such as WSi. Even for WSi gate material, however, the deposition temperature is limited to below 350°C because AuGeNi ohmic contacts are sintered at 400–450°C.

17. INTERCONNECTION

The interconnection technology used in GaAs LSIs is very different from that used in Si LSIs. In GaAs LSIs, connections are made to Schottky gate metal of Ti/Pt/Au and ohmic contact metal of alloyed AuGe/Ni or AuCr. The interconnection metal should maintain good contact conditions to these metals over long periods. Furthermore, GaAs LSIs operate at very high speed to exhibit the intrinsic high-performance properties of GaAs FETs. High-speed operation requires high current flow, which in turn requires very low resistance in the interconnection line. As a result, gold is used as an interconnection metal because of its good contact and low resistance.

Although Al interconnection is generally used in Si LSIs, it is not appropriate for GaAs LSIs. Several problems arise in Al interconnection. First, electromigration occurs at a high current density in GaAs LSIs. Second, RIE for Al uses chlorine-containing gases, such as CCl$_4$, BCl$_3$, and

$SiCl_4$, that also etch GaAs material. The direct deposition of Al to Au produces a purple plague. A barrier metal, such as Ti or Mo, is needed between Al and Au in order to prevent alloying. Then, a Ti/Pt/Au or Mo/Au interconnection layer would be appropriate for contacting the Al Schottky metal in GaAs LSIs.

Cross-sections of three kinds of interconnection structures are shown in Fig. 41. Although GaAs substrates are usually semi-insulating, it is preferable to lay the first level interconnection on SiO_2 or SiN film in order to avoid current leakage for long parallel lines. Figure 41a shows conventional technology in which a via hole with tapered walls is adopted for second-level interconnection. The size of this hole is usually 2–3 μm square. The second-level metal is Au with Ti or Mo and is defined using an ion-milling technique. Next, Figure 41b shows a completely planar interconnection that was initially developed by Rockwell using spacer lift-off technology (Welch *et al.*, 1979). The details of this interconnection will be described later.

The final part of Fig. 41c shows an air-bridged interconnection that is applicable to a digital IC, which is already used in monolithic microwave integrated circuits (MMIC). This technology is very effective in reducing the interconnection capacitance, particularly that of crossover. A semicircular air bridge is formed using the following process. Photoresist with a semicircular cross-section is formed over the first level metal and a Au bridge is created by electroplating. Finally, the semicircular photoresist is removed.

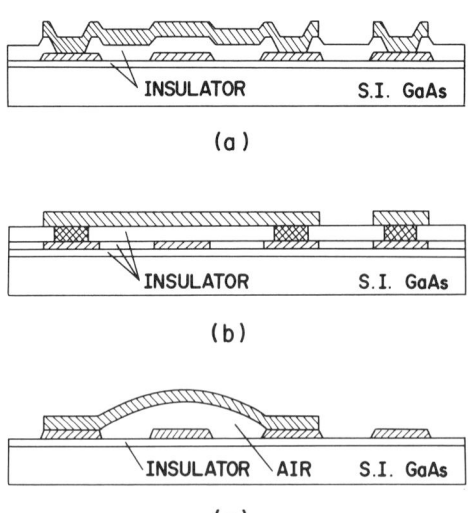

FIG. 41. Interconnection structures for GaAs LSIs using (a) conventional technology, (b) planar technology, and (c) air bridge technology.

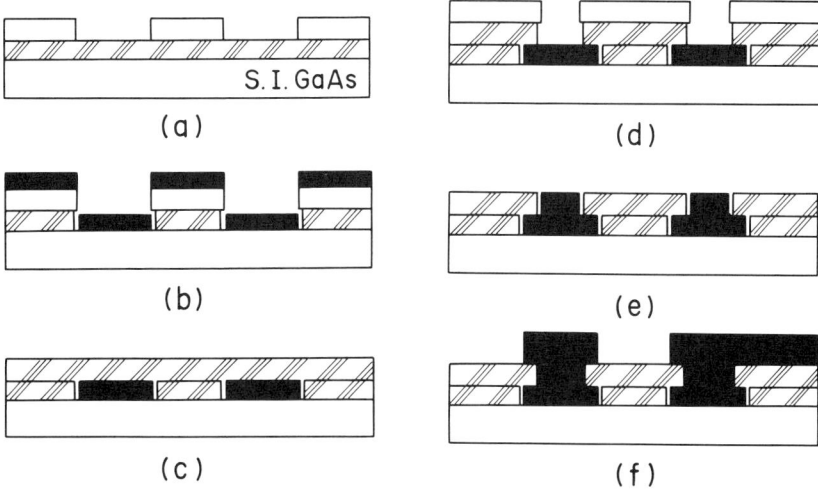

FIG. 42. Fabrication process for planar interconnection using spacer lift-off technology: (a) spacer-film deposition and photoresist-patterning; (b) spacer-film etching and metal evaporation; (c) metal lift-off and insulator film deposition; (d) via-hole opening; (e) via-metal embedding by lift-off; (f) second-level metal formation.

A problem with this technology, however, is that difficulty is encountered in accurately shaping the semicircular photoresist bridge.

A completely planar interconnection process is illustrated in Figs. 42a to 42f. In the figure, the first-level insulator below the first-level metal is omitted. The first step shown in Fig. 42a represents spacer-film deposition and the first-level interconnection photoresist patterning. Next, the spacer film is etched by RIE and metal is evaporated onto a wafer as shown in (b). Then, metal is removed by lift-off and an insulator film is deposited (c), and windows are opened in the insulator film (d). Finally, the same lift-off process is employed to make via-metal embedding in the via hole as shown in Fig. 41e. After the via metal embedding, the entire wafer surface becomes flat because the spacer is used. The second-level interconnection is made using an ordinary deposition and etching technique on a flat surface.

An advantage of the above technology is that it is free of the step-coverage problem because the via metals allow complete connection between the first- and second-level lines. A narrow but adequate gap between the spacer film and the first-level metal, or the insulator film and the via metal, is carefully adjusted during the SiO_2 or SiN layer etching process to ensure the lift-off. The gap is filled with an insulator or metal in the following process. Only a trace of the gap appears as a line on the flat surface of the second-level interconnection.

TABLE VI

INTERCONNECTION METAL LINE SHEET RESISTANCES FOR GaAs LSIs

	Thickness (μm)				Sheet resistance ($\Omega/\square$)	Resistance (Ω)
	Ti	Mo	Pt	Au		
First level	0.1	—	0.05	0.2	0.22	—
Second level	—	0.1	—	1.0	0.026	—
Via (2 μm square)	—	—	—	0.5	—	0.1

The results for interconnection in the fabrication process TEG mentioned in Section 7 are presented in Table VI. Ti or Mo is used to maintain good adhesion to the insulator in the first- and second-level metals. In addition, Pt prevents Ti and Au alloying. The interconnection resistance is almost entirely determined by the Au layer thickness. The sheet resistances of the first- and second-level metals are larger than bulk Au resistivity by factors of 1.4 and 1.1, respectively. The contact resistance of the via metal is small enough to apply to GaAs LSIs but is 20 to 30 times larger than the calculated value for the ideal case. This is because the effective contact area between

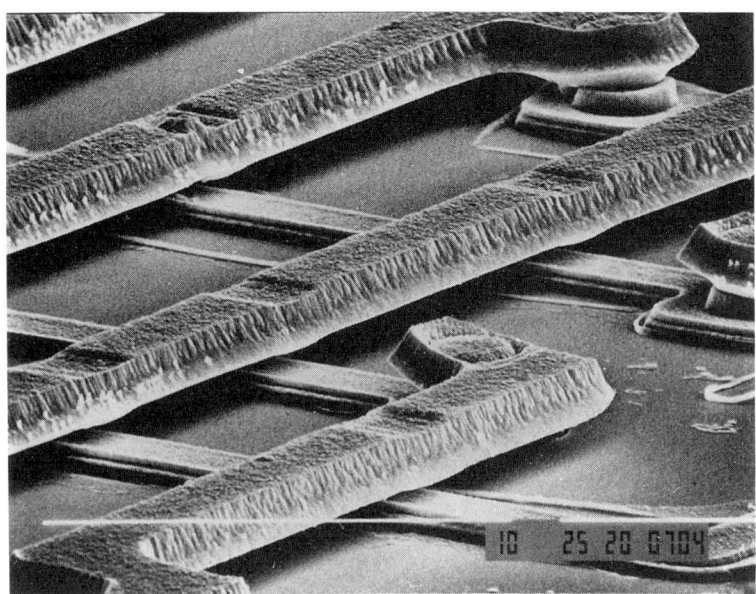

FIG. 43. Fabricated planar interconnection on GaAs SSI chip. Second-level gold metal thickness is 1 μm.

the via metal and the second-level metal is limited to a partial area. This may be improved by reducing the gap between the via metal and the hole.

Currently, the technology for creating a planar air bridge is available (Enoki et al., 1986), as shown in Fig. 43. A SiN film that fills the gap between two-level interconnection lines is plasma-etched to form an air bridge. The via metal acts as piers for supporting the second-level metal in the air. This bridge also has a low interconnection capacitance and a crossover capacitance that is similar to that of a semicircular bridge interconnection. With the planar air bridge, a one-by-two frequency divider could be operated at a frequency up to 11 GHz, which is 1.2 times higher than that of the SiN-filled interconnection. It should be noted that the surface of the chip must be passivated with a SiO_2 or SiN film even for the planar air-bridge interconnection.

VI. Assembly Techniques and Packaging

18. Dicing and Bonding

The assembly technology for a GaAs LSI is different from that of a Si LSI. Assembly and packaging technologies for very high-speed LSIs are also different from those for ordinal Si digital IC's operating at frequencies lower than 100 MHz. The assembly process and conditions are shown in Table VII. GaAs-wafer dicing with a diamond blade is very difficult as compared with Si-wafer dicing, so the spindle speed should be at least 30,000 rpm. In addition, the cutting speed should be very slow: 1 to 15 mm/sec for GaAs (Simada, 1985), as compared to 50 to 200 mm/sec for Si. Since GaAs crystal is brittle in comparison with Si, the cutting blade must rotate rapidly and move slowly.

A microphotograph of a diced GaAs wafer is shown in Fig. 44. The vertical lines show more, and more severe, chipping than the horizontal lines, which are parallel to the orientation flat of a $\langle 100 \rangle$ surface wafer. A hubless diamond blade 40 μm in thickness was used, and the groove width cut by the

TABLE VII

Assembly Technology and Conditions for GaAs LSIs

Process	Condition
Wafer dicing	Very low speed
Die bonding	Low temperature (less than 300°C)
Wire bonding	Ultrasonic bonder
Lid bonding	Low temperature (less than 300°C)

FIG. 44. Chipping phenomena on a ⟨100⟩ surface GaAs wafer cut with a diamond saw. The horizontal line is parallel to the orientation flat of the wafer.

blade is 50 μm. The chipping width is less than 15 μm in the worst direction. The loss width is sufficient for dicing because the bonding pads are usually located 50 μm apart from the edge of the dicing line. Die, wire, and lid bonding process are almost the same as in Si LSI assembly except for the allowable highest temperature. A GaAs Schottky barrier with nonrefractory metal begins to alloy with GaAs at temperatures of more than about 400°C. The alloying deteriorates Schottky characteristics and FET performances. The highest temperature should be lower than 300°C and the working time shorter than 30 min to keep barrier metals, such as Ti/Pt/Au, unchanged.

For die bonding, a AuSn solder is used after metallization on the back side of a wafer. An example of good die bonding using a AuSn solder is shown in Fig. 45. The solder sheet used is 20–30 μm thick and is molten at temperatures lower than 300°C. The microphotograph shows no extra-soldered areas and the applied solder stays precisely along the fringe of the chip (Shimada, 1985). Since the bonding pad is 100 μm square, the solder line width is less than 10 μm. Resin with Ag adhesive is also usable without back side metallization. Wire bonding using a Au whisker with an ultrasonic wire bonder is a sure and safe method for GaAs LSI.

19. PACKAGE

Packages for ultrahigh-speed operations must satisfy the following requirements:

FIG. 45. A GaAs bonded chip in a ceramic package using AuSn solder.

(1) Impedance matching between package leads and interconnection lines on the motherboard;
(2) Small stray capacitance to ground;
(3) Small coupling capacitance among the leads;
(4) Small lead resistance of metallized lines in ceramic material; and
(5) High-density pin location.

The cross-section of a sample package operating at superhigh frequencies with 24 pins (named SH-24) (Hirayama, 1984) is shown in Fig. 46. The ground level metal has a step in order to maintain a constant impedance lead with 50 Ω because the lead width is changed at the center. The via holes are used to connect the two ground levels. The lead length on the ceramic material is reduced to 3 mm for minimizing propagation delay time, transmission loss, and extra area. A typical pitch of leads is 1.27 mm.

The measured lead capacitance to ground and lead resistance are 0.68 pF and 100 mΩ, respectively. The stray capacitance between leads is as small as 0.05 pF, which is less than one-tenth that of conventional packages. The ratio of capacitances to the ground and to the neighboring lead is sufficiently large of 0.68/0.05 (=14); however, very little crosstalk occurs. A circuit must maintain sufficient noise immunity to keep complete operations.

Frequency characteristics of the insertion loss of packages are shown in Fig. 47. SH-24 is the package shown in Fig. 46, while UH-24 is a nonflat

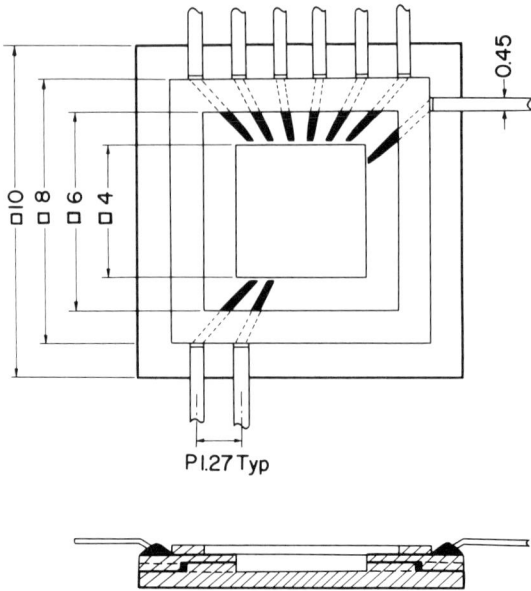

FIG. 46. A specified flat package for superhigh frequency (SHF) or a gigabit-rate digital IC with 24 pins using a stepped ground plane (SH-24). Upper part is the top view of the package with 50 Ω line leads. Lower part is the cross-section of the package. Ground level has a step to maintain 50 Ω line constant impedance at a point of center-metal width change.

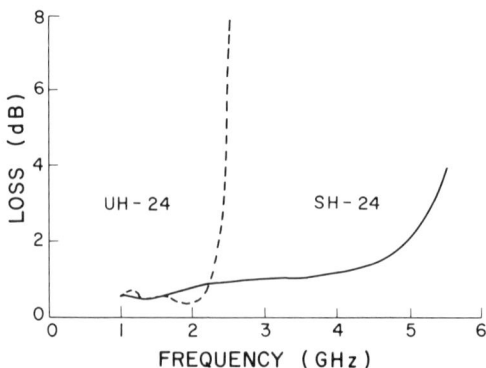

FIG. 47. Insertion loss characteristics of specified flat package, SH-24 (solid line) and nonflat package, UH-24 (dashed line).

package specially made for operation up to 2 GHz. A small loss up to 5 GHz in SH-24 is featured. The main reason for this improvement is the 50 Ω lead design, which uses a step ground.

VII. LSI Fabrication and Performance

20. UNIFORMITY AND CONTROLLABILITY OF THRESHOLD VOLTAGE

Two of the most significant problems in fabrication of GaAs LSIs are uniformity and controllability of the FET threshold voltage V_T. Circuit operation requires uniformity in circuit elements, especially for FETs. Furthermore, controllability means absolute threshold-voltage control of FETs.

GaAs is not as nearly perfect a material as Si. Dislocations exist at high densities of about 10^4 cm^{-2} and directly affect the FET threshold voltage in conventional LEC crystals, as mentioned in Section 6. FET threshold voltage nonuniformity has been a major problem arising from the existence of dislocations. Excellent uniformity (Yamazaki et al., 1984) was obtained by using In-doped dislocation-free (DF) LEC-grown crystals with etch pit densities (EPD) of less than 200 cm^{-2} as an LSI substrate. Thus, the absolute-value control of FET threshold voltage becomes an important problem in the present process technology.

All experimental results described in this section are based on the process evaluation TEG described in Section 7. The TEG was fabricated on the same wafer using the same technology with which a real LSI chip has been fabricated (Matsuoka et al., 1984). *In situ* process monitoring for the other results is also made for addition of the FET threshold voltage check. These results are used to estimate the operational circuit margin.

The threshold voltages of 1 μm gate SAINT FETs fabricated on low Cr-doped LEC wafers, ingot-annealed low Cr-doped LEC wafers, and In-doped DF LEC wafers were measured by using process-evaluation FETs. The results are shown in Fig. 48. The differences among those wafers are obvious. The DF wafer shows the smallest threshold voltage scattering. The standard deviation of EFETs is 20 mV over a 2-inch wafer. Over conventional and ingot-annealed LEC wafers, the standard deviations are 67 and 69 mV, respectively, as shown in Figs. 48a and 48b. The standard deviation of a DF wafer is about one-third that of conventional LEC wafers. Moreover, the ingot-annealed wafer shows better uniformity than the conventional LEC wafer, but some FETs have an unusually low threshold voltage. The reason for these low threshold voltages is not yet clear.

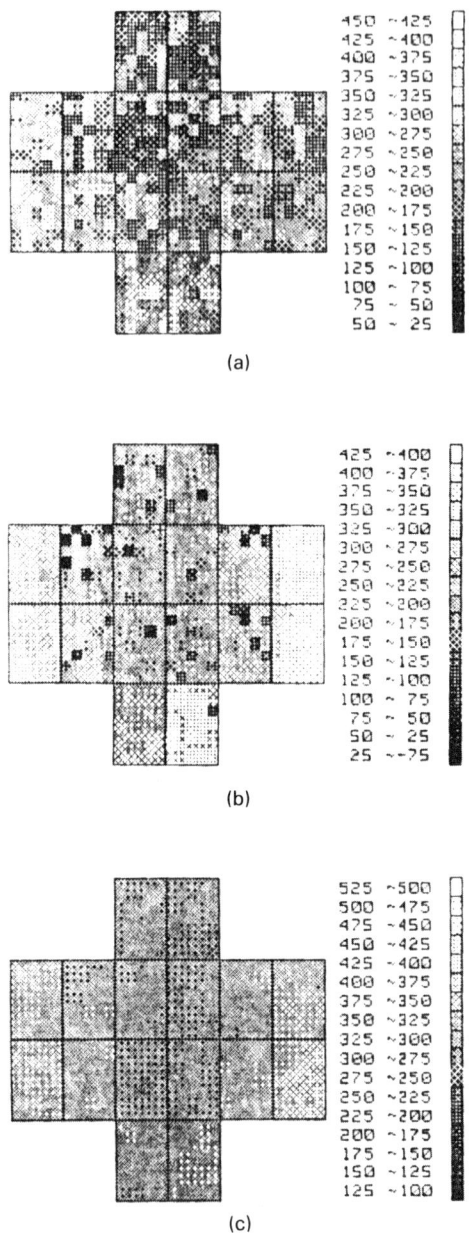

FIG. 48. Threshold voltage uniformity map of 1 μm gate SAINT FETs as a process evaluation TEG. The TEG adopts (a) a low Cr-doped conventional LEC, (b) a low Cr-doped long hour annealed LEC, and (c) In-doped very low EPD LEC crystals.

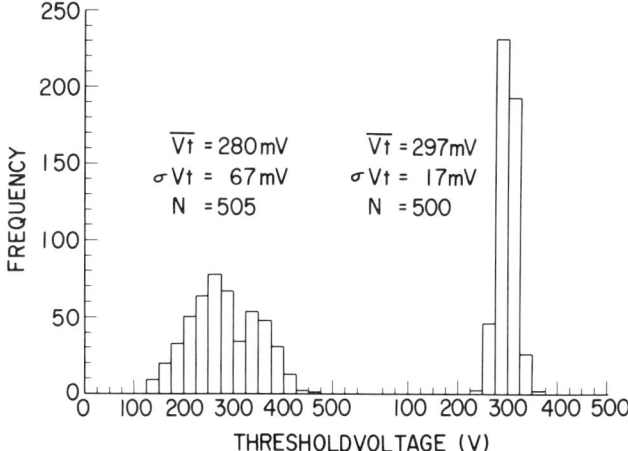

FIG. 49. Histograms of threshold voltage deviations of FETs fabricated on a low Cr-doped LEC wafer (left) and an In-doped very low EPD LEC wafer (right).

Histograms for EFETs fabricated on conventional and DF wafers are shown in Fig. 49. The standard deviations are 17 mV in the DF wafer and 67 mW in the conventional one. Over a small area of less than 100 mm^2, it is less than 10 mV for DF wafers.

All the data for threshold voltage nonuniformity shown above were obtained from FETs fabricated using a 10:1 projection-type stepper. In previous experiments, a contact aligner was used. Another main cause of threshold voltage nonuniformity is in-process scattering of the gate length. A contact aligner has larger scattering in photoresist patterning. Since a 1 μm gate FET has slight short-channel effect, gate-length scattering results in threshold-voltage scattering.

Threshold-voltage scattering is dependent both on the aligners used and on GaAs crystals, as shown in Fig. 50. These data represent the results from approximately 500 E- and D-type pairs of FETs separated by 8.5 μm gate-to-gate. The difference between (b) and (c) is due to the different mask aligners. Threshold voltage scattering is reduced by 10 mV by the use of a stepper. In Fig. 50b, the standard deviations for E- and DFETs are 74 and 104 mV, respectively. The difference between DF and conventional LEC wafers is shown in Figs. 50a and 50b. Presently, the technique using In-doped DF wafers and a stepper gives the best result for uniformity, with 20 and 24 mV of standard deviation for E- and DFETs, respectively.

A crystal was also evaluated using DCFL inverters. Figure 51 shows transfer characteristics of E/D DCFL inverters fabricated on (a) a conventional LEC, (b) an ingot-annealed LEC, and (c) a DF LEC. V_{out} is limited

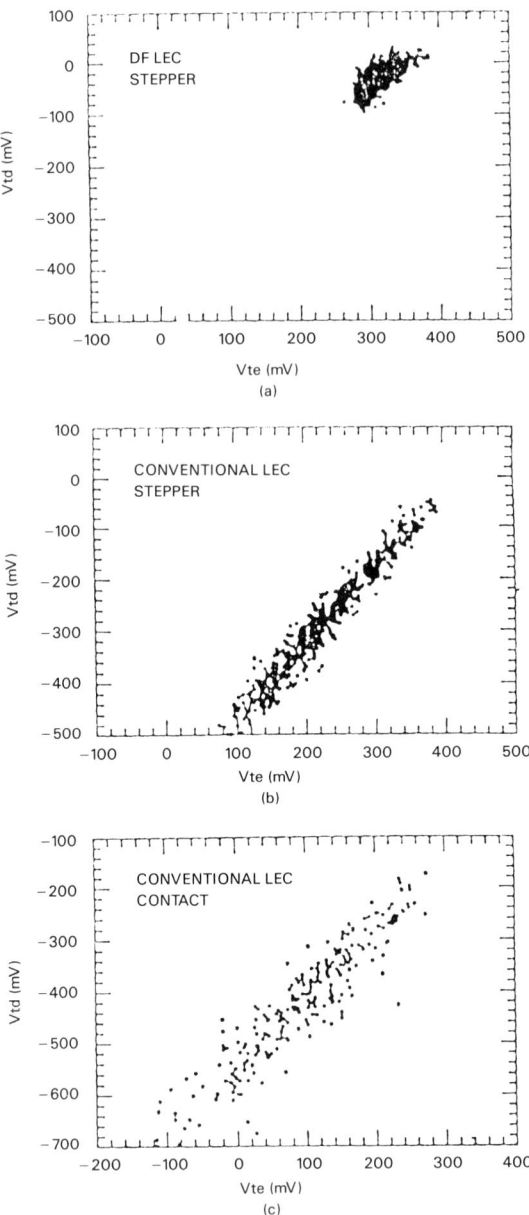

FIG. 50. Threshold voltage correlation for E and D type FET pair using (a) In-doped LEC and (b, c) conventional LEC crystal. Stepper photolithography was used in (a) and (b), and contact photolithography in (c). [From Ohmori (1984).]

5. GaAs LSI FABRICATION AND PERFORMANCE

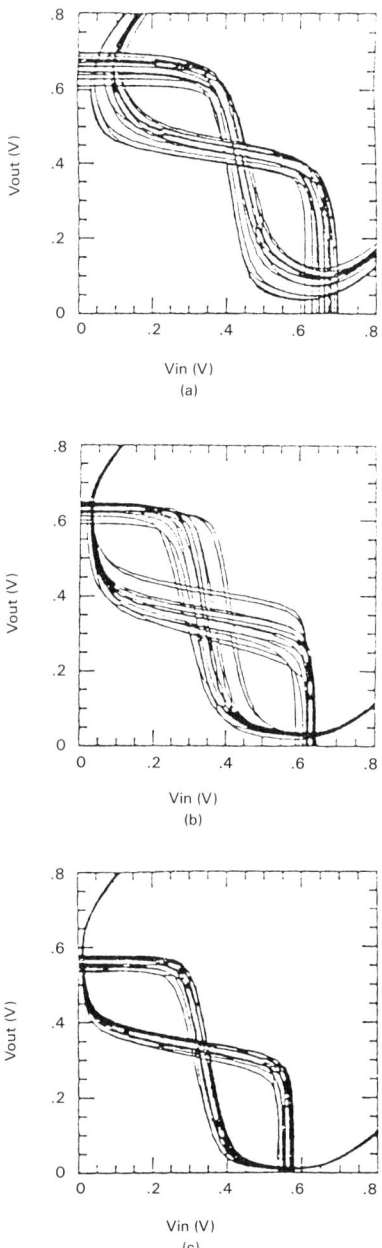

FIG. 51. DCFL circuit inverter characteristics for (a) conventional LEC, (b) long hour annealed LEC, and (c) In-doped LEC crystals.

by the input voltage of a Schottky gate at the next stage. Scattering of the switching voltage is small even in the conventional LEC, because the threshold voltages of E- and DFETs located near each other changes to the same direction. The ingot-annealed wafer shows very large scattering because of the unusually low threshold voltages of DFETs.

The low logic level V_L and high logic level V_H in transfer characteristics are different in all three of the wafers tested. The standard deviations for V_L and V_H are 10 and nearly 0 mV, respectively, in the DF LEC wafers. The nearly zero V_L scattering is an indication of high crystalline uniformity of the DF LEC wafer, and such a low scattering assures reliable operation of GaAs DCFL LSIs. It should be noted that the transconductance of SAINT FETs on DF wafers is not different from that on conventional dislocated wafers.

Threshold voltage control is the most important issue in LSI fabrication. Dependence of the threshold voltage on ^{29}Si ion dosage at an implantation energy of 67 keV is shown in Fig. 52. The cross-hatched area shows the scattering of threshold voltages derived from the results of 1 μm gate SAINT FETs fabricated on conventional, ingot-annealed, and DF LEC wafers. As a result, threshold voltage standard deviations for the three wafers are different from each other, but the mean threshold voltages for these wafers are plotted in the same area. Of course, the scattering is different, as mentioned above. The controllable regions for EFETs with a threshold voltage near 0.15 V and DFETs near -0.5 V are 100 and 150 mV, respectively. The reproducibility of the mean threshold voltage described above was also the same for Cr-doped HB crystals.

In-process monitoring of the threshold voltage was carried out by using 1 μm gate SAINT FETs. The measured results are shown in Fig. 53. Threshold voltages of FETs described in Section 7 were in-process-monitored

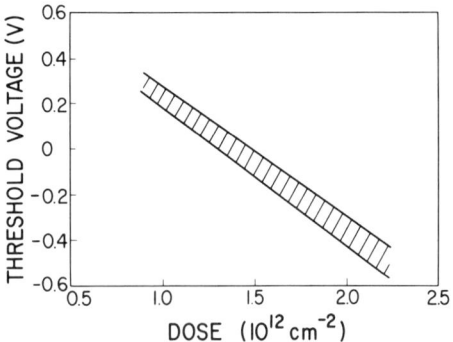

FIG. 52. Controllable threshold voltage for SAINT FETs using conventional, long hour annealed, and In doped LEC crystals.

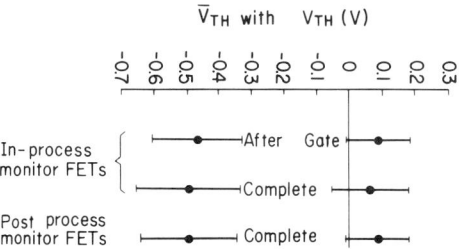

FIG. 53. Threshold voltage measurement results of FET threshold voltage for in-process and post-process monitor FETs using conventional LEC crystal. The bar length indicates the standard distribution of threshold voltage.

after gate formation and interconnection completion. The mean threshold voltages for EFETs and DFETs are lowered by less than 30 mV after the interconnection process. Post-process monitor FETs cannot be measured after gate formation and are identical to the FETs used in real LSIs. Moreover, the threshold voltage is almost the same as that of the in-process monitor FETs.

21. MEMORY LSIs

GaAs LSI technology in the research state experienced widespread applications in 1980, which was about 10 years later than Si LSIs. GaAs LSIs required the development of new fabrication processes and device structures. The mass production of 2–3 inch wafers has also been achieved. Highly uniform wafers become available by In-doping for GaAs LSI applications. Since low-temperature process techniques are required for GaAs LSI fabrication, plasma-enhanced CVD is used. Fine patterning has been established using a stepper and reactive ion etching (RIE). In addition, high-g_m MESFETs have been developed using self-aligned n^+-layer implantation technology, such as SAINT and WSi gate. The first version of the above-mentioned technologies, including interconnection, was developed from 1980 to 1982 for high-speed GaAs LSIs.

The first chip produced was a logic LSI of an 8×8 bit multiplier from Rockwell. The first memory LSI was an 8-bit static RAM from Thomson-CSF (Bert *et al.*, 1981). GaAs memory is limited to static random access memory (RAM) because a soft-pinchoff FET allows large leakage current in the nanoamp range. A dynamic RAM applied to Si memory requires a very small pinchoff current on the order of a picoamp. Therefore, GaAs MESFETs can be applied only to static RAMs.

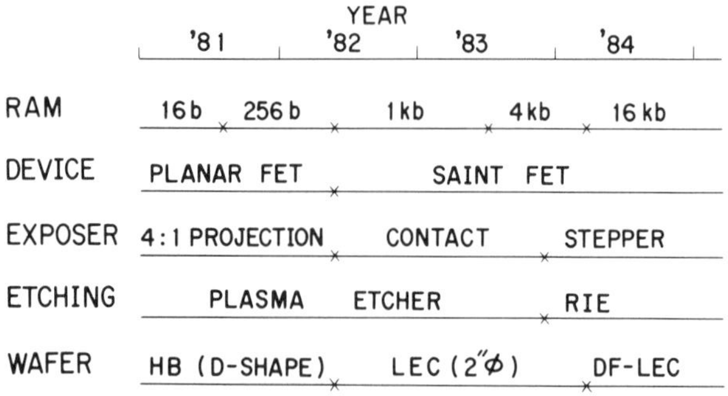

	YEAR			
	'81	'82	'83	'84
RAM	16 b 256 b	1 kb	4 kb	16 kb
DEVICE	PLANAR FET		SAINT FET	
EXPOSER	4:1 PROJECTION	CONTACT		STEPPER
ETCHING	PLASMA	ETCHER		RIE
WAFER	HB (D-SHAPE)	LEC (2″∅)		DF-LEC

FIG. 54. Research and development history of GaAs LSIs at NTT. DF-LEC represents In-doped wafer.

A typical technological flow of LSI memory development is shown in Fig. 54. The 16- and 256-bit RAMs triggered further development of LSI technology. During this stage, two-level interconnection technology was already established. Next, a 1-kbit RAM, the first LSI RAM, was fabricated using a 2-inch-diameter LEC wafer and the n^+ self-aligned FET technology (SAINT). The integration density was limited by the pattern rule using a contact mask aligner and a plasma etcher, and the design rule for line/space was 3/3 μm. As the next step, the integration density was improved by four timers by using a 10 to 1 reduction stepper and RIE. In this case, a design rule of 1.5/1.5 μm was adopted for line and space. By using both (3/3 μm and 1.5/1.5 μm) types of technology, 4-kbit SRAMs (Hirayama et al., 1984; Idda et al., 1984) were sequentially fabricated. Finally, the memory cell area was reduced by two-thirds from 4030 to 1350 μm². Metallic amorphous silicon gate FET (MASFET) (Suzuki et al., 1984) was also applied to 4 kbit static RAM (Suzuki et al., 1985).

Next, a 16-kbit SRAM was successfully fabricated on an In-doped DF wafer (Hirayama et al., 1986) by adopting the 1.5 μm design rule. A

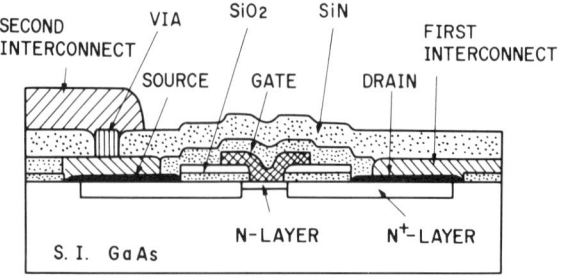

FIG. 55. Developed GaAs 16-kbit RAM with a cross-sectional view of planar integration.

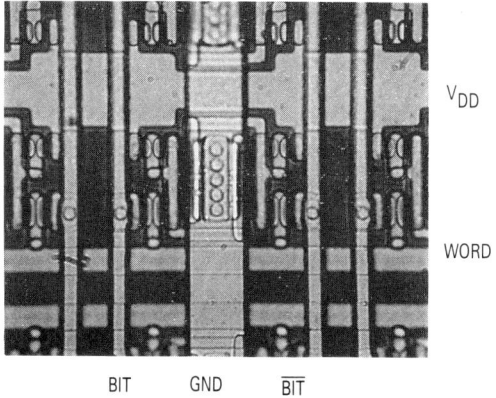

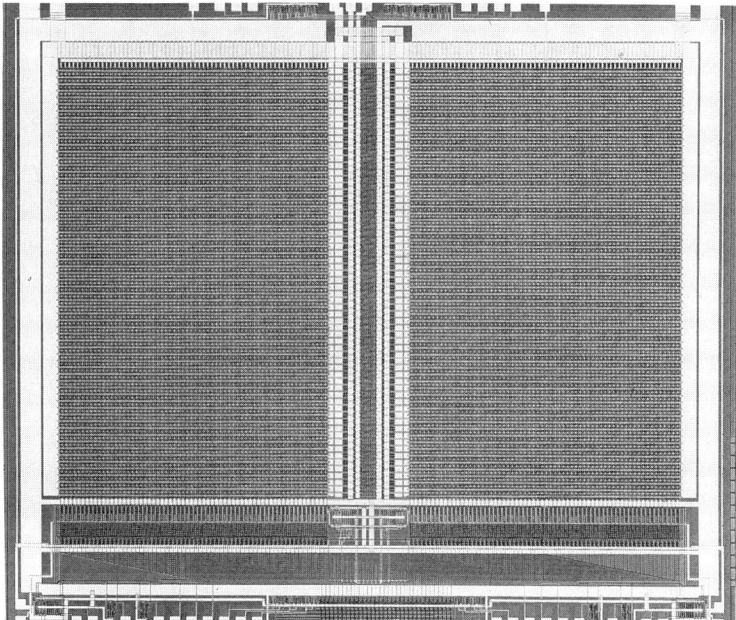

FIG. 56. Microphotographs of (a) a memory cell using SAINT FETs and (b) a memory chip integrating a total of 102,000 FETs and diodes.

cross-sectional view of an LSI using SAINT and planar interconnection is shown in Fig. 55. Although the area on the sputtered SiO_2 and the gate metal is not flat, the second-level interconnections can be easily overlaid without disconnection. The SiN film for spacer lift-off and for the insulator is deposited at 200°C, which is a sufficiently low temperature after Schottky gate formation.

A fabricated memory cell microphotograph is shown in Fig. 56a. As seen in the figure, all FET gates are vertically aligned. Two driver FETs with the widest gates (12 μm) are at the center of the memory cell and are located under the second-level metal of the ground. In addition, the ground line is completely isolated from the gate metal, which sits on a SiO_2 step. The DC supply line and word line in a cell are oriented horizontally using the first-level metal. Moreover, bit lines and a ground line are arranged vertically with the second-level metal.

A microphotograph of the 16 kbit SRAM chip is shown in Fig. 56b. The memory cell array is divided into two parts to reduce the word line voltage drop. The minimum measured access time was 4.1 nsec with a dissipation power of 1.46 W, using the package measurement system. The monitor FET threshold voltages on this chip are 0.186 V for the EFET and -0.393 V for the DFET. Standard deviations for these FETs were 9 and 14 mV for the EFET and DFET, respectively.

The most advanced RAM technology to date is a high-density integration using tri-level interconnection (Kato et al., 1985a) as shown in Fig. 57a. A large g_m (transconductance) of 260 mS/mm has been obtained for the EFETs. A large g_m value is achieved in a MASFET by using both BP-SAINT, described in Section 15, and shallow n-layer implantation at an acceleration energy of 30 keV. A novel static RAM cell layout dramatically reduced the cell area to 455 μm^2, as shown in Fig. 57b. To achieve such a compact cell in addition to tri-level interconnection, the electron-beam delineation of $0.8 \times 1.3 \mu$m via holes has been developed for direct contact to gates and ohmic metals. In addition, ohmic metals are also applied to inner cell interconnection.

Complementary JFETs technology for an extremely low-power static RAM (Zuleeg et al., 1984) was developed with the circuit design. $^{29}Si^+$ and $^{24}Mg^+$ were implanted for n^+/n and p^+/p regions, respectively. A junction gate was then formed by the n^+/p and p^+/n junctions. Only a memory cell stack utilized the complementary circuitry with a power dissipation of 100 nW/cell.

22. Logic LSIs

Logic LSI fabrication technology is not significantly different from that for memory LSIs. The sheet resistance of the first-level interconnection is larger than that of the second-level interconnection, which occasionally presents a problem. This means that the wide line width of the first-level metal obstructs the high density integration of logic circuits, whose interconnection area is much larger than that of memory circuits.

The largest scale of logic circuit made to date is a 16 × 16 bit parallel multiplier using WSi gate self-aligned FETs (Nakayama et al., 1983). The

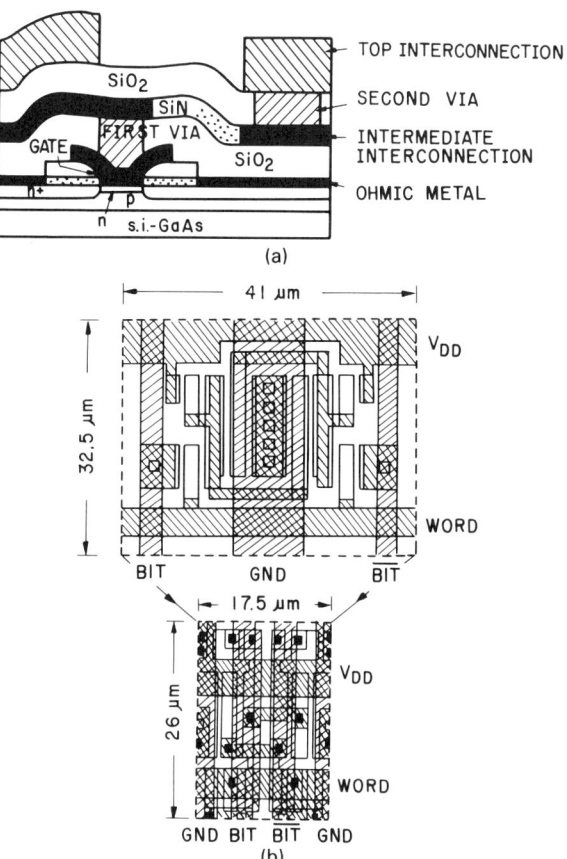

FIG. 57. High-density RAM technology using three-level interconnection and direct contact to Schottky barrier area: (a) cross-section and (b) reduction of memory cell area to one-third.

technology used in the LSI was 2 μm gate length, 2 × 2 μm contact hole, and 2 μm line width. In this case, a DCFL circuit with 3168 NOR gates was applied to an LSI with a 4 × 4 mm chip size. The multiplication time was 10.5 nsec with an associated power dissipation of 952 mW. The mean value of the FET threshold voltages was 0.002 V for the EFET and −0.779 V for the DFET, and typical transconductance was 170 mS/mm for the EFETs.

Half-micron-gate FETs (Yamasaki *et al.*, 1984) were successfully adopted for LSIs. BP-SAINT FETs were applied to an 11 GHz frequency divider, a 1 GHz frequency prescaler with 5 mA of operation current (Takada *et al.*, 1985a), and a 2 Gb/sec throughput digital time switch LSI for communications systems. A 4-channel digital time switch LSI having a 2.0 Gb/sec throughput and 500 Mb/sec channel data rate was then developed using a 0.5 μm gate-length BP-SAINT (Takada *et al.*, 1985b). The fabrication process of the 0.5 μm gate FETs began with 1 μm width photoresist patterning with a tri-level resist using a stepper. After descumming, SiO_2 etching, bottom resist patterning, and undercutting, the Schottky gate contact length to the GaAs became 0.5 μm. The threshold voltage standard deviation was as small as 49 mV for 0.5 μm gate FETs. This was due to the high uniformity of dislocation-free wafers as well as the reproducible BP-SAINT process. The mean threshold voltage was 0.18 V, which was appropriate to low power source coupled FET logic (LSCFL) circuits. Here, the FET transconductance was 160 mS/mm.

23. Preliminary Yield and Reliability

It is felt that the GaAs process has reached a stage of development for providing a satisfactory circuit yield. To date, NTT has reported laboratory yield data for several MSI–LSI chips. In general, however, the yield discussed is for mass production, which has large process tolerances. Despite this fact, laboratory yield data can sometimes be used to estimate improvements in the fabrication process.

The reported preliminary yields obtained in NTT laboratories are shown in Fig. 58 (Kato *et al.*, 1985b). All of these data were obtained from SSI-LSIs with the source coupled FET logic (SCFL) circuit configuration. Circuits

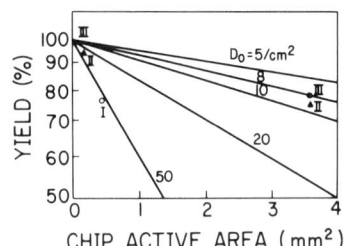

Version	Process	Lg	Lithography	n-Layer	Wafer
I	SAINT	0.8 μm	Contact+EB	67 keV	Conventional LEC
II	BP-SAINT	0.55 μm	Stepper	67 keV	Low EPD LEC
III	BP-SAINT	0.5 μm	Stepper	30 keV	Low EPD LEC

Fig. 58. Preliminary chip yield of a test run at NTT using SAINT. The stepper contributed to a photoresist defect reduction. [From Kato *et al.* (1985b).]

with greater complexity naturally have lower yields. To clarify the improvement for various defects including GaAs crystal defects, the Poisson distribution $Y = e^{-D_0 A}$ is also shown by solid lines, where D_0 is the randomly distributed defect density of any type, and A is the active chip area. The defects are caused by dust, scratches, nonuniformity in threshold voltage, or other local abnormalities in the device fabrication process. A process having a small D_0 value can be regarded as advanced.

Improvements from version I to version II are based on the application of DF wafers and a stepper. The DF wafer effectively reduces the scattering in FET threshold voltage and the resistance in the ion-implanted layers. The stepper also reduces photoresist defects that cause photoresist striping in the contact print. Furthermore, no significant differences are observed between versions II and III.

Reliability of GaAs LSIs has not yet been sufficiently investigated, because GaAs LSIs are still in the early stage of development. The gate voltage, gate width, and threshold voltage of digital LSIs are different from those of discrete FETs. However, the reliability values for gate metal and ohmic contacts in LSIs are similar to other FETs and microwave circuits.

Discrete FETs for a microwave amplifier were tested to examine the gate-metal structure. Reliability in microwave power FETs was also investigated because of severe operational conditions. For example, a 500 Å Ti/4500 Å Al/GaAs structure was tested in a cumulative failure distribution (Wada and Chino, 1983). In this structure, the overlaid Ti metal suppressed electromigration of the Al layer. Although the use of Au/refractory metal gates is one approach for improving electromigration endurance, this metallization may cause failures, such as an increase in current leakage induced by the Au and GaAs reaction. As a result, the reliability of a Au/refractory metal gate at the operation bias point of GaAs LSIs should be substantiated.

The reliability of ohmic contacts for 400 Å Pt/1000 Å Au/500 Å Ge/50 Å Ni/GaAs was tested in relation to first-level interconnection metallization of 5000 Å Au/500 Å Pt/250 Å Ti. The inserted 500 Å TiN film served as an effective barrier to degradation for 300 h at 250°C in air (Remba *et al.*, 1985). The TiN film was deposited by reactive sputtering.

The final reliability factor to be tested is radiation hardness. A half-micron-gate BP-SAINT FET was tested in gamma radiation. Samples were packaged in a ceramic DIP with AuSn solder and Au wire bonding. A FET has a Au/Pt/Ti/GaAs gate metal with a 0.55 μm gate length and 10 μm width. The gamma ray source was ^{60}Co and the total dose was 10^8 rad (Ogawa *et al.*, 1985).

The results are shown in Fig. 59. From the figure, it can be seen that the total gamma ray dose reduces transconductance and increases series resistance. In addition, a threshold voltage shift in the positive direction is

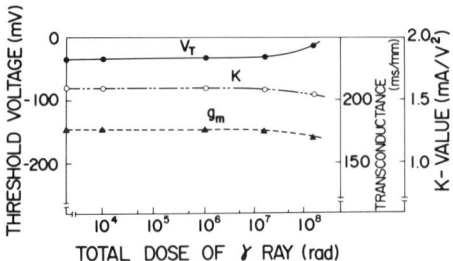

FIG. 59. Radiation hardness characteristics of one-half micron gate-length SAINT FETs for gamma rays.

observed. It is estimated that this deterioration is caused by damage to the GaAs bulk crystal. These results are quite similar to those with other gate-material FETs.

VIII. Summary and Future Trends

Research on GaAs LSI technology began in 1980, about 10 years later than the start of Si LSI technology. This new field of technology has rapidly advanced over the past six years. Before LSI technology development, basic study has been carried out on microwave devices for MESFETs, diodes, transfer electron devices (TEDs), and impact-avalanche transit-time (IMPATT) diodes. Furthermore, the technology for crystal growth, epitaxy, ion implantation, and photolithography has also been developed. However, the current state of GaAs LSI technology must be improved regarding dimensions, quality, uniformity, and reliability. Nevertheless, some technologies in GaAs LSIs have advanced to the same level of those in Si LSIs, such as photoresist patterning, pattern layout, and circuit simulation.

Next, crystal growth has effectively evolved from the HB method to the LEC method for 2-inch round wafers. However, 3-inch wafers are now being used for device processes, as of 1983. The other major event in crystal growth was the development of dislocation free crystal with In-doped materials. By the use of this crystal, uniformity was dramatically improved. In the future, more nearly perfect crystals having the same diameter as the Si wafer will be required for mass production. To achieve this, GaAs on Si wafers grown by MOCVD or MBE will provide integration of GaAs and Si LSIs on the same chip.

The fabrication technology for GaAs LSIs was developed using selective ion implantation, self-aligned n^+-layer formation, and two-level interconnection. These steps were achieved by applying a 1.5 micron design rule. In addition, high-density integration with a factor of three has been developed using a three-level interconnection. Moreover, GaAs EFETs with more than 300 mS/mm transconductance and less than 20 mV threshold-voltage standard deviation will enable ultrahigh-speed operation for GaAs LSIs.

Highly reliable and reproducible technology for GaAs LSIs must be ensured, including the protection circuits. GaAs LSIs have been confirmed to be highly durable elements under severe conditions such as radiation. Finally, the application field will certainly be widened with further advances in GaAs technology.

REFERENCES

Asai, K., Ino, M., Kurumada, K., Kawasaki, Y., and Ohmori, M. (1981). *Int. Symp.* GaAs *and Related Compound; Inst. Phys. Conf. Ser.* **63**, 533.
Asbeck, P. M., Miller, D. L., Anderson, R. J., and Eisen, F. H. (1984a). *IEEE Electron Device Lett.* **EDL-5**, 310.
Asbeck, P. M., Lee, C. P., and Chang, M. F. (1984b). *IEEE Trans. Electron. Devices* **ED-31**, 1377.
Bert, G., Morin, J. P., Nuzillat, G., and Arnold, C. (1981) GaAs *IC Symp. Tech. Dig.*, p. 27.
D'avanzo, D. C. (1982) *IEEE Trans. Microwave Theory and Techniques* **MTT-30**, 955.
Dingfew, W. U., and Heime, K. (1982). *Electron. Lett.* **18**, 940.
Eden, R. C., and Welch, B. M. (1977). *IEEE Trans. Electron Devices* **ED-24**, 1209.
Eden, R. C., Welch, B. M., and Zucca, R. (1978). *IEEE J.* **SC-13**, 419.
Enoki, T., Yamasaki, K., Osafune, K., and Ohwada, K. (1985). *Extended Abstracts of the 17th Conf. on Solid State Devices and Materials*, p. 413.
Enoki, T., Yamasaki, K., Osafune, K., and Ohwada, K. (1986). *Electron. Lett.* **22**, 68.
Greiling, P. T., Krumm, C. F., Ozdemir, F. S., Hackett, L. H., and Lohr, Jr., R. F. (1978). *Annual Device Research Conf.*, paper MP-A1.
Hendel, R. H., Pei, S. S., Kiehl, R. A., Tu, C. W., Feuer, M. D., and Dingle, R. (1984). *IEEE Electron Device Lett.* **EDL-5**, 406.
Higashisaka, A., Ishikawa, M., Katano, F., Asai, S., Furutsuka, T., and Takayama, Y. (1983). *Extended Abstracts of the 15th Conf. on Solid State Devices and Materials*, p. 69.
Hirayama, M. (1984). *National Convention Record on IECE Japan*, 2-341.
Hirayama, M., Ino, M., Matsuoka, Y., and Suzuki, M. (1984). *ISSCC Tech. Dig.*, p. 46.
Hirayama, M., Togashi, M., Kato, N., Suzuki, M., Matsuoka, Y., and Kawasaki, Y. (1986). *IEEE Trans.* **ED-33**, 104.
Huber, A. M., Morillot, G., and Linh, N. T. (1979). *Appl. Phys. Lett.* **34**, 858.
Idda, M., Yamazaki, H., Kato, N., and Ohmori, M. (1984). *The 16th Conference on Solid State Devices and Materials Late News*, p. 30.
Imanura, K., Yokoyama, N., Ohnishi, T., Suzuki, S., Nakai, K., Nishi, H., and Shibatomi, A. (1984). *Japan. J. Appl. Phys.* **23**, L342.
Ino, M., Hirayama, M., Ohwada, K., and Kurumada, K. (1982). GaAs *IC Symp. Tech. Dig.*, p. 2.
Ishii, Y., Ino, M., Idda, M., Hirayama, M., and Ohmori, M. (1984a). GaAs *IC Symp. Tech. Dig.*, p. 121.
Ishii, Y., Miyazawa, S., and Ishida, S. (1984b). *IEEE Trans.* **ED-31**, 1051.
Ishikawa, H., Kusakawa, H., Suyama, K., and Fukuta, M. (1977). *ISSCC Tech. Dig.*, p. 200.
Iversen, W. R. (1983). *Electronics* **56**, 97.
Kato, N., Yamasaki, K., Asai, K., and Ohwada, K. (1983). *IEEE Trans. Electron Devices* **ED-30**, 663.

Kato, N., Matsuoka, Y., Ohwada, K., and Moriya, S. (1984). *IEEE Electron Device Lett.* **EDL-4,** 417.
Kato, N., Hirayama, M., Asai, K., Matsuoka, Y., Yamasaki, K., and Ogino, T. (1985a). *IEDM Tech. Dig.*, p. 90.
Kato, N., Takada, T., Yamasaki, K., and Hirayama, M. (1985b). *Extended Abstracts of the 17th Conference on Solid State Devices and Material*, p. 417.
Kroemer, H. (1982). *Proc. IEEE* **70,** 13.
Kuroda, S., Mimura, T., Suzuki, M., Kobayashi, N., Nishiuchi, K., Shibatomi, A., and Abe, M. (1984). GaAs *IC Symp. Tech. Dig.*, p. 125.
Lee, F. S., Eden, R. C., Long, S. I., Welch, B. M., and Zucca, R. (1980). *Int. Conf. Circuit. Computers.*, p. 697.
Lee, C. P., Welch, B. M., and Fleming, W. P. (1981). *Electron. Lett.* **11,** 407.
Lee, C. P., Vahrenkamp, R., Lee, S. J., Shen, Y. D., and Welch, B. M. (1982). *Proc. of* GaAs *IC Symp.*, p. 169.
Lee, F. S., Kaelin, G. R., Welch, B. M., Zucca, R., Shen, E., Asbeck, P., Lee, C. P., Kirkpatrick, C. G., Long, S. I., and Eden, R. C. (1982). *IEEE J.* **SC-17,** 638.
Liechti, C. A. (1984). *IEDM Tech. Dig.*, p. 13.
Magee, T., xxxxxx, J., Lee, K., xxxxxx, S., Ormond, R., Batter, R. J. and Evans, Jr., C. A. (1980). *Appl. Phys. Lett.* **37,** 447.
Matsuoka, Y., Ohwada, K., and Hirayama, M. (1984). *IEEE Trans.* **ED-31,** 1062.
Mimura, T., Hiyamizu, S., Fujii, T., and Nanbu, K. (1980). *Japan. J. Appl. Phys.* **19,** L225.
Mimura, T., Joshin, K., Hiyamizu, S., Hikosaka, K., and Abe, M. (1981). *Japan. J. Appl. Phys.* **20,** L598.
Miyazawa, S., Mizutani, T., and Yamazaki, H. (1982). *Japan J. Appl. Phys.* **21,** L542.
Miyazawa, S., Ishii, Y., Ishida, S., and Nanishi, Y. (1983). *Appl. Phys. Lett.* **43,** 853.
Mizutani, T., Kato, N., Ishida, S., Osafune, K., and Ohmori, M. (1980). *Electron. Lett.* **16,** 315.
Mizutani, T., Honda, T., Ishida, S., and Kawasaki, Y. (1982). *Solid State Electron.* **25,** 885.
Muraguchi, M., Ohwda, K., and Hirayama, M. (1984). *Proc. 6th Symp. Dry Process., Tokyo*, p.37.
Nakayama, Y., Suyama, K., Shimizu, K., Yokoyama, N., Shibatomi, A., and Ishikawa, H. (1983). *ISSCC Tech. Dig.*, p. 48.
Nanishi, Y., Yamazaki, H., Mizutani, T., and Miyazawa, S. (1981). GaAs *and Related Compound Symp.; Inst. Phys. Conf. Ser.* **63,** 7.
Nanishi, Y., Ishida, S., Honda, T., Yamazaki, H., and Miyazawa, S. (1982). *Japan. J. Appl. Phys.* **21,** L335.
Nishi, H., Onodera, H., Kawata, H., Yokoyama, N., and Shibatomi, A. (1983). *Annual Device Research Conf.*, paper IV A-5.
Nottoff, J. K., and Zuleeg, R. (1975). *IEDM Tech. Dig.*, 624.
Ogawa, K., Nawata, K., Matsuoka, Y., and Simada, K. (1985). *Japan Appl. Phys. Fall Meeting*, 2P-B-6.
Ohmori, M. (1983). *Conf. Solid State Devices and Materials, Tokyo*, p. 61.
Ohmori, M. (1984). *11th Int. Symp.* GaAs *and Related Compounds, Biarritz*.
Ohnishi, T., Yamaguchi, Y., Onodera, T., Yokoyama, N., and Nishi, H. (1984). *Extended Abstracts of 16th Int. Conf. on Solid State Devices and Materials*, p. 391.
Ohnishi, T., Onodera, T., Yokoyama, N., and Nishi, H. (1985). *IEEE Electron Device Lett.* **EDL-9,** 172.
Remba, R. D., Suni, I., and Nicolet, M.-A. (1985). *IEEE* **EDL-6,** 437.
Sadler, R. A., and Eastman, L. F. (1983). *Appl. Phys. Lett.* **43,** 865.
Shah, N. J., Pei, S. S., Tu, C. W., and Tiberio, R. C. (1986). *IEEE Trans. Electron Devices* **ED-33,** 543.

Simada, K. (1985). (Private communication.)
Spicer, W. E., Chye, P. W., Skeath, P. R., Su, C. Y., and Lindau, I. (1979). *J. Vac. Sci. & Technol.* **16,** 1422.
Stall, R., Wood, C. E. C., Board, K., and Eastman, L. F. (1979). *Electron. Lett.* **15,** 800.
Suzuki, M., Murase, K., Asai, K., and Kurumada, K. (1983). *Japan J. Appl. Phys.* **22,** L709.
Suzuki, M., Murase, K., and Hirayama, M. (1984). *Extended Abstract of the 16th Conference on Solid State Devices and Materials*, p. 387.
Suzuki, M., Murase, K., Kato, N., Togasi, M., and Hirayama, M. (1986). Paper *Tech. Group on Solid State Devices, IEEE, Japan, 85-11*, p. 29.
Takada, T., Saito, S., Kato, N., and Idda, M. (1985a). *Electron. Lett.* **21,** 731.
Takada, T., Shimazu, Y., Yamasaki, K., Togashi, M., Hoshikawa, K., and Idda, M. (1985b). *IEEE Microwave and Millimeterwave Monolithic Circuits Symp. Tech. Dig.*, p. 22.
Toyoda, N., Mochizuki, M., Mizoguchi, T., Nii, R., and Hojo, A. (1981). *Int. Symp. GaAs and Related Compounds*, p. 521.
Van Tuyl, R. L. (1974). *IEEE J. Solid-State Circuits* **SC-9,** 269.
Van Tuyl, R. L., and Liechti, C. A. (1974). *ISSCC Tech. Dig.*, p. 114.
Van Tuyl, R. L., Liechti, C. A., Lee, R. E., and Gowen, E. (1977). *IEEE J.* **SC-12,** 485.
Wada, Y., and Chino, K. (1983). *Extended Abstract of ECS*, **83-2,** 463.
Welch, B. M., and Eden, R. C. (1977). *IEDM Tech. Dig.*, p. 205.
Welch, B. M., Shen, Y. D., Zucca, R., and Eden, R. C. (1979). *IEDM Tech. Dig.*, p. 493.
Yamane, Y., Makimura, T., and Matsuoka, Y. (1984). *Nat. Conv. Record. IECE of Japan*, pp. 2–20.
Yamane, Y., Imanura, Y., Iwadate, K., and Hirayama, M. (1985). *Proc. 12th Int. Symp. GaAs and Related Compounds*, p. 493.
Yamasaki, K., Asai, K., Mizutani, T., and Kurumada, K. (1982a). *Electron. Lett.* **18,** 119.
Yamasaki, K., Asai, K., and Kurumada, K. (1982b). *Proc. 14th Conf. on Solid State Devices*, p. 381.
Yamasaki, K., Yamane, Y., Kurumada, K. (1982c). *Electron. Lett.* **18,** 592.
Yamasaki, K., Asai, K., and Kurumada, K. (1982d). *IEEE Trans. Electron Devices* **ED-29,** 1772.
Yamasaki, K., Kato, N., Matsuoka, Y., and Ohwada, K. (1982e). *IEDM Tech. Dig.*, p. 166.
Yamasaki, K., Kato, N., and Hirayama, M. (1984). *Electron. Lett.* **20,** 1029.
Yamazaki, H., Honda, T., and Miyazawa, S. (1981). *Electron Lett.* **17,** 817.
Yamazaki, H., Honda, T., Ishida, S., and Kawasaki, Y. (1984). *Appl. Phys. Lett.* **45,** 1109.
Yokoyama, N., Mimura, T., Fukuta, M., and Ishikawa, M. (1981). *ISCC Dig. of Tech. Papers*, p. 218.
Yokoyama, N., Ohnishi, T., Onodera, H., Shinoki, T., Shibatomi, A., and Ishikawa, H. (1983). *ISSCC Dig. of Tech. Papers*, p. 44.
Yuan, H. T. (1982). *GaAs IC Symp. Tech. Dig.*, 100.
Yuan, H. T., Mclevige, W. V., Shih, H. D., and Hearn, A. S. (1984). *ISSCC Dig. Tech. Papers*, 42.
Zucca, R., Welch, B. M., Eden, R. C., and Long, S. I. (1980). *IEEE Trans.* **ED-27,** 1109.
Zuleeg, R., Notthoff, J. K., and Troeger, G. L. (1984). *IEEE* **EDL-5,** 21.

Index

A

Annealing. *See* Encapsulant; Furnace annealing; Rapid thermal annealing

C

Circuit configurations, 190
  BFL, 195
  CCFL, 188
  CDFL, 188, 196
  HSCFL, 198
  LBFL, 196
  LSCFL, 198
  SCFL, 188, 198
  SDFL, 187, 190
Circuit elements of LSI
  capacitor, 188
  diodes, 187
  interconnection line, 188
  resistor, 188
Circuit simulation program, 170
  ASTAP, 170
  SPICE, 170
  statistical analysis, 170
  statistical circuit simulation, 206
Custom LSI
  full-custom LSI, 209
  gate array, 210, 215
  macro-cell array, 210, 215
  master-slice, 215
  semi-custom LSI, 209
  standard cell, 215
Cutoff frequency, 168

D

Device analysis, 160
  drift velocity overshoot, 167
  intrinsic response time (tint), 164
  macroscopic FET analysis, 160
  Monte Carlo simulation, 167
  particle simulation, 166
  transient characteristics, 164
  two-dimensional Monte Carlo simulation, 167
Dicing, 283–284
  chipping, 284
Die bonding, 284
Diffusion
  Cr outdiffusion, 29
  dopant diffusion, 29, 33, 38–39, 42–45
Diode current, 277
Drift mobility, 251
Dry etching, 111–114
HCl cleaning, 127
  hydrogen plasma treatment, 113
  ion-beam milling, 114
  surface cleaning, 113
  via hole opening, 114
  WSi_x, 112
XPS spectra of GaAs surface just after RIE, 128

E

Encapsulant
  AlN, 6
  double-layer, 9
  material requirement, 4–5
  Si_3N_4, 6
  SiO_2, 5–6
  SiO_xN_y
    annealing, 15–22
    characterization, 11–15
    composition, 12–13
    deposition, 10–11
    film stress, 14–15
    historical background, 8
    rapid thermal annealing, 35–36, 39
    refractive index, 12
  see also Furnace annealing
Equivalent circuits

gate capacitances, 168, 179
gate resistance Rg, 185
parasitic capacitances, 186

F

Fan-in, 247
Fan-out, 247
FET(s)
  extrinsic, 185
  intrinsic, two-region model, 180
  long gate, 251
  normally-off, 170
  normally-on, 170
  short source-drain, 173
  simple physical model, 171
  see also Self-aligned FET(s)
FIB (focused ion beam) implantation
  compositional disordering, AlGaAs/GaAs, 82–85
  electrical characteristics, GaAs
    activation efficiency, 77–78
    electron mobility, 77–78
  induced damage, GaAs, Raman scattering spectroscopy, 74–77
  lateral spread, implanted atom, 71–73
  optical properties, GaAs, photoluminescence spectroscopy, 78–79
  pattern doping, 71–73
  positional alignment, 65, 86–87
  semi-insulating of GaAs, 79–82
FIB implantation doping MBE (FIBI-MBE)
  contamination, during growth interruption, 89-90
  crystal quality
    carrier concentration profile, 87–89
    effect of, growth condition, 92–93
  crystal structure for HEMT, 91–92
  growth system, 85–87
FIB implanter
  implanter system, 64–65
  throughput, 93–94
Flip-flop
  D-type (D-F/F), 221
  T-type (T-F/F), 217
Frequency divider(s), 216
  D-type flip flop, 221
  dynamic, 219
  modules, 221
  static, 217
  toggling frequency, 217
  T-type flip flop, 217
Furnace annealing, 3–25, 240, 254
  capped, 5–10, 15–22
  controlled atmosphere, 23–25
see also Encapsulant

G

GaAs substrate, 238–245
  Cr concentration, 242f, 243
  defect density, 289
  denuded zone, 245
  dislocation pit, 245
  etch pit density (EPD), 244
  horizontal Bridgeman (HB), 238
  In-doped, 237
  liquid-encapsulated Czochralski (LEC), 238, 242
  specification for LSI, 238
  striation, 243
  undoped, 244
  uniformity, 242
  wafer selection, 239

H

Hall mobility, 242
HBT, 237
HEMT, 237

I

IC, 234–238
  complexity, 235f
  LSI, 236
  commercial production, 237
Interconnection technology, 279–283, 296
  air-bridged, 280
  Al, 279
  contact resistance, 282
  conventional, 280
  Au, 279
  planar, 280
  planar air bridge, 283
  sheet resistance, 282
  spacer film, 281
  trilevel, 296
  via metal, 281

Ion implantation, 1–59
  activation efficiency, 241
  active layer formation, 253–255
    shallow implantation, 254
    through implantation, 254–255
    see also Encapsulant; Furnace annealing; Rapid thermal annealing
  ion bombardment for isolation, 263
  selective ion implantation, 251
  self-aligned implantation, 252
    see also Self-aligned FET(s)
Isolation, 249, 262–263
  mesa isolation, 262
  ion bombardment, 263
  sidegating effect, 262–263

K

K-value, 116, 129–130, 140

L

LDD (lightly doped drain) technology, 127–131
  fabrication process, 128
  FET characteristics, 125, 127
  n^+-region insertion, 129
Leakage current, 240
Liquid metal (LM) ion source
  Au-Si-Be alloy, 67–68
  beam intensity profile, 70–71
  configuration, 65–66
  lifetime, 69
  mass spectrum
    Au-Si-Be, 67
    Pd-Ni-Si-Be-B, 69
  Pd-Ni-Si-Be-B alloy, 68–69
Logic LSI, 297

M

Maskless implanter. See FIB implanter
Memory, 202
  cell, 204
  pattern, 208
  SRAM, 202
MESFET
  array, 242
  structures, 251–253
  mesa-epitaxial FET, 251–252
  planar FET, 251–252
  see also Self-aligned FET(s); Self-aligned MESFET
Metal insulator metal (MIM), 277

N

N_G factor, 249

O

Ohmic contact(s), 99–104
  alloyed, 260
  band diagram, 260
  contact resistance, 101, 260–262
  Ni-AuGe, 100–101
  nonalloyed, 103, 262
  preferential lateral diffusion of n^+-layer dopant, 151–153
  pulse annealed, 102
  thermal stability, 262
  transmission line method (TLM), 260
  see also Piezoelectric effect

P

Package(s), 226
  dual in-line, 226
  flat, 226
  for ultrahigh speed, 284–287
  frequency characteristics, 285
Piezoelectric effect, 147–151
  FET characteristics for (100) and (111) substrate, 152
  V_T shift, 147, 149, 151
  V_T variation in 3-inch wafer, 148
  see also Orientation effect

R

RAM, 293–296
  1, 4 kbit, 294
  16 kbit, 294
    In-doped wafer, 294
Rapid thermal annealing
  annealing system, 25–26
  capless annealing, 27
  channel implant, 27–34
  controlled atmosphere, 56–57
  defect
    deep level, 52–53

INDEX

slip line, 52
surface dissociation, 53
device application, 54–56
general, 25–27
high dose n-type implant
  nonalloyed ohmic contact, 39–41
  S implant, 37–38
  Se implant, 36–37
  Si implant, 35–36
  Sn implant, 38
  thermal stability, 38–39
p-type implant
  Be implant, 51
  Mg implant, 49–51
  Zn implant, 42–29
  see also Encapsulant; Furnace annealing
Refractory metal gate technology, 115–119
  Au/TiN/WSi gate, 117
  Au/WSiN gate, 118
  K-value, 116, 129, 130
  MOCVD n^+-region, 116
  through-AIN implantation, 116
  TiW gate, 115
  W/WSi bilayer film, 118
Reliability, 299
  radiation hardness, 299

S

SAINT (self-aligned implantation for n^+-layer technology), 119–123, 173–174, 208, 236
  a-Si-Ge-B gate, 121–122
  gate-source external series resistance, 121
  MAS FET, 121
  pattern inversion technique, 122
  process sequence, 120
Schottky gate technology, 104–111, 255–257
  Al, 104
  amorphous Si-Ge-B, 110, 259
  barrier height, 106–107, 257–259
  gate material, 255–256
  ideality factor, 107
  Mo, 105
  patterning techniques, 255–257f
  Pt, 104–105
  surface cleaning, 257
  $TaSi_2$, 110
  Ti-Pt-Au, 104
  TiW,, 106
  TiWSi, 106

W, 105
W-Al, 109
WN, 108
WSi, 106–107
Self-aligned FET(s), 236, 252–253, 263–271
  electrical performance
    gate capacitance, 269–270
    orientation dependence, 270–271
    source resistance, 269–271
  fabrication technology, 263–269
    flat gate SAINT, 268
    refractory gate self-aligned technology, 263–264
    refractory metal gate, 236
    SAINT, 236
    self-aligned implantation for n^+-layer technology (SAINT), 264–269
  see also SAINT
Self-aligned MESFET(s), 114–131, 236
  refractory metal gate, 236
  see also Self-aligned FET(s)
Sense amplifier, 205
Sheet resistance, 279
  Si implanted layer, 279
Short-channel effects, 131–145
  buried p-layer, 141–144
  control of n'-layer-gate-gap, 135–137
  K-value, 140
  L_{off}, 138
  n^+-layer lateral stretch, 138-149
  n^+ selective epitaxial layer, 144–145
  N_G value, 138–139, 144, 249
  reduction in n^+-layer thickness, 137–138
  shallow channel layer with higher doping concentration, 132–135
  V_T-shift, 133–134, 144
Signal swing voltage, 247
Spectroscopy (spectra)
  photoluminescence, 78–79
  Raman scattering, 74–77
  XPS, 128
Submicron gate length FET, 272–276
  buried p-layer FET, 272–274
    propagation delay time, 275
    short channel effects, 272–276
  Ge/GaAs source/drain FET, 272–276
Subthreshold characteristics, 249
  factor N_G, 249

short channel effects, 249
subthreshold current, 249
SWAT (side-wall assisted self-alignment technology), 123–127
  cross-sectional SEM view, 125
  fabrication procedure, 124
  FET characteristics, 125, 127
  Side wall, 124

T

Test element group (TEG), 245–251
  Circuit TEG, 247
    inverter, 247
    ring oscillator, 247
  process TEG, 247
    basic recognition, 247
    circuit element, 247
    crossover interconnection, 249
    interconnection, 247
Testing technology, 223
  coaxial type probe card, 224
  coupled coaxial probe (CCP), 224
  high-frequency probe card, 223
Thermal stability, 240
Threshold voltage (V_T), 243–244, 247, 287–293
  controllable region, 292
  controllability, 287, 292
  in-process monitoring, 292
  scattering, 289–291
  standard deviation, 244
  uniformity, 287–293
    In-doped DF LEC, 287–292
    ingot annealed, 287–291
    low Cr-doped LEC, 287–291
    small area, 289
Transfer characteristics, 289, 291
Transmission line method (TLM), 251
  see also Ohmic contact(s)

Y

Yield, 298

Contents of Previous Volumes

Volume 1 Physics of III–V Compounds
C. Hilsum, Some Key Features of III–V Compounds
Franco Bassani, Methods of Band Calculations Applicable to III–V Compounds
E.O. Kane, The $k \cdot p$ Method
V.L. Bonch-Bruevich, Effect of Heavy Doping on the Semiconductor Band Structure
Donald Long, Energy Band Structures of Mixed Crystals of III–V Compounds
Laura M. Roth and Petros N. Argyres, Magnetic Quantum Effects
S.M. Puri and T.H. Geballe, Thermomagnetic Effects in the Quantum Region
W.M. Becker, Band Characteristics near Principal Minima from Magnetoresistance
E.H. Putley, Freeze-Out Effects, Hot Electron Effects, and Submillimeter Photoconductivity in InSb
H. Weiss, Magnetoresistance
Betsy Ancker-Johnson, Plasmas in Semiconductors and Semimetals

Volume 2 Physics of III–V Compounds
M.G. Holland, Thermal Conductivity
S.I. Novkova, Thermal Expansion
U. Piesbergen, Heat Capacity and Debye Temperatures
G. Giesecke, Lattice Constants
J.R. Drabble, Elastic Properties
A.U. Mac Rae and G.W. Gobeli, Low Energy Electron Diffraction Studies
Robert Lee Mieher, Nuclear Magnetic Resonance
Bernard Goldstein, Electron Paramagnetic Resonance
T.S. Moss, Photoconduction in III–V Compounds
E. Antončik and J. Tauc, Quantum Efficiency of the Internal Photoelectric Effect in InSb
G.W. Gobeli and F.G. Allen, Photoelectric Threshold and Work Function
P.S. Pershan, Nonlinear Optics in III–V Compounds
M. Gershenzon, Radiative Recombination in the III–V Compounds
Frank Stern, Stimulated Emission in Semiconductors

Volume 3 Optical of Properties III–V Compounds
Marvin Hass, Lattice Reflection
William G. Spitzer, Multiphonon Lattice Absorption
D.L. Stierwalt and R.F. Potter, Emittance Studies
H.R. Philipp and H. Ehrenreich, Ultraviolet Optical Properties
Manuel Cardona, Optical Absorption above the Fundamental Edge
Earnest J. Johnson, Absorption near the Fundamental Edge
John O. Dimmock, Introduction to the Theory of Exciton States in Semiconductors
B. Lax and J.G. Mavroides, Interband Magnetooptical Effects

CONTENTS OF PREVIOUS VOLUMES

H.Y. Fan, Effects of Free Carries on Optical Properties
Edward D. Palik and George B. Wright, Free-Carrier Magnetooptical Effects
Richard H. Bube, Photoelectronic Analysis
B.O. Seraphin and H.E. Bennett, Optical Constants

Volume 4   Physics of III–V Compounds

N.A. Goryunova, A.S. Borschevskii, and D.N. Tretiakov, Hardness
N.N. Sirota, Heats of Formation and Temperatures and Heats of Fusion of Compounds $A^{III}B^{V}$
Don L. Kendall, Diffusion
A.G. Chynoweth, Charge Multiplication Phenomena
Robert W. Keyes, The Effects of Hydrostatic Pressure on the Properties of III–V Semiconductors
L.W. Aukerman, Radiation Effects
N.A. Goryunova, F.P. Kesamanly, and D.N. Nasledov, Phenomena in Solid Solutions
R.T. Bate, Electrical Properties of Nonuniform Crystals

Volume 5   Infrared Detectors

Henry Levinstein, Characterization of Infrared Detectors
Paul W. Kruse, Indium Antimonide Photoconductive and Photoelectromagnetic Detectors
M.B. Prince, Narrowband Self-Filtering Detectors
Ivars Melngailis and T.C. Harman, Single-Crystal Lead-Tin Chalcogenides
Donald Long and Joseph L. Schmit, Mercury-Cadmium Telluride and Closely Related Alloys
E.H. Putley, The Pyroelectric Detector
Norman B. Stevens, Radiation Thermopiles
R.J. Keyes and T.M. Quist, Low Level Coherent and Incoherent Detection in the Infrared
M.C. Teich, Coherent Detection in the Infrared
F.R. Arams, E.W. Sard, B.J. Peyton, and F.P. Pace, Infrared Heterodyne Detection with Gigahertz IF Response
H.S. Sommers, Jr., Macrowave-Based Photoconductive Detector
Robert Sehr and Rainer Zuleeg, Imaging and Display

Volume 6   Injection Phenomena

Murray A. Lampert and Ronald B. Schilling, Current Injection in Solids: The Regional Approximation Method
Richard Williams, Injection by Internal Photoemission
Allen M. Barnett, Current Filament Formation
R. Baron and J.W. Mayer, Double Injection in Semiconductors
W. Ruppel, The Photoconductor-Metal Contact

Volume 7   Application and Devices
PART A

John A. Copeland and Stephen Knight, Applications Utilizing Bulk Negative Resistance
F.A. Padovani, The Voltage-Current Characteristics of Metal-Semiconductor Contacts
P.L. Hower, W.W. Hooper, B.R. Cairns, R.D. Fairman, and D.A. Tremere, The GaAs Field-Effect Transistor
Marvin H. White, MOS Transistors

CONTENTS OF PREVIOUS VOLUMES

G.R. Antell, Gallium Arsenide Transistors
T.L. Tansley, Heterojunction Properties

PART B

T. Misawa, IMPATT Diodes
H.C. Okean, Tunnel Diodes
Robert B. Campbell and Hung-Chi Chang, Silicon Carbide Junction Devices
R.E. Enstrom, H. Kressel, and L. Krassner, High-Temperature Power Rectifiers of $GaAs_{1-x}P_x$

Volume 8 Transport and Optical Phenomena

Richard J. Stirn, Band Structure and Galvanomagnetic Effects in III–V Compounds with Indirect Band Gaps
Roland W. Ure, Jr., Thermoelectric Effects in III–V Compounds
Herbert Piller, Faraday Rotation
H. Barry Bebb and E.W. Williams, Photoluminescence I: Theory
E.W. Williams and H. Barry Bebb, Photoluminescence II: Gallium Arsenide

Volume 9 Modulation Techniques

B.O. Seraphin, Electroreflectance
R.L. Aggarwal, Modulated Interband Magnetooptics
Daniel F. Blossey and Paul Handler, Electroabsorption
Bruno Batz, Thermal and Wavelength Modulation Spectroscopy
Ivar Balslev, Piezooptical Effects
D.E. Aspnes and N. Bottka, Electric-Field Effects on the Dielectric Function of Semiconductors and Insulators

Volume 10 Transport Phenomena

R.L. Rode, Low-Field Electron Transport
J.D. Wiley, Mobility of Holes in III–V Compounds
C.M. Wolfe and G.E. Stillman, Apparent Mobility Enhancement in Inhomogeneous Crystals
Robert L. Peterson, The Magnetophonon Effect

Volume 11 Solar Cells

Harold J. Hovel, Introduction; Carrier Collection, Spectral Response, and Photocurrent; Solar Cell Electrical Characteristics; Efficiency; Thickness; Other Solar Cell Devices; Radiation Effects; Temperature and Intensity; Solar Cell Technology

Volume 12 Infrared Detectors (II)

W.L. Eiseman, J.D. Merriam, and R.F. Potter, Operational Characteristics of Infrared Photodetectors
Peter R. Bratt, Impurity Germanium and Silicon Infrared Detectors
E.H. Putley, InSb Submillimeter Photoconductive Detectors
G.E. Stillman, C.M. Wolfe, and J.O. Dimmock, Far-Infrared Photoconductivity in High Purity GaAs
G.E. Stillman and C.M. Wolfe, Avalanche Photodiodes

CONTENTS OF PREVIOUS VOLUMES

P.L. Richards, The Josephson Junction as a Detector of Microwave and Far-Infrared Radiation
E.H. Putley, The Pyroelectric Detector–An Update

Volume 13 Cadmium Telluride

Kenneth Zanio, Materials Preparation; Physics; Defects; Applications

Volume 14 Lasers, Junctions, Transport

N. Holonyak, Jr. and M.H. Lee, Photopumped III–V Semiconductor Lasers
Henry Kressel and Jerome K. Butler, Heterojunction Laser Diodes
A. Van der Ziel, Space-Charge-Limited Solid-State Diodes
Peter J. Price, Monte Carlo Calculation of Electron Transport in Solids

Volume 15 Contacts, Junctions, Emitters

B.L. Sharma, Ohmic Contacts to III–V Compound Semiconductors
Allen Nussbaum, The Theory of Semiconducting Junctions
John S. Escher, NEA Semiconductor Photoemitters

Volume 16 Defects, (HgCd)Se, (HgCd)Te

Henry Kressel, The Effect of Crystal Defects on Optoelectronic Devices
C.R. Whitsett, J.G. Broerman, and C.J. Summers, Crystal Growth and Properties of $Hg_{1-x}Cd_x Se$ Alloys
M.H. Weiler, Magnetooptical Properties of $Hg_{t-x}Cd_x Te$ Alloys
Paul W. Kruse and John G. Ready, Nonlinear Optical Effects in $Hg_{t-x}Cd_x Te$

Volume 17 CW Processing of Silicon and Other Semiconductors

James F. Gibbons, Beam Processing of Silicon
Arto Lietoila, Richard B. Gold, James F. Gibbons, and Lee A. Christel, Temperature Distributions and Solid Phase Reaction Rates Produced by Scanning CW Beams
Arto Lietoila and James F. Gibbons, Applications of CW Beam Processing to Ion Implanted Crystalline Silicon
N.M. Johnson, Electronic Defects in CW Transient Thermal Processed Silicon
K.F. Lee, T.J. Stultz, and James F. Gibbons, Beam Recrystallized Polycrystalline Silicon: Properties, Applications, and Techniques
T. Shibata, A. Wakita, T.W. Sigmon, and James F. Gibbons, Metal-Silicon Reactions and Silicide
Yves I. Nissim and James F. Gibbons, CW Beam Processing of Gallium Arsenide

Volume 18 Mercury Cadmium Telluride

Paul W. Kruse, The Emergence of $(Hg_{t-x}Cd_x)Te$ as a Modern Infrared Sensitive Material
H.E. Hirsch, S.C. Liang, and A.G. White, Preparation of High-Purity Cadmium, Mercury, and Tellurium
W.F.H. Micklethwaite, The Crystal Growth of Cadmium Mercury Telluride
Paul E. Petersen, Auger Recombination in Mercury Cadmium Telluride
R.M. Broudy and V.J. Mazurczyck, (HgCd)Te Photoconductive Detectors
M.B. Reine, A.K. Sood, and T.J. Tredwell, Photovoltaic Infrared Detectors
M.A. Kinch, Metal-Insulator-Semiconductor Infrared Detectors

CONTENTS OF PREVIOUS VOLUMES

Volume 19   Deep Levels, GaAs, Alloys, Photochemistry

G.F. Neumark and K. Kosai, Deep Levels in Wide Band-Gap III–V Semiconductors
David C. Look, The Electrical and Photoelectronic Properties of Semi-Insulating GaAs
R.F. Brebrick, Ching-Hua Su, and Pok-Kai Liao, Associated Solution Model for Ga-In-Sb and Hg-Cd-Te
Yu. Ya. Gurevich and Yu. V. Pleskov, Photoelectrochemistry of Semiconductors

Volume 20   Semi-Insulating GaAs

R.N. Thomas, H.M. Hobgood, G.W. Eldridge, D.L. Barrett, T.T. Braggins, L.B. Ta, and S.K. Wang, High-Purity LEC Growth and Direct Implantation of GaAs for Monolithic Microwave Circuits
C.A. Stolte, Ion Implantation and Materials for GaAs Integrated Circuits
C.G. Kirkpatrick, R.T. Chen, D.E. Holmes, P.M. Asbeck, K.R. Elliott, R.D. Fairman, and J.R. Oliver, LEC GaAs for Integrated Circuit Applications
J.S. Blakemore and S. Rahimi, Models for Mid-Gap Centers in Gallium Arsenide

Volume 21   Hydrogenated Amorphous Silicon
Part A

Jacques I. Pankove Introduction
Masataka Hirose, Glow Discharge; Chemical Vapor Deposition
Yoshiyuki Uchida, dc Glow Discharge
T.D. Moustakas, Sputtering
Isao Yamada, Ionized-Cluster Beam Deposition
Bruce A. Scott, Homogeneous Chemical Vapor Deposition
Frank J. Kampas, Chemical Reactions in Plasma Deposition
Paul A. Longeway, Plasma Kinetics
Herbert A. Weakliem, Diagnostics of Silane Glow Discharges Using Probes and Mass Spectroscopy
Lester Guttman, Relation between the Atomic and the Electronic Structures
A. Chenevas-Paule, Experiment Determination of Structure
S. Minomura, Pressure Effects on the Local Atomic Structure
David Adler, Defects and Density of Localized States

Part B

Jacques I. Pankove, Introduction
G.D. Cody, The Optical Absorption Edge of a-Si:H
Nabil M. Amer and Warren B. Jackson, Optical Properties of Defect States in a-Si:H
P.J. Zanzucchi, The Vibrational Spectra of a-Si:H
Yoshihiro Hamakawa, Electroreflectance and Electroabsorption
Jeffrey S. Lannin, Raman Scattering of Amorphous Si, Ge, and Their Alloys
R.A. Street, Luminescence in a-Si:H
Richard S. Crandall, Photoconductivity
J. Tauc, Time-Resolved Spectroscopy of Electronic Relaxation Processes
P.E. Vanier, IR-Induced Quenching and Enhancement of Photoconductivity and Photoluminescence
H. Schade, Irradiation-Induced Metastable Effects
L. Ley, Photoelectron Emission Studies

CONTENTS OF PREVIOUS VOLUMES

Part C

Jacques I. Pankove, Introduction
J. David Cohen, Density of States from Junction Measurements in Hydrogenated Amorphous Silicon
P.C. Taylor, Magnetic Resonance Measurements in a-Si:H
K. Morigaki, Optically Detected Magnetic Resonance
J. Dresner, Carrier Mobility in a-Si:H
T. Tiedje, Information about Band-Tail States from Time-of-Flight Experiments
Arnold R. Moore, Diffusion Length in Undoped a-Si:H
W. Beyer and J. Overhof, Doping Effects in a-Si:H
H. Fritzche, Electronic Properties of Surfaces in a-Si:H
C.R. Wronski, The Staebler-Wronski Effect
R.J. Nemanich, Schottky Barriers on a-Si:H
B. Abeles and T. Tiedje, Amorphous Semiconductor Superlattices

Part D

Jacques I. Pankove, Introduction
D.E. Carlson, Solar Cells
G.A. Swartz, Closed-Form Solution of I-V Characteristic for a-Si:H Solar Cells
Isamu Shimizu, Electrophotography
Sachio Ishioka, Image Pickup Tubes
P.G. LeComber and W.E. Spear, The Development of the a-Si:H Field-Effect Transitor and Its Possible Applications
D.G. Ast, a-Si:H FET-Addressed LCD Panel
S. Kaneko, Solid-State Image Sensor
Masakiyo Matsumura, Charge-Coupled Devices
M.A. Bosch, Optical Recording
A.D'Amico and G. Fortunato, Ambient Sensors
Hiroshi Kukimoto, Amorphous Light-Emitting Devices
Robert J. Phelan, Jr., Fast Detectors and Modulators
Jacques I. Pankove, Hybrid Structures
P.G. LeComber, A.E. Owen, W.E. Spear, J. Hajto, and W.K. Choi, Electronic Switching in Amorphous Silicon Junction Devices

Volume 22 Lightwave Communications Technology
Part A

Kazuo Nakajima, The Liquid-Phase Epitaxial Growth of InGaAsP
W.T. Tsang, Molecular Beam Epitaxy for III-V Compound Semiconductors
G.B. Stringfellow, Organometallic Vapor-Phase Epitaxial Growth of III--V Semiconductors
G. Beuchet, Halide and Chloride Transport Vapor-Phase Deposition of InGaAsP and GaAs
Manijeh Razeghi, Low-Pressure Metallo-Organic Chemical Vapor Deposition of $Ga_xIn_{t-x}As_yP_{t-y}$ Alloys
P.M. Petroff, Defects in III–V Compound Semiconductors

Part B

J.P. van der Ziel, Mode Locking of Semiconductor Lasers
Kam Y. Lau and Amnon Yariv, High-Frequency Current Modulation of Semiconductor Injection Lasers
Charles H. Henry, Spectral Properties of Semiconductor Lasers

Yasuharu Suematsu, Katsumi Kishino, Shigehisa Arai, and Fumio Koyama, Dynamic Single-Mode Semiconductor Lasers with a Distributed Reflector

W.T. Tsang, The Cleaved-Coupled-Cavity (C^3) Laser

Part C

R.J. Nelson and N.K. Dutta, Review of InGaAsP/InP Laser Structures and Comparison of Their Performance

N. Chinone and M. Nakamura, Mode-Stabilized Semiconductor Lasers for 0.7–0.8- and 1.1–1.6-μm Regions

Yoshiji Horikoshi, Semiconductor Lasers with Wavelengths Exceeding 2 μm

B.A. Dean and M. Dixon, The Functional Reliability of Semiconductor Lasers as Optical Transmitters

R.H. Saul, T.P. Lee, and C.A. Burus, Light-Emitting Device Design

C.L. Zipfel, Light-Emitting Diode Reliability

Tien Pei Lee and Tingye Li, LED-Based Multimode Lightwave Systems

Kinichiro Ogawa, Semiconductor Noise-Mode Partition Noise

Part D

Federico Capasso, The Physics of Avalanche Photodiodes

T.P. Pearsall and M.A. Pollack, Compound Semiconductor Photodiodes

Takao Kaneda, Silicon and Germanium Avalanche Photodiodes

S.R. Forrest, Sensitivity of Avalanche Photodetector Receivers for High-Bit-Rate Long-Wavelength Optical Communication Systems

J.C. Campbell, Phototransistors for Lightwave Communications

Part E

Shyh Wang, Principles and Characteristics of Integratable Active and Passive Optical Devices

Shlomo Margalit and Amnon Yariv, Integrated Electronic and Photonic Devices

Takaaki Mukai, Yoshihisa Yamamoto, and Tatsuya Kimura, Optical Amplification by Semiconductor Lasers

Volume 23 Pulsed Laser Processing of Semiconductors

R.F. Wood, C.W. White, and R.T. Young, Laser Processing of Semiconductors: An Overview

C.W. White, Segregation, Solute Trapping, and Supersaturated Alloys

G.E. Jellison, Jr., Optical and Electrical Properties of Pulsed Laser-Annealed Silicon

R.F. Wood and G.E. Jellison, Jr., Melting Model of Pulsed Laser Processing

R.F. Wood and F.W. Young, Jr., Nonequilibrium Solidification Following Pulsed Laser Melting

D.H. Lowndes and G.E. Jellison, Jr., Time-Resolved Measurements During Pulsed Laser Irradiation of Silicon

D.M. Zehner, Surface Studies of Pulsed Laser Irradiated Semiconductors

D.H. Lowndes, Pulsed Beam Processing of Gallium Arsenide

R.B. James, Pulsed CO_2 Laser Annealing of Semiconductors

R.T. Young and R.F. Wood, Applications of Pulsed Laser Processing

Volume 24 Applications of Multiquantum Wells, Selective Doping, and Superlattices

C. Weisbuch, Fundamental Properties of III–V Semiconductor Two-Dimensional Quantized Structures: The Basis for Optical and Electronic Device Applications

H. Morkoc and H. Unlu, Factors Affecting the Performance of (Al, Ga)As/GaAs and (Al, Ga)As/InGaAs Modulation-Doped Field-Effect Transistors: Microwave and Digital Applications

N.T. Linh, Two-Dimensional Electron Gas FETs: Microwave Applications

M. Abe et al., Ultra-High-Speed HEMT Integrated Circuits

D.S. Chemla, D.A.B. Miller, and P.W. Smith, Nonlinear Optical Properties of Multiple Quantum Well Structures for Optical Signal Processing

F. Capasso, Graded-Gap and Superlattice Devices by Band-gap Engineering

W.T. Tsang, Quantum Confinement Heterostructure Semiconductor Lasers

G.C. Osbourn et al., Principles and Applications of Semiconductor Strained-Layer Superlattices

Volume 25 Diluted Magnetic Semiconductors

W. Giriat and J.K. Furdyna, Crystal Structure, Composition, and Materials Preparation of Diluted Magnetic Semiconductors

W.M. Becker, Band Structure and Optical Properties of Wide-Gap $A^{II}_{1-x}Mn_xB^{VI}$ Alloys at Zero Magnetic Field

Saul Oseroff and Pieter H. Keesom, Magnetic Properties: Macroscopic Studies

T. Giebultowicz and T.M. Holden, Neutron Scattering Studies of the Magnetic Structure and Dynamics of Diluted Magnetic Semiconductors

J. Kossut, Band Structure and Quantum Transport Phenomena in Narrow-Gap Diluted Magnetic Semiconductors

C. Riqaux, Magnetooptics in Narrow Gap Diluted Magnetic Semiconductors

J.A. Gaj, Magnetooptical Properties of Large-Gap Diluted Magnetic Semiconductors

J. Mycielski, Shallow Acceptors in Diluted Magnetic Semiconductors: Splitting, Boil-off, Giant Negative Magnetoresistance

A.K. Ramdas and S. Rodriquez, Raman Scattering in Diluted Magnetic Semiconductors

P.A. Wolff, Theory of Bound Magnetic Polarons in Semimagnetic Semiconductors

Volume 26 III–V Compound Semiconductors and Semiconductor Properties of Superionic Materials

Zou Yuanxi, III–V Compounds

H.V. Winston, A.T. Hunter, H. Kimura, and R.E. Lee, InAs-Alloyed GaAs Substrates for Direct Implantation

P.K. Bhattacharya and S. Dhar, Deep Levels in III–V Compound Semiconductors Grown by MBE

Yu. Ya. Gurevich and A.K. Ivanov-Shits, Semiconductor Properties of Superionic Materials

Volume 27 High Conducting Quasi-One-Dimensional Organic Crystals

E.M. Conwell, Introduction to Highly Conducting Quasi-One-Dimensional Organic Crystals

I.A. Howard, A Reference Guide to the Conducting Quasi-One-Dimensional Organic Molecular Crystals

J.P. Pouget, Structural Instabilities

E.M. Conwell, Transport Properties

C.S. Jacobsen, Optical Properties

J.C. Scott, Magnetic Properties

L. Zuppiroli, Irradiation Effects: Perfect Crystals and Real Crystals

CONTENTS OF PREVIOUS VOLUMES

Volume 28 **Measurement of High-Speed Signals in Solid State Devices**

J. Frey and D. Ioannou, Materials and Devices for High-Speed and Optoelectronic Applications
H. Schumacher and E. Strid, Electronic Wafer Probing Techniques
D. Auston, Picosecond Photoconductivity
J. Valdmanis, Electrooptic Measurement Techniques
R. Jain and J. Wiesenfeld, Direct Optical Probing of Integrated Circuits and High-Speed Devices
G. Plows, Electron Beam Probing
A. M. Weiner and R. B. Marcus, Photoemissive Probing